WATTS MY LINE?

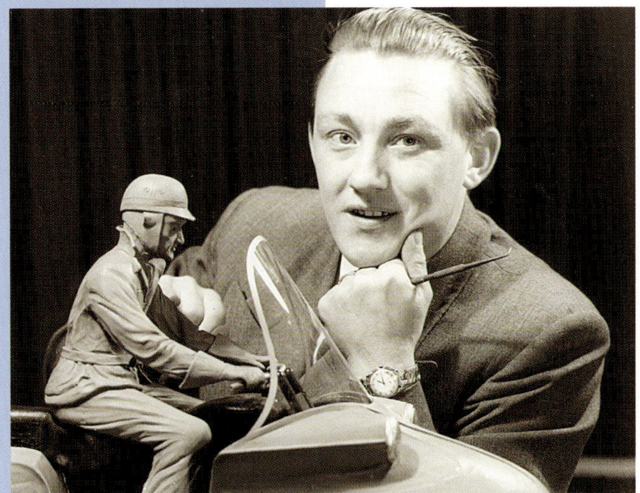

THE LIFE AND WORK OF EDITORIAL ARTIST, LAWRIE WATTS

REDLINE BOOKS

Dedicated to Lawrie and his wife Eileen,
and also my own wife Christine, for her tolerance of my frequent absences.

ACKNOWLEDGEMENTS

To three fellow members of 'The Club' (which is for executive members of the motor cycle industry) who saved me pulling my remaining hair out when I hit another brick wall, especially during the search for information on 'The Ones That (almost) Got Away '.

Norman Hyde, Triumph and Edward Turner information; for putting me on the track of the Norton 500 four and much else... **Keith Blair**, for his enthusiasm in entering the chase for people who had long since disappeared off my radar in connection with 'The Ones That (almost) Got Away', but were still on his. **Nick Jefferies**, for identifying a Rotax 250 in Mick Walker's 'European Racing Motorcycles'. Oops! It was there, but it shouldn't have been because it was a road engine, not a racer! To cap it all, nobody could remember a thing about it...

Steve Purcell, for his permission to use large extracts from his history 'The Corby Iron and Steel Works', and for the invaluable information on a byegone era.

Mick Woollett, for his thoughtful advice, 'Never write a book, it'll take over your life, not worth the effort...'. Then, after many years, I took the plunge, he helped and encouraged whenever I came up against another brick wall....

Fred Swift and **Harry Whitehouse** for pointing me towards **Charlie Edwards**. Charlie for his generous time and impressive memory of Joe Craig and the ill-fated Norton four.

Peter Williams for his efforts, ultimately unsuccessful, to identify a mysterious Heenan and Froude dyno, and for confirmation of some anecdotes told about his father, Jack, (which Peter promised not to emulate).

Bill Bennett, **John Hancox** and **Alan Baker** for their generous tributes to their old and valued friend Lawrie.

Glynne Bowsher for his help with Thrust 2 and Thrust SSC.

Lawrie's daughters, **Julie Sewell** and **Rosemary Powell**, and their respective spouses Paul and Daniel for invaluable assistance with finding, identifying and printing photographs from the family archives.

Richard Bailey, for generously printing the archive pictures of Percy Pilcher.

Jim Reynolds, for being Jim-ever-helpful with contact numbers, anecdotes, more contacts and encouragement when needed.

David Garside, for his encyclopaedic knowledge of the Wankel engine and for lending his personal papers as a guide into the world of the trochoids.

Peter Hooper, who followed his brilliant father, Bernard Hooper, in developing the stepped piston engine from its beginnings in motorcycles to its present state-of-the art in unmanned air vehicles (UAV), and putting all that knowledge at the author's disposal.

Thanks to **Autocar, Flight Magazine, The Motor, Farmer's Weekly, Farmer and Stockbreeder, Mechanical Farming, Hollis Publishing and Mortons Motorcycle Media** for their help and assistance.

To one and all, my heartfelt thanks!
David Dixon

www.redlinebooks.co.uk REDLINE BOOKS

Honda CB750 engine with CR Keihin carburettors.

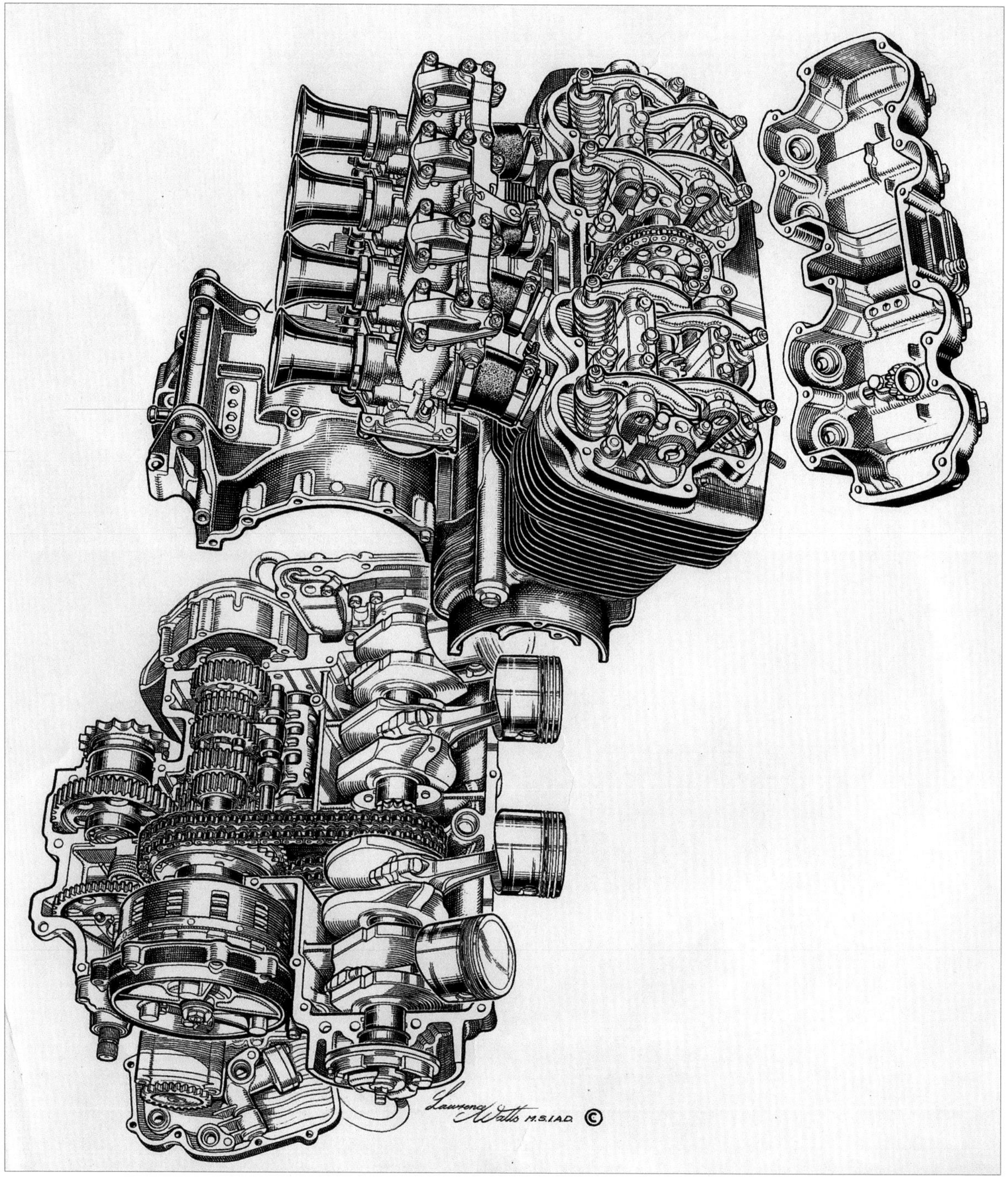

Contents:

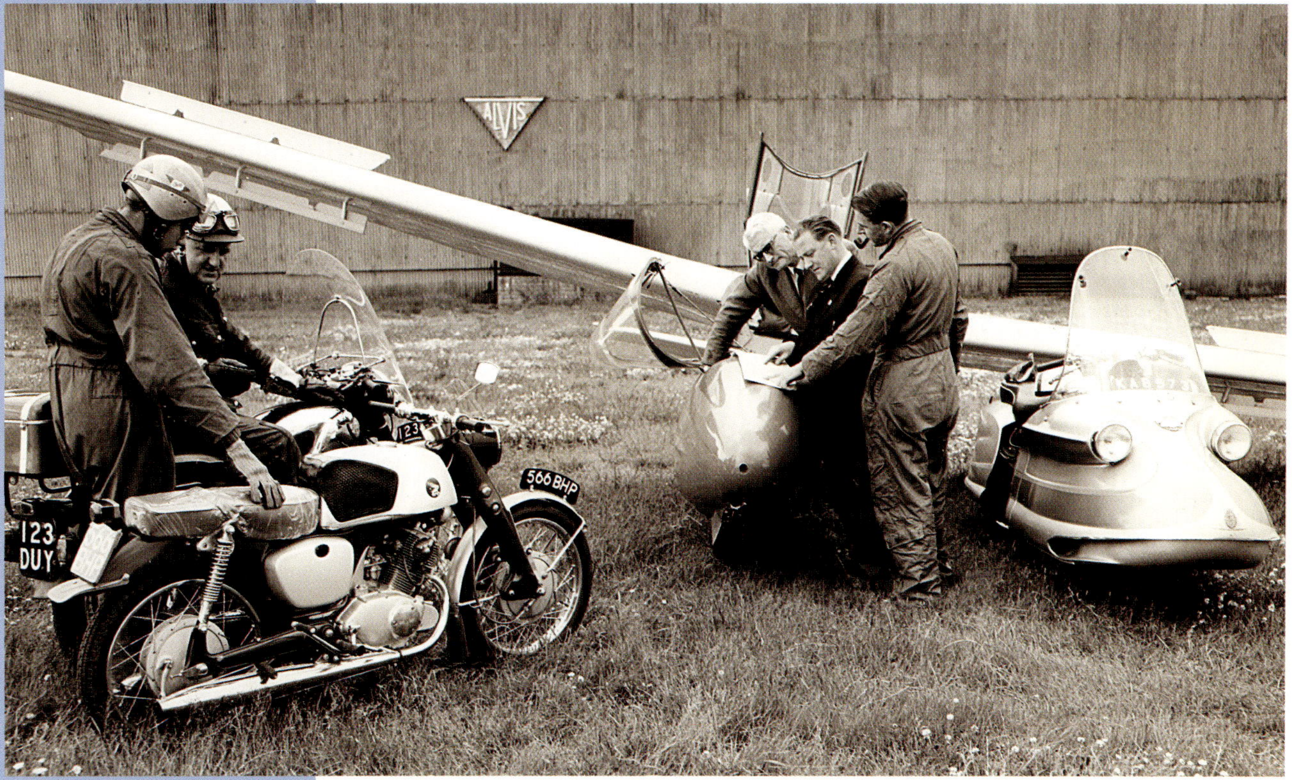

REDLINE BOOKS

Richard Noble's Thrust

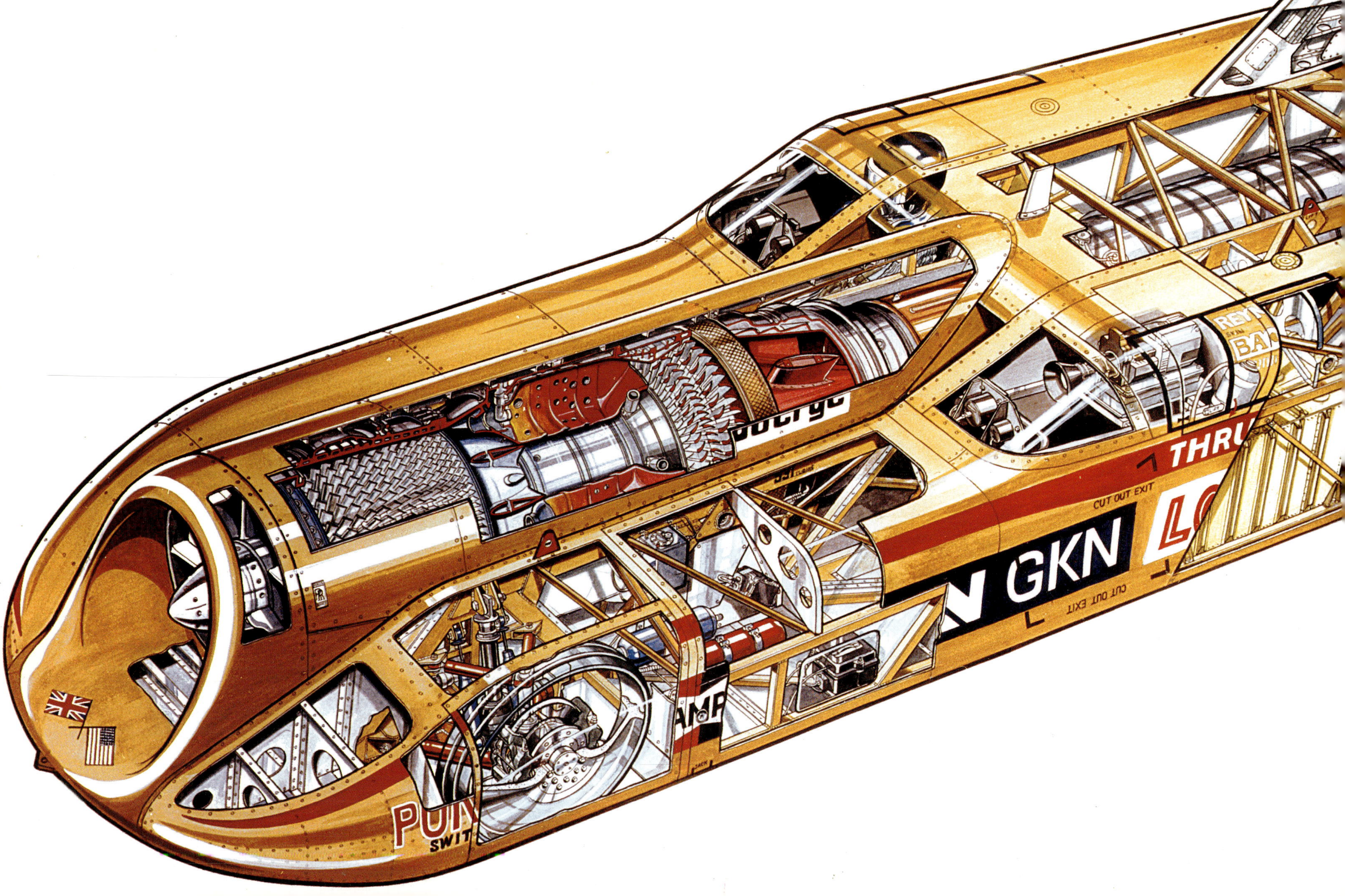

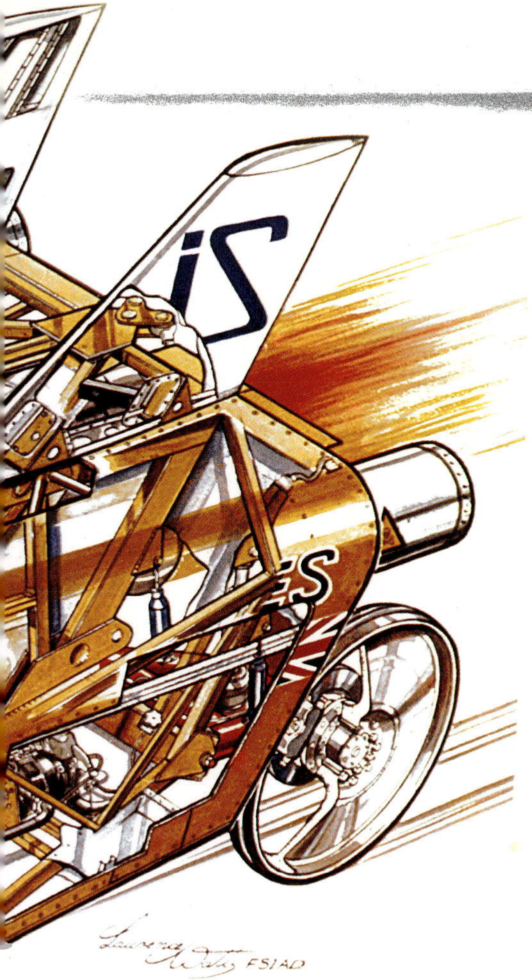

Foreword

I am writing this in my study in Hampshire and on the right hand wall is a Loctite poster of the Lawrie Watts cutaway of my 1983 land speed record car Thrust2. Thrust2 and I went over 600mph 11 times and I often look at Lawrie's cutaway with great affection. Of course it looks absolutely right – but I never realised that to achieve that Lawrie had had to draw each Rolls-Royce Avon turbine blade as a separate unit in order to ensure that the perspective was right and that he could get the correct spacing. In this book he states, 'The repetition nearly drove me round the twist.' Take it from me, this is an understatement – it would have driven any normal artist to fondly contemplate a lifetime of interior decoration!
Lawrie came down to Fishbourne on the Isle of Wight to draw the 35,000hp Thrust2 for Motor Magazine. At that point the car was far from finished and designer John Ackroyd remembers Lawrie sitting quietly against the wall of our small workshop, totally absorbed in the huge job before him. As he says in the book, 'Soon as I get started, I seem to go into some kind of trance.' The Thrust2 cutaway was a great help to us – explaining the car to would-be sponsors and engineers alike. It also gave us a good idea of how the final product would look!
Although Lawrie doesn't know this, he and I go back many many years. When I was a kid, I was locked away in one of those boarding schools with an unhealthy routine based on sport and Latin. Relief came every Thursday with my copy of The Eagle kid's comic. It would usually arrive rolled up and the outside would be damaged – but that didn't matter because Dan Dare and the Mekon took second place to the centre spread which was often one of Lawrie's magic drawings. This was truly wonderful material – aircraft, cars, ships at a time of tremendous post war development. Not only did the pictures live in the sense that the artist truly understood the activity and context, but they gave you a clear understanding of the product's purpose and how it worked.
These cutaway drawings are simply works of art. I have always gone to Lawrie because the alternative seems to be cold CAD type drawings, which fail to inspire. It's only by reading this book that simple people like me get to understand why: CAD drawings leave no room for the artist's interpretation and statement. CAD is a like a balance sheet – Lawrie's work is like a book. In the chapter on techniques you will find your first lessons on perspective and interpretation. And you'll find those special touches - the half open rear window on the Montego and the reflection in the ERF cab mirror.
But there is much much more – this book is a truly wonderful pictorial history of the development of aircraft, cars, motorcycles and even farm machinery over the last century. Not only do you get beautiful illustrations – but also a clear understanding of how the subject worked. Always wanted to know about the Wilson gearbox? - see page 101!
The book depicts the tremendous age of engineering creativity right from Percy Pilcher, whom I like to think flew before the Wright brothers, and much of it seems to have been centred around Coventry, that powerhouse of manufacturing where you can still find the skills and engineering talent to make almost anything. Name key local manufacturers like Daimler, Armstrong Whitworth, Jaguar, Land Rover and Lawrie either worked for them or with them! Make a special point of looking up the Jaguar XJ-13.
Take the trouble to read this book. In it, David Dixon, journalist and long time friend of Lawrie, presents the lifetime talent and work of a very exceptional designer and artist. I promise you will learn something on every page.

Richard Noble
Greywell
May 2006.

Bill Bennett Technical Artist

'Lawrie was my 'opposition', he worked for publications which were in direct competition with those for which I worked. Deadly rivals in fact! However, during most of our 40+ working lives we grew to know one another and developed a healthy respect for each other's work and in so doing also became friends. Lawrie's work was 'different'; he evolved his own distinctive style which could be recognised without looking at the signature, the meticulous attention to detail, the way he drew onlookers into the picture, the way his drawing told a story even though his brief had been to merely 'draw the thing'. Some artists would have done just that, not Lawrie! It gave me pleasure to study his interpretation of a subject which I'd also drawn. Sometimes this set me thinking 'Why did he do it that way?'. Most times the penny would drop and I'd appreciate what he'd wanted you to see.

Lawrie is not just an artist, he's a designer too. A fine example is his Royal Enfield-powered Dreamliner. It looked just right, most others went over the top, and was way ahead of its time.

Did he influence my drawing? Being quite a bit older, I probably started earlier, so when he came along I felt that I was well established. Nevertheless, analysing his work made me evaluate afresh what I did so, in that respect, he made me look to my laurels. Artists enjoyed greater camaraderie than writers ever did, we looked after one another, we had a greater appreciation of one another's skills and, for that reason, supported one another more than journalists did. That's still true today. I never shared a job with Lawrie though I did work with most of the other 'opposition' artists. Although it's years since we retired from full time work, Lawrie and I still exchange Christmas cards. How I enjoy his! This is typical Lawrie. Every Christmas he draws a new one, usually a scene around Olde Coventry, frequently with a car or motorcycle of that period on snow-covered roads. I look forward to those cards, and then examine them minutely because each one is a little masterpiece, with his trademark details. Even in retirement, he can't resist the temptation to tell a story through his drawings. To my mind that's the hallmark of a great artist.

I'm proud to have known him, as a man, as a friend, but most of all as a great artist. I'm sure that this book will give great enjoyment to many people'

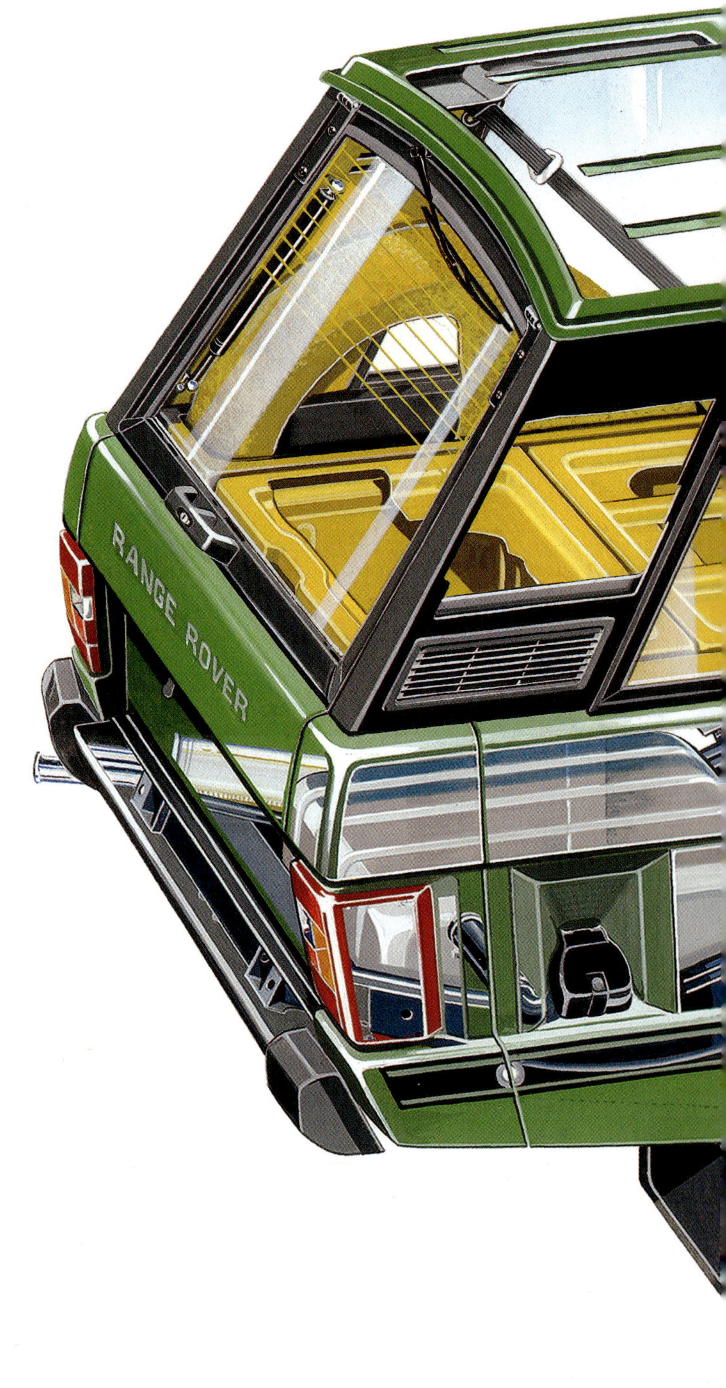

Alan Baker Technical Editor of 'The MotorCycle' from 1953 to 1958.

'Lawrie has always shown a wonderful mechanical understanding with a good idea of how to show the subject far better than any other artist I've worked with. At the annual TT races in the Isle of Man, when I wrote the 'Technical Aspects of the TT' from 1953 to 1958, numerous drawings were required and I found that Lawrie and I spoke the same language. Of course it was called 'work' covering the TT, but to us motorcycle enthusiasts it was actually a hard-working holiday. Between all the rushing around, we managed to squeeze in some leisure time and Lawrie forced himself to come along and enjoy himself too. I soon learned that Lawrie could turn his knowledge to anything. He had a happy knack of reducing the most complex piece of machinery into a simplified drawing, as the great Phil Irving once said, 'simple and direct'. This characterised his design work too, as exemplified in his Dreamliner motorcycle. As with most successful designs the concept was simple, the execution was thorough and earned for Lawrie the backing of his employer Associated Iliffe Press and the wholehearted co-operation of the Royal Enfield company. The only problem was that the concept was so far ahead of its time that the paying public failed to back it in sufficient numbers. Lawrie was also responsible for me sampling gliding, which enabled me to make comparisons with powered flight, another of his many interests.'

The Range Rover

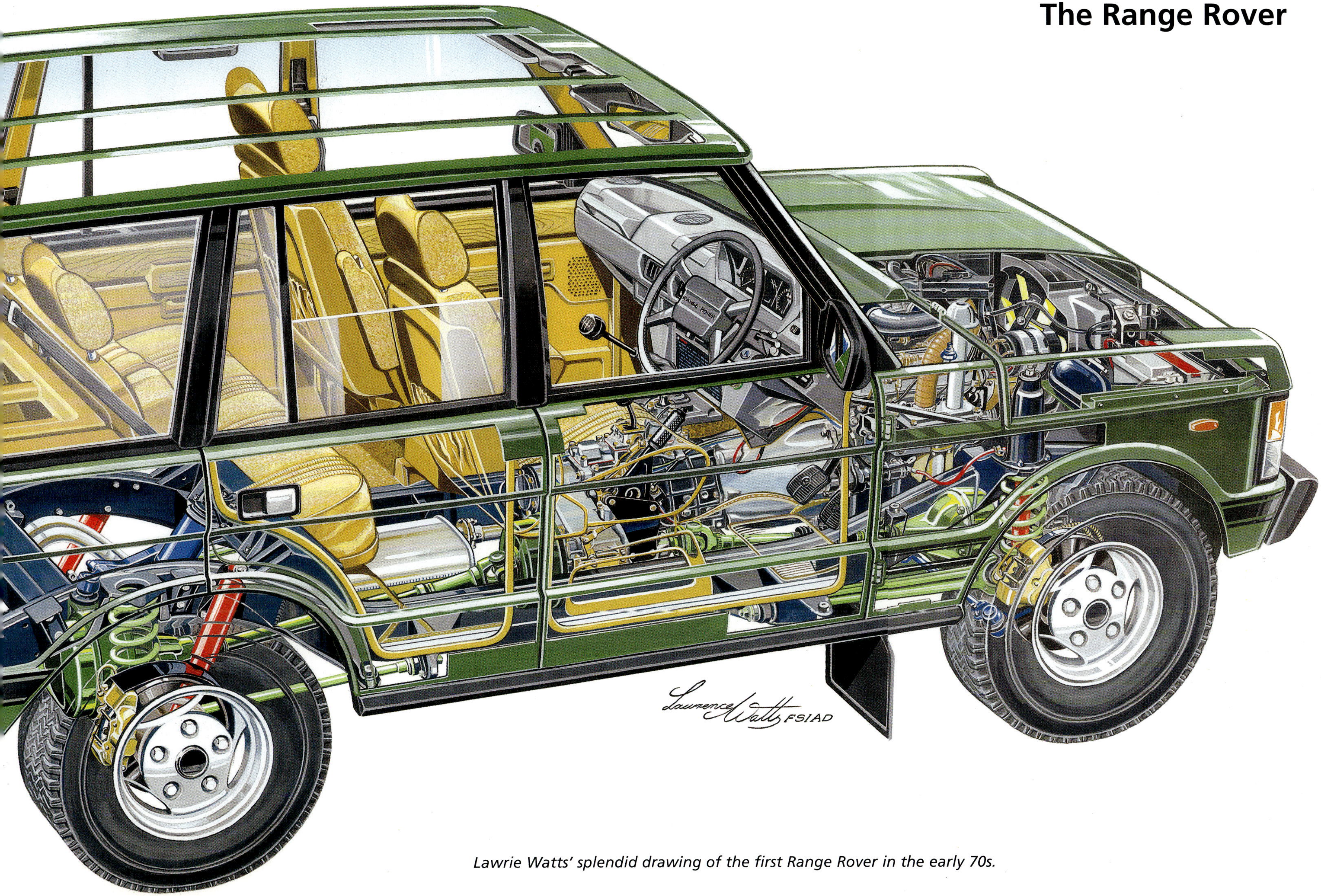

Lawrie Watts' splendid drawing of the first Range Rover in the early 70s.

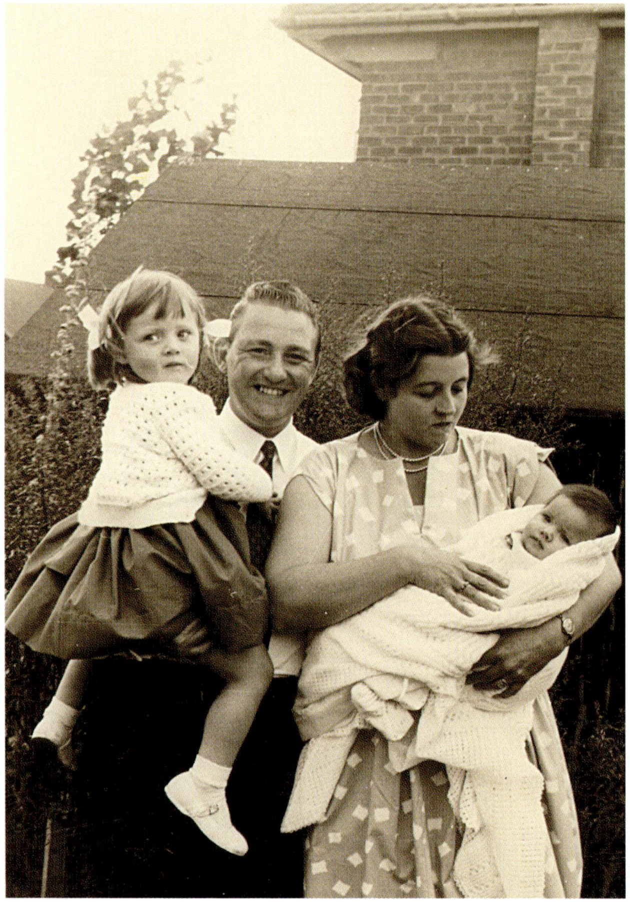

Happy families! Loll holding elder daughter Julie; Eileen with Rosemary.

Right: Little Loll in Auntie's arms on a family outing. Far right: Dad (Arthur Watts) in his Standard 9 chauffeuring young Master Watts.

Below: Great Uncle John with his vendor cart in Nottingham. John was the first to sell electric light to the public.

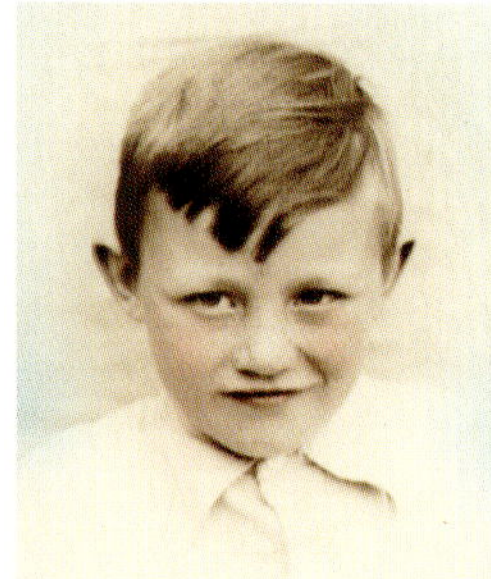

Left: Will I get away with this? Plotting, aged 6.

Below: Pram testing! Pass = bag of sweets. Fail = walk home.

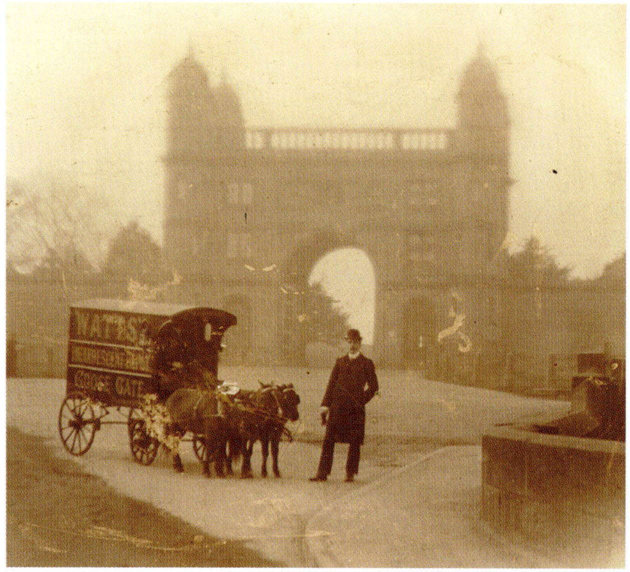

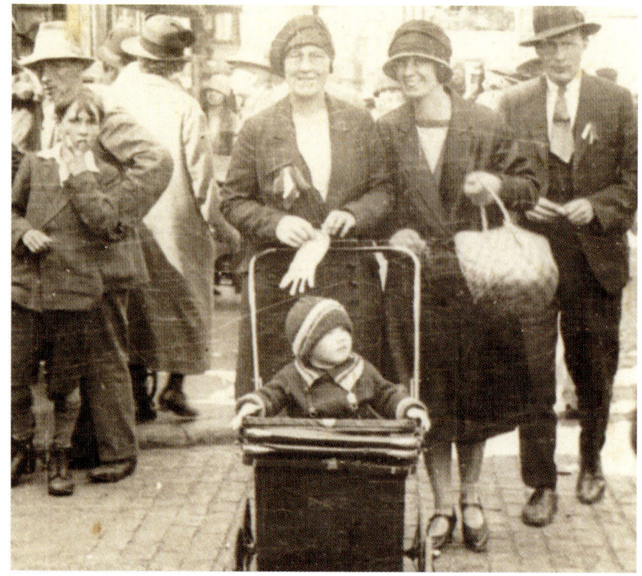

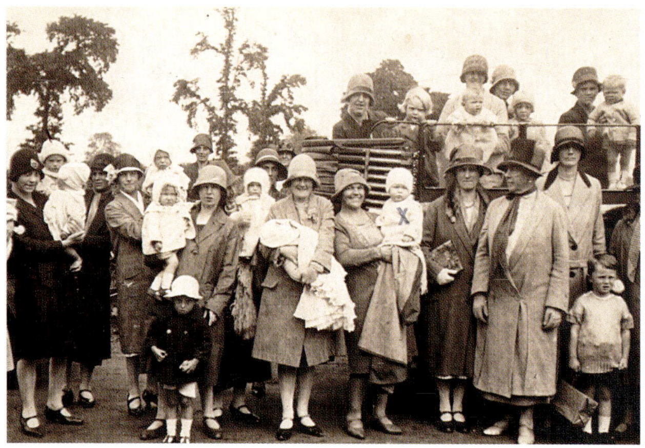

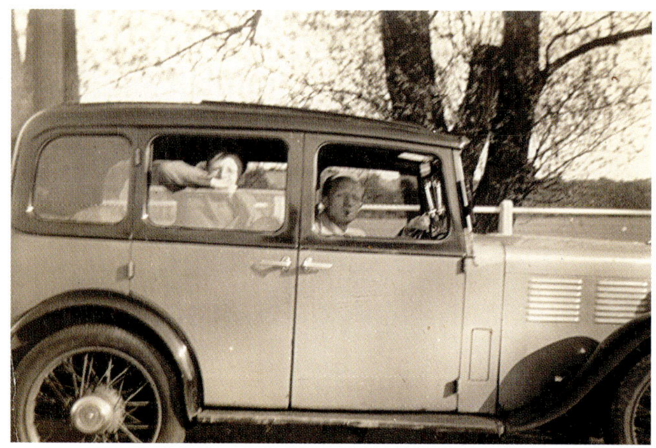

Introduction

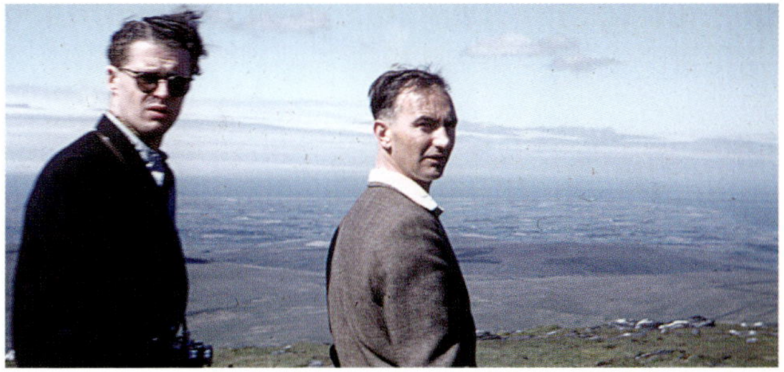

Alan Baker and the late Bill Banks hard at work in the Isle of Man.

The Challenge

'Nobody's ever written about me, not a dickey bird,' remarked the chap across the table. 'Why is that?' I asked. 'Probably because I've had my light pushed under a bushel over the years,' he responded. After a minute or two's reflection, he continued, 'I think I've got a story to tell... could you write it?' My instinctive answer was, 'Of course!' He looked suitably pleased. Reaching into my pocket, I extracted a postcard, 'Would it fit on this?' Slowly, he got to his feet, leant across the table - his five feet four inches at full stretch - and smacked me smartly on top of my head. All around, conversation ceased, forks wavering between mouth and plate, as startled eyes glanced in our direction. Carefully adjusting his bow tie, he resumed eating his dinner... The message went home. Ten years later, I got around to writing his story, on a rather larger card, for you to enjoy.

David Dixon.

My head-patter was friend and ex-colleague Lawrie Watts, editorial artist of 'The Motor Cycle', and the occasion was technical editor Vic Willoughby's 80th birthday party. We had worked together more than 30 years previously for the magazine, popularly known as the 'Blue 'Un'; its blue masthead distinguished it from rival, 'Motor Cycling', which had a green masthead. Birthday boy Vic arrived already the worse for wear, with a piratical eye patch over one eye, numerous facial scars and bruises. Apparently, he'd been feeling frisky after a recent triple by-pass operation and had run for a bus, failed to see a protruding pavement slab, and went base over apex along Enfield's High Street. Which is why he'd begged a lift with ex-sports editor Mick Woollett, now an honourary 'Blue Un' member because he transferred from the defunct 'Motor Cycling' in 1967 and was generous in his time caring for Vic, who lived nearby.

Wine and conversation resumed flowing among Vic's old colleagues around a table in the Manor Inn Hotel at Newlands Corner, near Guildford. Founded in 1903, 'The Motor Cycle' was the premier motor cycle publication of its day, the jewel in the crown of owner Associated Iliffe Press. Technical Editor until his retirement in 1978, Vic had succeeded Alan Baker in 1959. Alan was also at the table, along with Bill Banks, former chief of the photographic department, and a superb action photographer in his day. Other staff who could make the venue, such as road tester and off road specialist, Peter Fraser, and features writer and tester, Michael Evans, and myself were the 'junior' members of a select coterie. And then there was the man of many talents, Lawrie - technical artist, pilot, designer, aerodynamicist, inventor, known mainly to those who recognised his signature on countless drawings in 'The Motor Cycle', 'Autocar', 'The Motor', 'Flight', and 'Mechanical Farming'. As a freelance in later life, he had been commissioned by the Castrol oil company to do a sectioned drawing of Richard Noble's world land speed record cars, and by Richard Branson for sectioned drawings of his trans-Atlantic and trans-Pacific hot air balloons. So, yes, Lawrie did have a story to tell as one of this country's leading technical illustrators. Like many another back room boy, his light had burned beneath a bushel for too long.

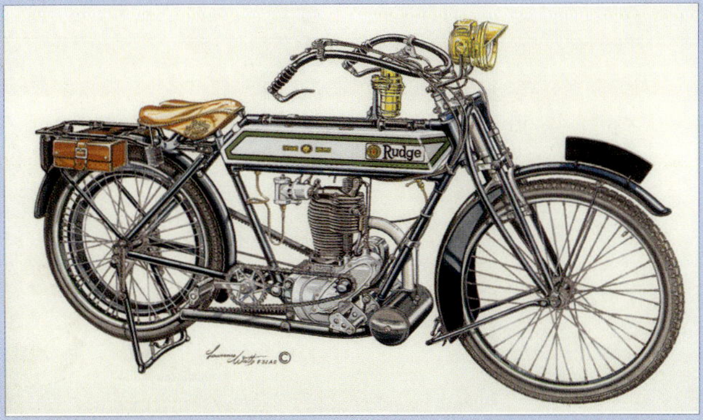

Lawrie painted this 1912 Rudge in watercolours, without any preparatory pencil work. It was not commissioned. He did it for his own satisfaction. Bob Currie, then editor of 'The Classic Motor Cycle', subsequently used the painting in his magazine.

Mini Watts

Mum, Ada, Foo Foo the dog and Dad, Arthur.

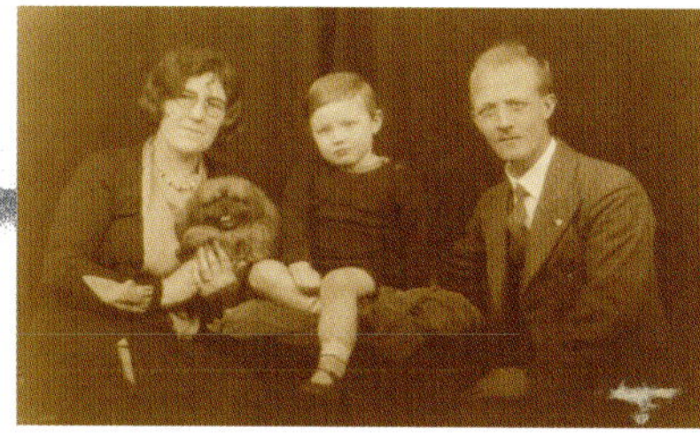

The early years

Born 20th August, 1927, Lawrie says he weighed 10lbs, '....which was a lot of money then. My mother must have thought that that was enough, because I remained an only child. My father and grandad were top flight engineers, grandad received many awards for his inventions. That he was versatile showed in the nature of the inventions - machine tool improvements, gas mantle design, electric light roof fittings and, eventually, toilet paper manufacture. The family claimed that grandad started at the top, then worked his way to the bottom. One of the awards was a large lathe presented by the government during the first world war for grandad's design work on the expanding breech end of naval guns. I spent many happy hours treadling it for him, while watching with great interest what he was doing. He died when I was 17. But I suffered a double blow. Within a week my beloved grandma also died. A devoted and loving couple, they clearly could not live without one another. Grandad was also a member of the Salvation Army and when he died I remember the band came to play outside his house. Sadly, a week later they were back again when Grandma died. It was not all bad news because grandad's lathe came to me and I soon converted it to electric power, adding it to my collection, which comprised a small lathe, milling and grinding machine, cutting machines, a pedestal drill and other bits and bobs of machinery.

'Wheels always fascinated me, aerodynamics and flying have played a great part in my life, which is why motorcycle, cars and aeroplanes are so close to my heart. I don't think I was much good at anything apart from drawing and painting at Radford village school. However, the head mistress did decorate her study with my work, saying to my mum one day that 'young Lawrence will be a great artist'. After I moved up to the juniors, life changed. I came under the charge of Mr Walker, a master who was an historian and artist. His class room was a mosaic of colour, which helped my attitude to learning, and his encouragement produced the best out of me. I was very grateful for this and was planning to thank him when the school was destroyed during the blitz on Coventry. When Mr Walker saw the extent of the damage, and realised the fate of his life's work, he had a heart attack and died on the spot.'

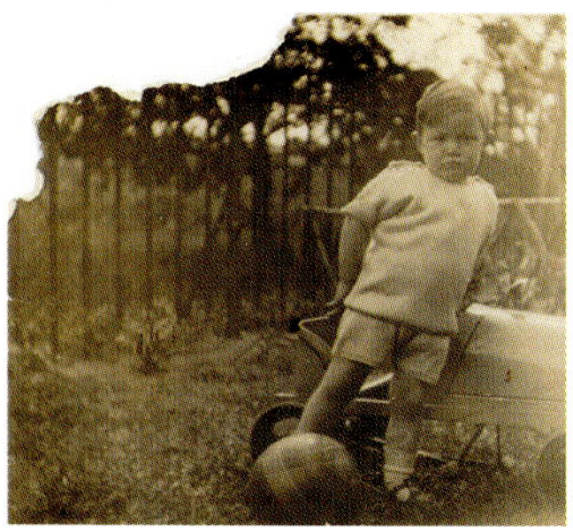

With my first car - and football, in case I failed my test.

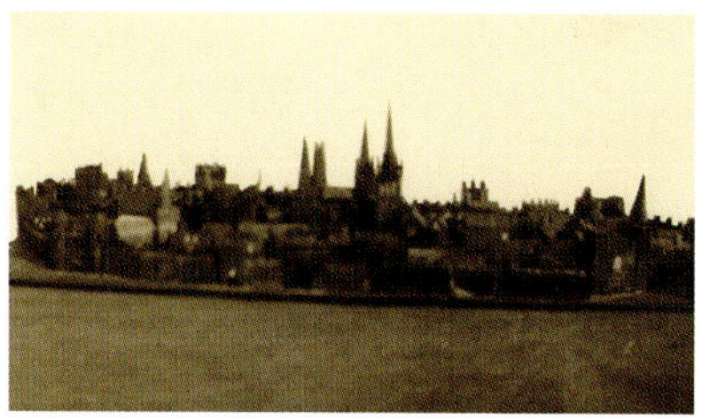

Model of Coventry as it was in Medieval times, made at Radford school, under the charge of Mr Walker.

What have I done now?.

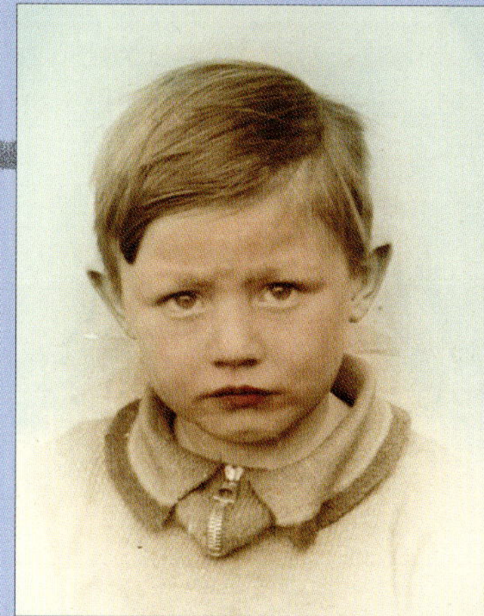

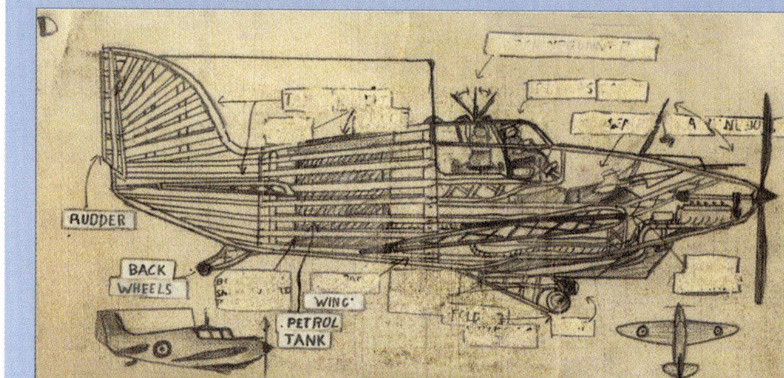

Next time you see a little boy drawing aircraft, or burying one in Plasticine, or its equivalent, and then cutting it open to expose the hidden object, pay attention. It may be another Lawrie Watts. As a small boy, all he ever wanted to do was draw cutaway aeroplanes, such as those produced by chief artist Max Miller, which were already appearing in 'Flight' magazine. So the nine year old Loll drew a very realistic single seat aircraft with annotated directions. That he aspired to the more difficult task of drawing the insides of the aircraft, or anything else which took his fancy, came as a natural progression from his early experiments with Plasticine-covered objects

Close shave! Shrapnel from a bomb nicked Loll's ear.

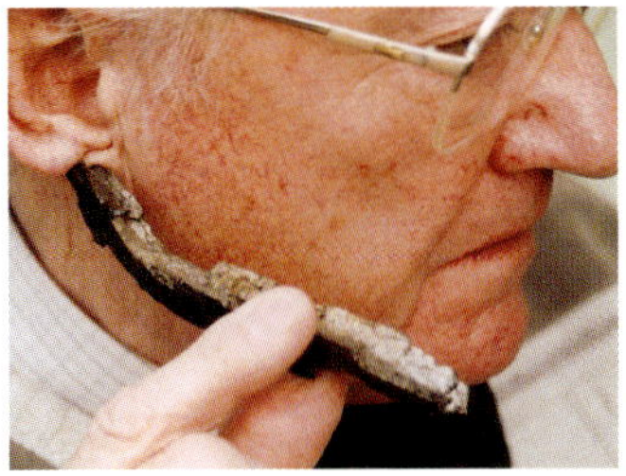

Roy and Jim

'At senior school, my two close friends were Roy and Jim. An epileptic, Jim would always come to me to hold him when he had a fit in class; I was glad to be able to comfort him and we remained good friends until one day during the bombing, my father and I went looking for Jim, whose home was close to the Daimler factory, the target of the German bombers and where my dad was in charge of the tool room. Although Jim's house was intact, there was no sign of him, he must have still been in the Anderson air raid shelter in the garden. So Dad and I hurried home, carrying his heavy tool box which he had retrieved from the wreckage of the factory. Just as we arrived home a second wave of aircraft dropped their deadly cargo onto the burning works. Just as we got home, a bomb fell close by and a piece of shrapnel clipped my ear before smashing into the pantry wall, where we had dived for shelter. Mum patched me up. In hindsight, the pantry was not the best place to have taken cover because above it was a 600 gallon water tank. Had that been hit, it might have made us a little wet! Later, I removed the shrapnel from the wall, in fact I still have it today.

'In the morning, I went to check on my school friends. Though his house had been badly damaged by incendiary bombs, Roy was unharmed. So we set off to find Jim. His house had been severely damaged and where the Anderson shelter had stood there was now a gaping hole in the ground. Then Roy shouted, 'I've found his boot!' We picked it up. Inside was a foot. Stunned, we reverently laid it back on the ground. In a daze, we walked slowly away, unable to speak. We went to our homes, passing other horrors on the way. One grew up quickly in those days.'

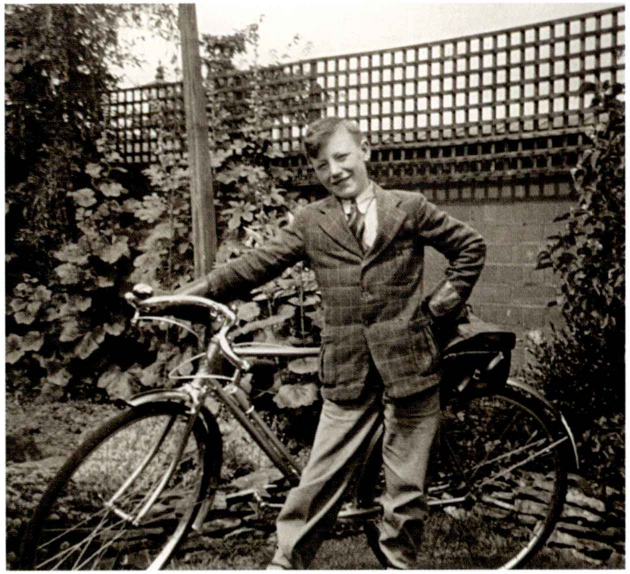

Loll with his celebrated Coventry Eagle Silver Sports bike. It (almost) survived an encounter with a bus.

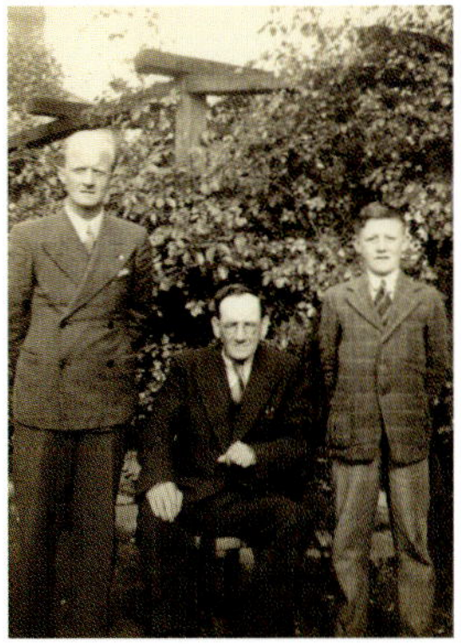

Three generations of Watts (left); Arthur (Lawrie's dad) Grandad and Lawrie

To work at 14

Soon it was time for Loll to leave school and start work. So, at the tender age of 14, he found a job in the body shop at Armstrong Siddeley cars making wooden parts for the Whitley bomber. The factory was at Parkside, two miles from the home he has lived in for the last 40 years or so. Commuting on a bicycle from his parents' home at Radford, the other side of the city, was no hardship; he'd been riding his bicycle to and from school, including going home for lunch, doing 28 miles a day on his favourite (and only) transport, a lovely, and exclusive, Coventry Eagle Silver Sports. But the harsh winters, which were a feature of the early 'forties, did cause him to think of powered wheels, if only he could afford them. Hardly surprising, then, that he fell in love with motor bikes. Few people could afford cars in those days, even fewer could run them because of wartime restrictions on fuel. So motor bikes, with or without sidecars, were the preferred means of transport. Young Loll needed no stronger incentive to earn money and buy his own set of powered wheels. Power and a 125cc two stroke James were not necessarily synonymous, but it took the sweat out of pedalling and got him to and from work. More importantly, it was reliable and economical. To spread his meagre petrol ration, Loll used his half a gallon (mixed with lubricating oil) and rode until the fuel ran out. Then he walked! Obviously, he chose his routes carefully so that the bike climbed the uphill bits, leaving the downhill return to foot power. 'Ridiculous, really, but it was my solution to wartime petrol rationing,' he explained years later.

Riding in the dark

After a hard 12 hour day in the factory, Loll opted for evening classes at Coventry Technical College to compensate for leaving school at 14. But Herr Hitler made this a somewhat hazardous choice because night time was when he blitzed Coventry, the UK's automotive and aircraft industries centre. As Loll recalled; 'My parents could have been worried silly by me being blown sky high on my 9pm ride from evening class, but they never said anything other than to offer love and encouragement - as they always had. A totally different problem was self-preservation in the black out. Street lighting was forbidden, not even a battery lamp on my bike, so developing the art of seeing in the dark, combined with my ability to make phenomenal avoidances, became life-saving prerequisites. Really, it became a matter of groping my way home, hoping to arrive in one piece and in time to dive into the pantry before the heavy bombing started. And this happened most nights. If we were not being bombed, then they bombed Birmingham, just 20 miles down the road. But you never knew who was going to get it; we were on the Luftwaffe flight path to Birmingham, so even if we were not that night's target, the buggers still flew over us, causing us to dive like meerkats into our pantry. I have to say, it was not a pleasant experience even though, after a time, you became used to it.'

Eileen has supported all of her husband's aerial activities. Here she poses by the tutor gilder.

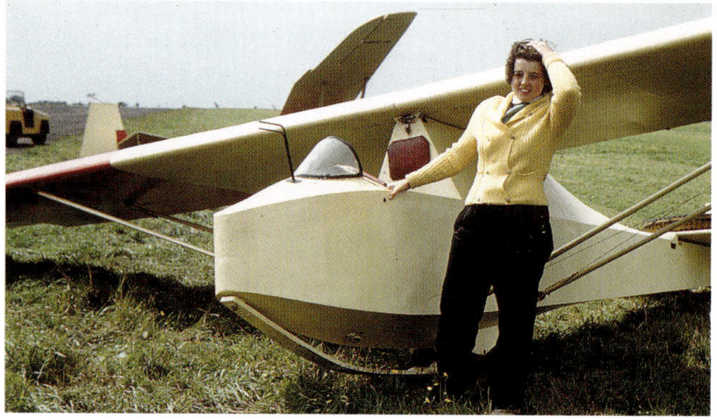

Eileen

'It was not long before I first saw the girl I was to fall in love with,' remarked Loll. 'She was just 14 when we met. We had to wait until she was 21 before we could get married. Eileen has stood by me in everything I have undertaken. I love her to bits but I have never felt good enough for her. Both of our daughters, Julie and Rosemary, have become successful artists in their own field of art; Rosemary as a wildlife artist, Julie has a degree in three-dimensional furniture design, she's also a talented watercolourist. My hobby was making model aeroplanes. One day when working at Armstrong Siddeley I had a brain wave; model making material was not easily available but in the saw mill I had noticed scrap pieces of 1mm thick plywood. Just the job for me! So I devised what I thought was a clever way of getting it out of the works past the gate men without being noticed. Having found a suitable piece, I sandwiched it between two sheets of newspaper and, when it was time to leave with all the other lads, they would all be reading their newspapers as they walked through the gates. So my little wheeze would not appear to be unusual. Except for one detail, the wind. I was the only one not battling to control a flailing newspaper, mine was as stiff as a board. A plywood board. How I got away with it I'll never know, but I did. So off to night school nursing the stiff newspaper which I left with my bike, expecting it to be there when I collected my bike after class. The bike was, but some light fingered 'friend' had pinched the board! Cycling home, I was concentrating on keeping my narrow-section sports tyres out of the tramlines when I had to pass between a parked bus and a tram line. Suddenly, the bus driver threw open his door and leapt from the cab. The door hit me hard on the shoulder, causing me to somersault onto the cobbled road in front of the bus. I landed just in time to see a car run over my front wheel! Never mind that it could have been me, I was distraught that my beautiful bike had been badly abused through no fault of mine, or the bike's. But it caused me to think deeply, why should this have happened? My conscience led to conclude that the accident had something to do with me nicking a piece of wood. So, first thing next morning, I told my boss what I'd done. When he asked why I wanted the plywood, I told him the truth. To my amazement, he suggested that I could bring my templates and cut them out on the band saw during lunch hour, and put the pieces in my empty lunch box. So it does pay to be honest, even though it took a bus to wake me up! That hard earned lesson has stayed with me ever since.'

Techniques

L.Watts, Editorial Artist

'On a recommendation from Max Miller, then chief editorial artist and one of my earliest heroes, (he had seen some of my work during a visit to Armstrong Whitworth when I was head of my section), the Managing Director of Associated Iliffe Press, Geoffrey Smith, came to interview me in Coventry,' recalled Lawrie of his progress up the ladder. Which is how, just 11 days short of his 22nd birthday, he joined the Coventry office of Associated Iliffe Press. 'I remember it well. It was at 9.15am on 9th August 1949. Technical art at that time was usually a one-dimensional view of machinery drawn in one of two ways; expertly, or not. From my seven years in the aviation and car industry after leaving school, I had been drawing components which required a high level of skill in order to show how they worked. So I'd served my apprenticeship well, and learnt my craft under very difficult conditions while absorbing an enormous amount of knowledge about the aircraft and automotive industries. My draughtsmanship had been honed to a high level and this is what attracted the Iliffe offer. Little did they know that this had been my aim all along! I was now with the biggest publisher of technical periodicals in the UK, possibly the world. I was in heaven!'

10 Lines led to CAD

Without yesterday there could be no today, so no CAD (Computer Aided Design).

Without realising it, the technique which ultimately became CAD was evolved over a long period during the late stages of the war. Daimler's 3-D shapes, which became the forerunner of today's Computer Aided Design (CAD), were evolved from the '10 lines technique'. In this, 30ft drawing boards, which were the full length of the design department, made from sheets of aluminium alloy, were sprayed with a special duck egg blue paint. Ten lines were drawn vertically in ink every ten inches. Superimposed over these were ten horizontal lines, also at ten inch intervals, to provide a grid of 10 inch squares, similar to a giant graph. Onto this the new car and its various key components were drawn, the squares allowing each component to be located accurately to its position in the overall full scale design.

Loll tells how the design process evolved. 'The first decision was where the front wheel would be, we called it the axle centre zero. From zero (the axle centre), we measured down 10 inches to establish the ground level. We then went six squares up, then two down. From the front axle centre, we counted four squares forward, then backwards 20. That established the elevation. We did the same from the centre line of the car at one end, to the extreme outside, which would be four squares (ie: 40 inches). The other end of the drawing would be the rear view of the car, calculated the same way. The centre line was the datum to the elevation. Then the car would be drawn, from little sketches already made, working up to full size. We used pencil, because you could rub out without affecting the ink lines. You were now able to locate any point in space because you could give a co-ordinate reading from the elevation.'

As to the modern development of CAD, Lawrie's views are typically forthright. 'CAD is best at producing a commercial end product intended for production and replication. The shortcoming for my type of drawing is that CAD cannot look around corners, can't emphasise because its lines are constant thickness, nor can it create the illusion of perspective. It certainly can't identify materials or textures or show welded seams. CAD can't paint nor can it vary weight of line for shape, nor can it produce different techniques. God willing, I'm optimistic of being in gainful employment for the foreseeable future!'

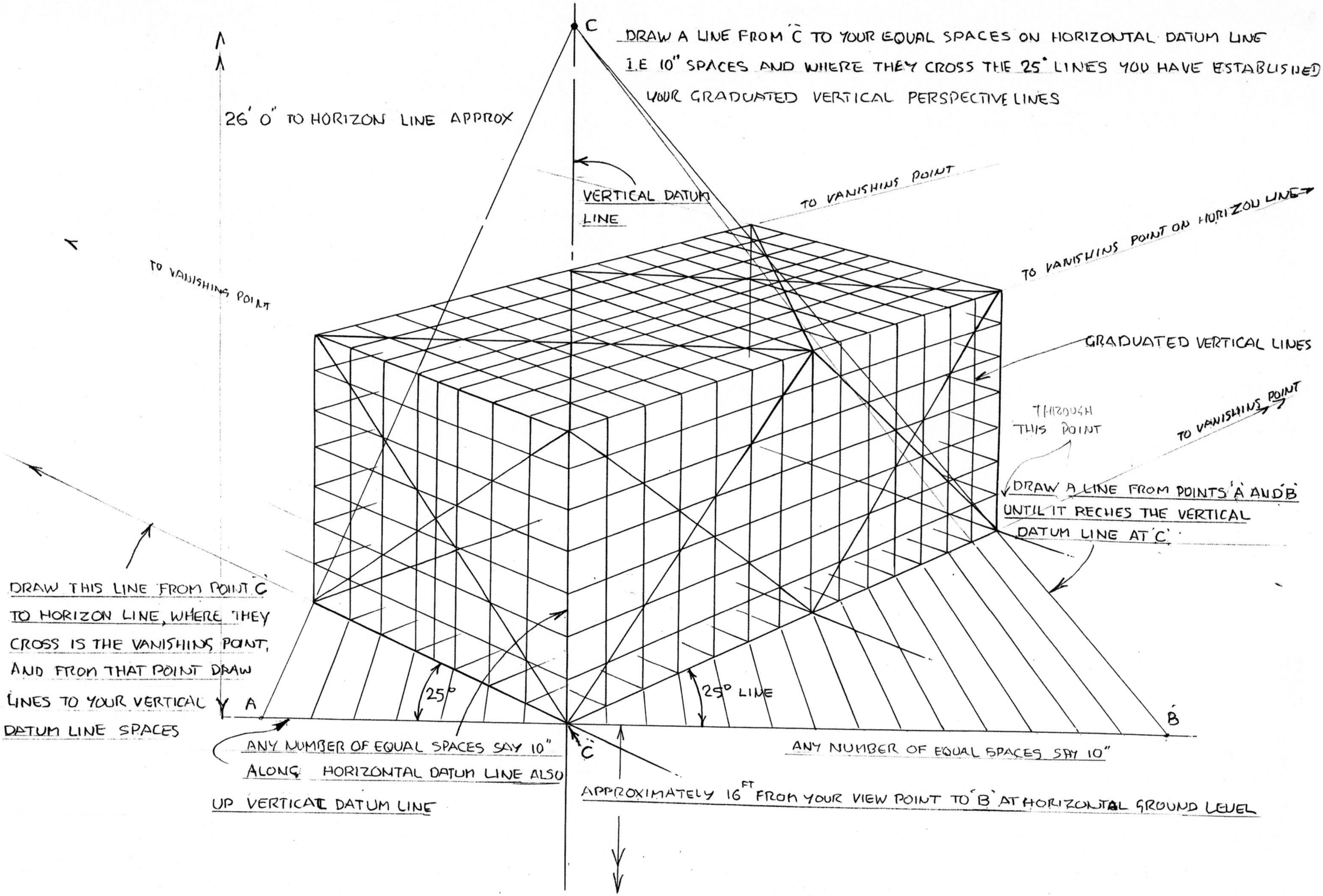

Drawing a motorcar. A 3-dimensional cuboid in perspective.

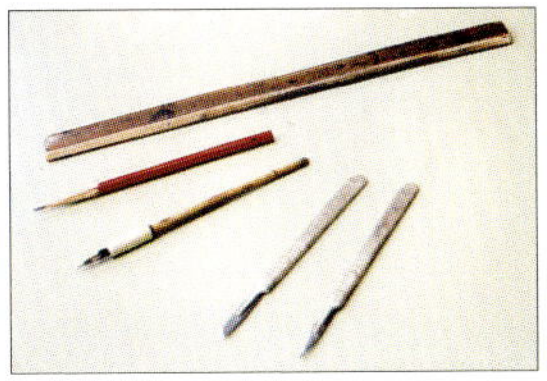

Under the skin

How could an artist show what was under the skin? This is how Lawrie worked:

1. He visited the factory to gain information on vital components.
2. Working sketches were made to aid the main drawing back at base.
3. That left an imprint on his mind exactly how everything worked.
4. At his home studio, he would spend maybe two or three weeks drawing it all together.

If the project was still being built, to ensure that the drawing was completed simultaneously with the press launch, he would keep checking on progress and update his drawing accordingly, such as with Richard Noble's Thrust, 'A huge project, and very challenging,' recalled Lawrie.

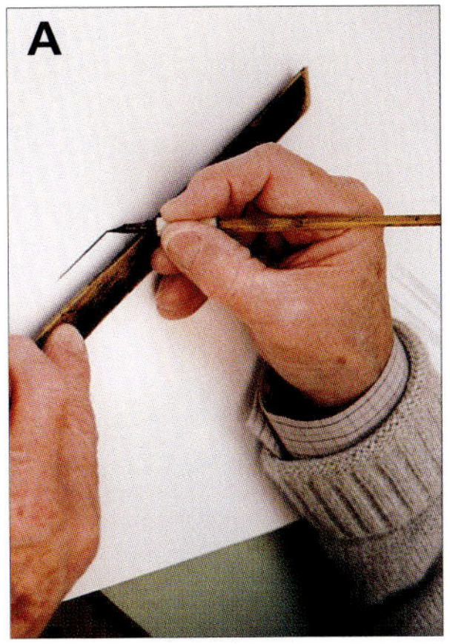

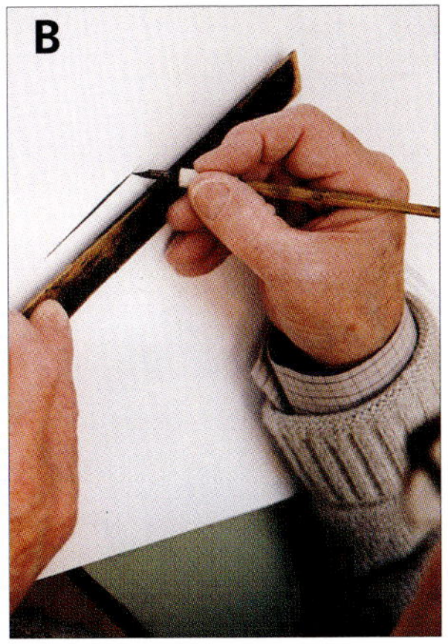

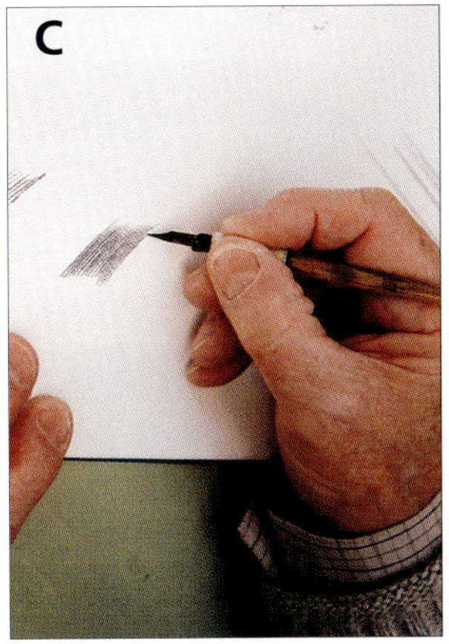

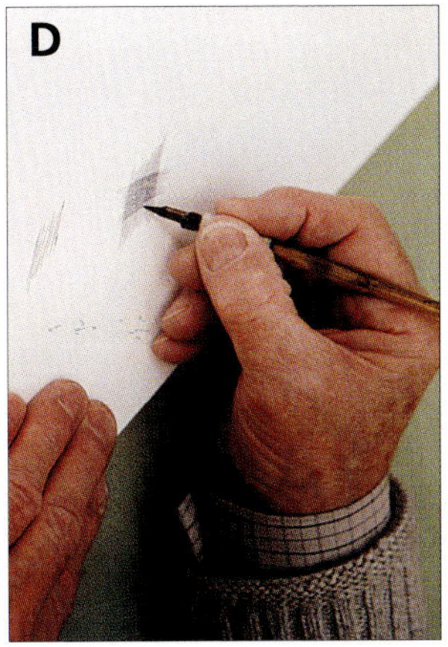

Evolving techniques

Hallmarks of any Watts engine drawing are the subtle vertical, hatched and circular shading lines. The vertical ones indicate the direction of travel (such as with a piston), the circular shading shows revolving parts (a rotary disc) and the hatched lines (with a fine nib) are to provide texture, such as in two stroke engine transfer ports. With the increasing availability of colour reproduction, Lawrie became more ambitious and introduced colour for the first time, in technical drawings. He used a hard pencil, which avoided smudging, instead of a pen, to draw the subject lightly. It was then water coloured on dry paper. Details were added over the dry paper and 'knife worked' for highlights.

He explains why he goes to all this trouble, inserting the tiniest amounts of detail to ensure absolute accuracy. 'Nothing is more boring for me than doing drawings of engines, which is similar to showing lots of spare parts. I want to tell a story which communicates to the viewer and that's why my drawing has evolved much differently to anyone else's.' Lifting up a drawing of a 125cc MZ racing motorcycle engine from the early 1960s, he continues; 'Here's an example of what I was saying. This could be a

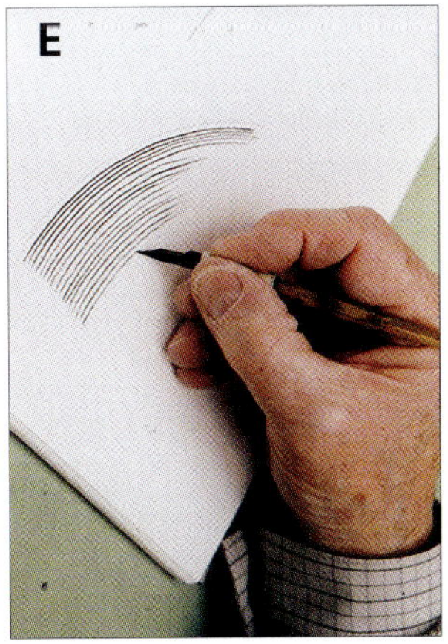

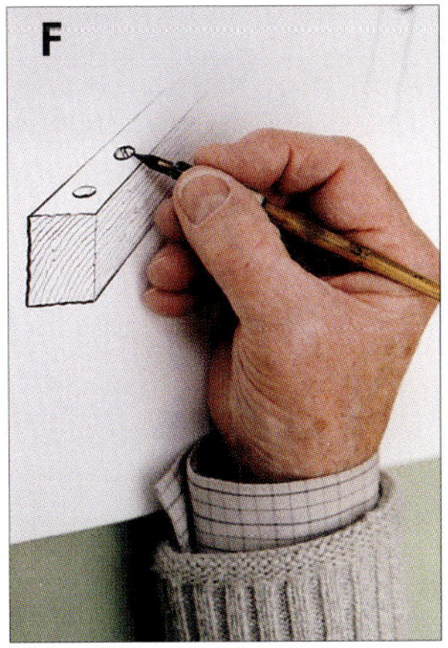

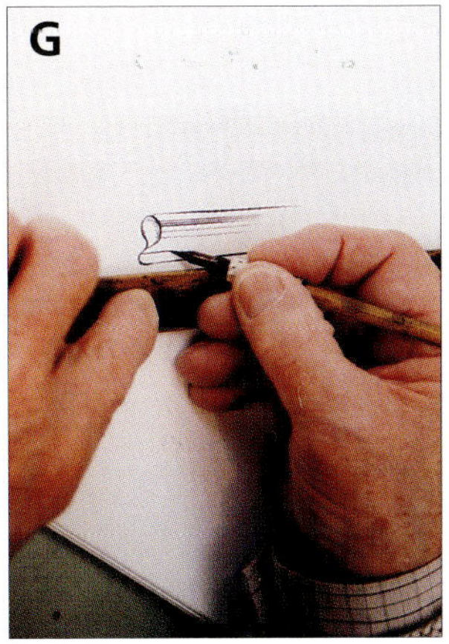

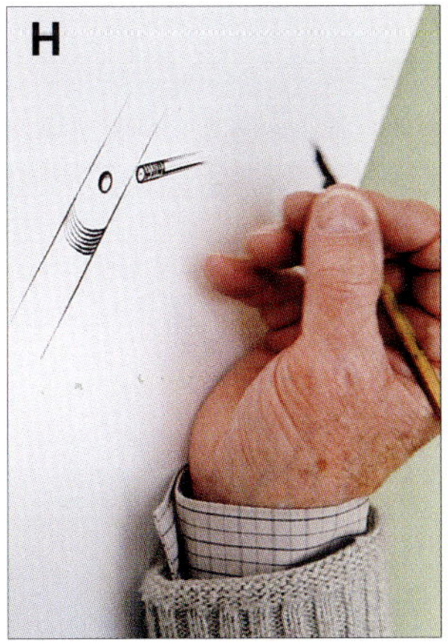

A *Drawing a tapered line with increasing width. This is done by gradually increasing the pressure on the nib.*
B *Then less pressure.*
C *Illustrating rotary action as on a two-stroke disc-valve.*
D *Hatched lines used on porting.*
E *Disc almost complete, shortening the lines gives a shiny impression.*
F *A timber texture also showing perspective. Showing screw heads completes the study in wood, minimum lines, maximum effect.*
G *Shading a metal tube to provide a metallic finish.*
H *Shading helps provide shape, and shows the tube to be hollow.*

Loading the pen before drawing is a precision operation - too much and it will blot.

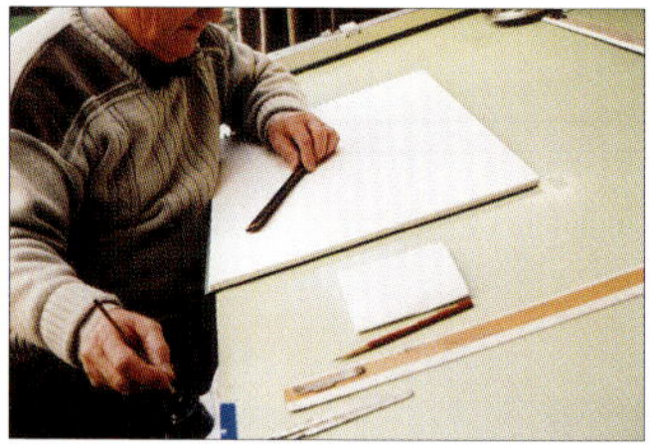

stretched out, one-piece, a very pretty drawing superimposed on another, but it doesn't tell you anything. You already know there's a left hand side and a right hand side to a crankshaft - anyone knows that! - but what I wanted to do was to show what's inside it to the best advantage. Slowly, I developed this technique so that you can see how the crank locates in this end cover, by positioning the cover at an angle to the crankshaft. Then I can build it up slowly, not as a pretty picture, but as a piece of engineering information. The shading indicates what's happening to the component when the engine is running. For instance, the crankshaft has radial lines, the piston has vertical lines, and the disc valve in there is lifted up quite nicely by the radial shading on the inside of the crankcase. The lines on the shading also show the direction of travel - the vertical shading on the piston shows its direction of travel. But if I'd drawn the piston without shading, the cylinder bore without its vertical shading, and the transfer ports without their shading, they would not have stood out as they do now. Instead, they would have been lifeless, just like in a spare parts list. I invented this technique, and got into trouble like I usually did. My boss claimed 'they'll never wear that....'. I defended my new technique by saying that the drawing matched the written article, thereby making it easier for the reader to follow. For instance, one of the most challenging things to draw, transfer ports in a two stroke engine, are here shaded and cross hatched with lines of varying thickness to provide a three-dimensional effect - which I could not have achieved by any other means.'

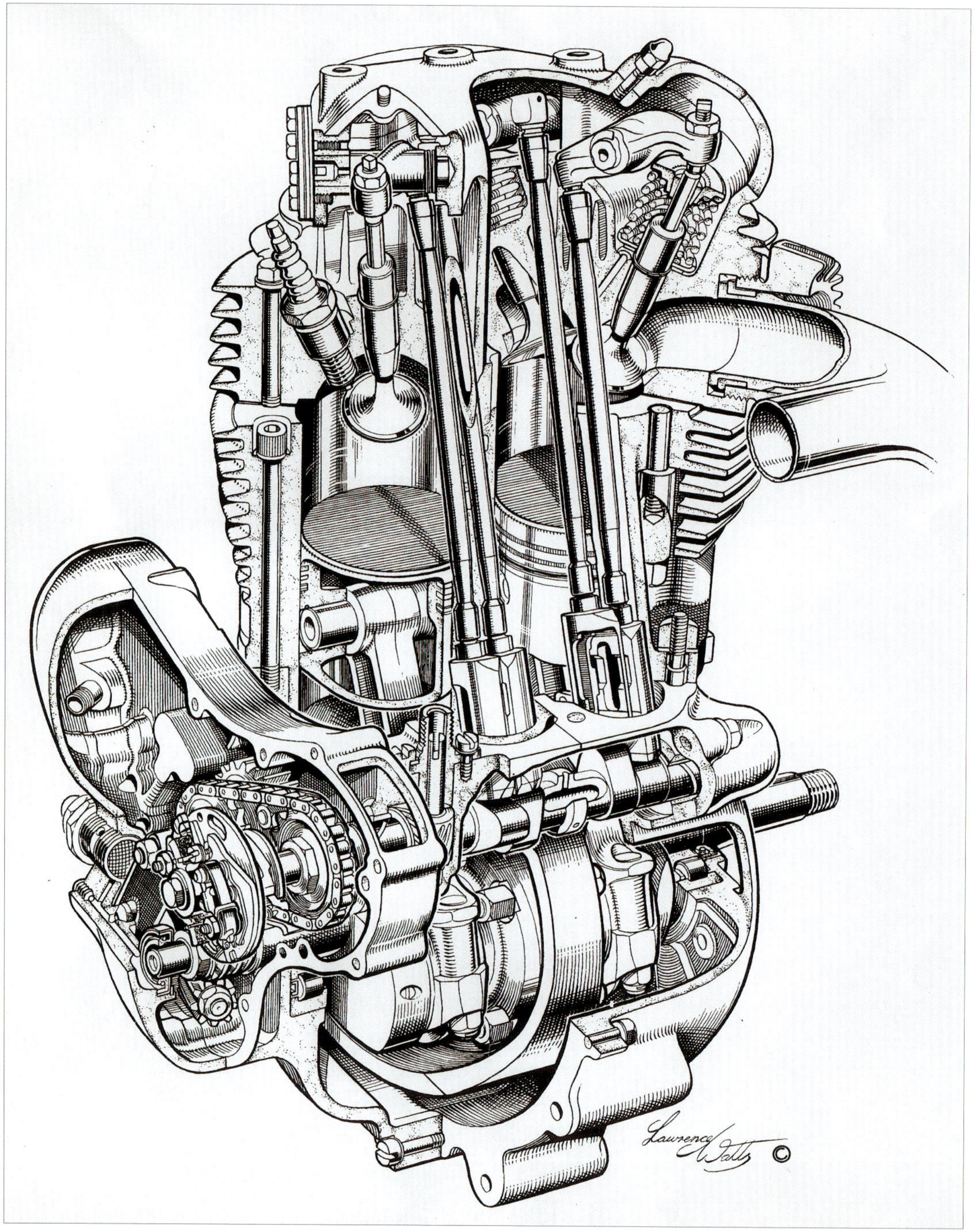

Norton 850cc Commando engine.

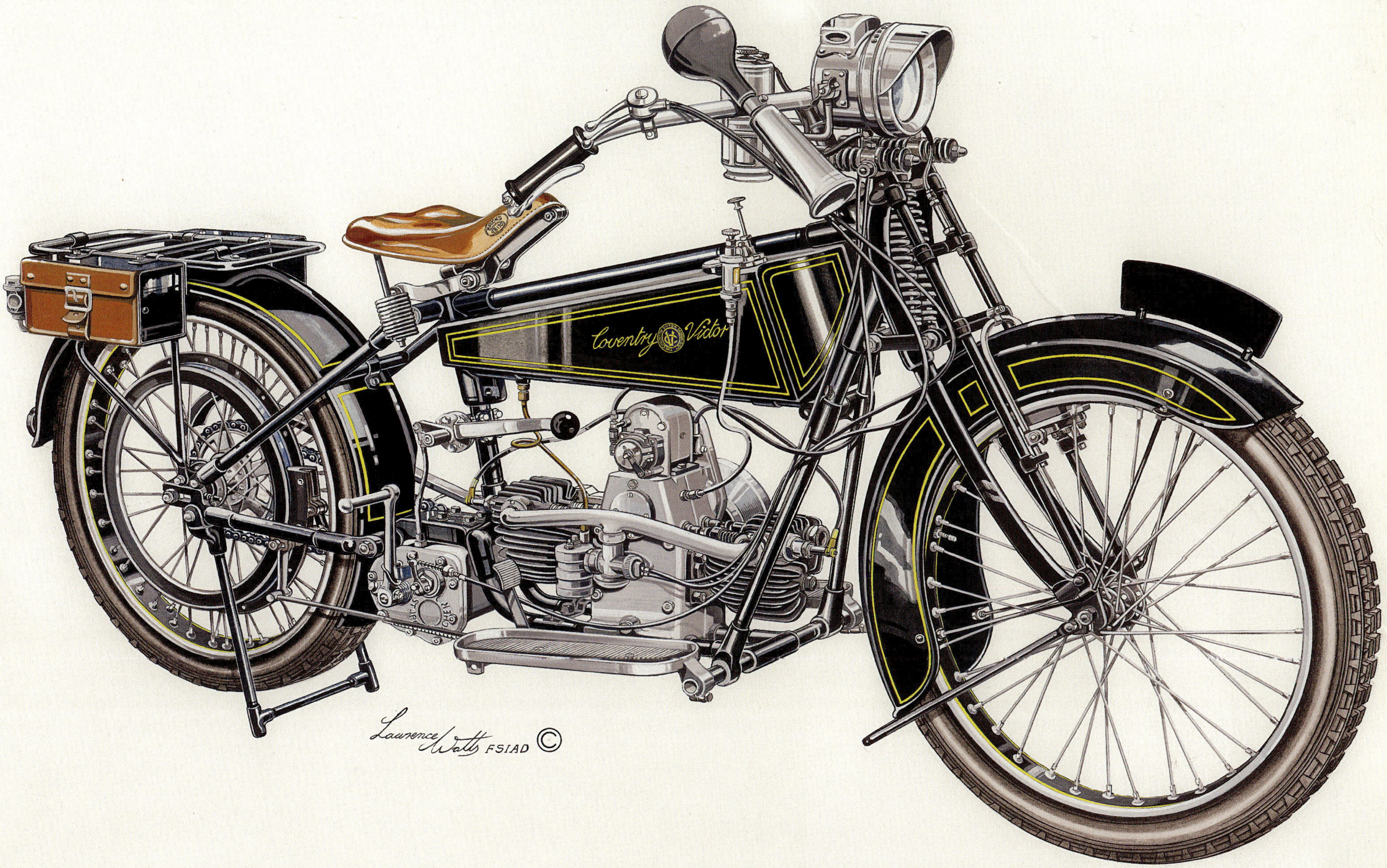

Another watercolour without any pencil work, this time a 1921 Coventry-Victor from the Coventry Transport Museum is the chosen model.

Below: 250cc MZ racing two stroke twin cylinder. Lawrie has split the crankshaft to show its construction. Note also the rotary disc valves controlling the ingoing fuel mixture.

Avoid cocking up

What determines when a drawing is finished? Is it when the artist becomes tired, after a week, or several in some cases, of working on the same subject day after day? Or is it when everything is included? One large brush stroke could kill the job, so you have to discipline yourself. Or is it when certain things are not included, for effect? As Loll explains, 'I don't think I have ever felt like I've reached a point when I thought to myself first thing in the morning - Oh no, not this wretched thing again... I'll just do this, and this, and finish it off... It's not like that at all. What usually happens, is that when I'm within an ace of finishing a drawing, I'm never quite sure whether it is finished or not. This is when I've reached what I call the cocking up point, I get myself into a real tight tizz over it... STOP IT! Don't do any more... But I keep doing a bit more... I'm ever so glad when someone distracts me, then I can leave it alone for a while. When I look at it the next time, I think, can't do anything more to it... if I do, I'll mess it up completely... that's when I leave it. Finito!'

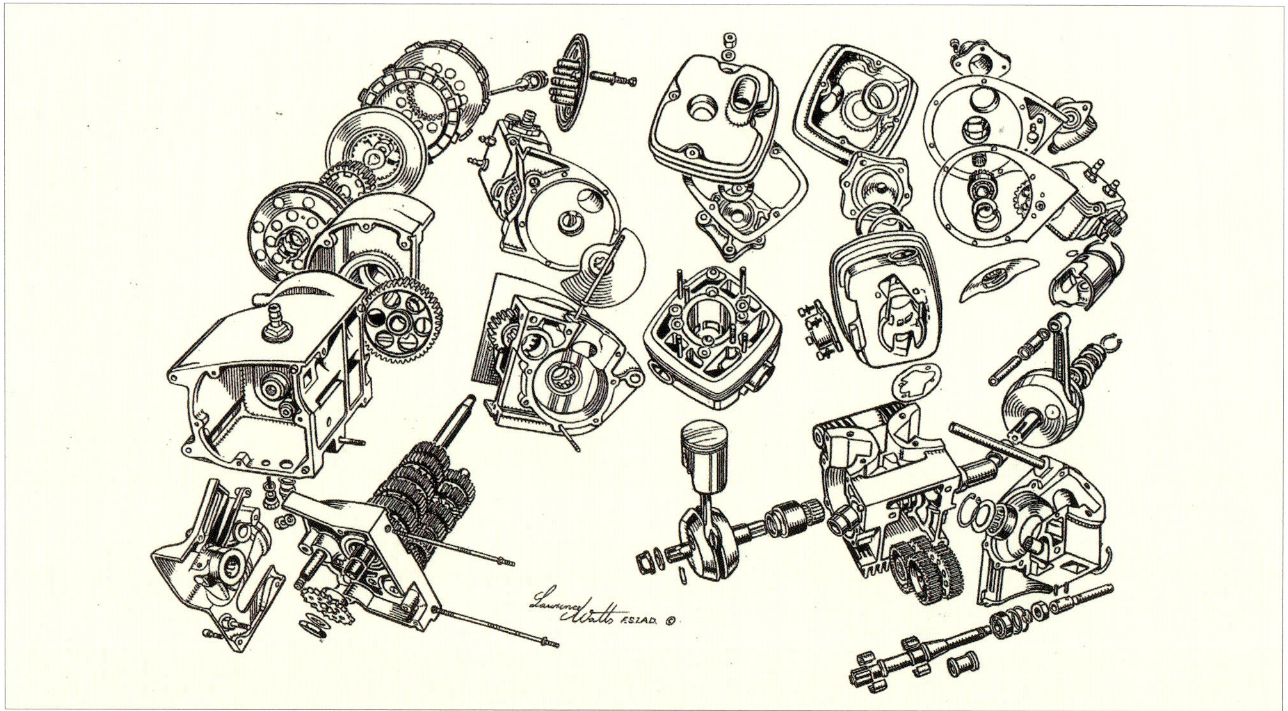

So much detail

In a different context, 'tis said that 'the devil's in the detail...' So it is with Lawrie Watts' drawings. Renowned for the mass of detail included, he misses nothing, going out of his way to include seemingly insignificant things which many of his contemporaries would dismiss in order to get the drawing finished. So how does he do it? His technique is based on an almost-perfect photographic memory. 'I seem to go into a kind of trance, so deeply involved do I become when doing sketches of a new project, it's as if my mind is taking photos of the most difficult bits, helping me to get it down on paper so that it can be completed back at base.'

Why does he do this, when it involves so much more work? As he explains, 'My aim is to tell a story through the drawing. I'm not interested in just doing another drawing, it must look interesting and vibrant. If it's to succeed you must attract the onlooker. Which is why unlike most, if not all, of my contemporaries, I add human interest by putting a person in context into the drawing to which the viewer's eye is automatically drawn. In the case of farming subjects, I include animals and humans. After all that's what you find in real life, people and animals going about their business. Otherwise a drawing could appear sterile, lifeless. Where it's not possible to include people or animals, for instance in a drawing showing mechanical components, I love to include tiny details to attract the eye of the beholder. This, as you've remarked, seems to have become my trade mark!'

Among the archives

Most of Loll's work has been commissioned either by his employer Iliffe (subsequently absorbed into IPC), or independently, and they usually kept the original works. However, he has recovered some originals, or made good photocopies, and he took me on a conducted tour of these, each with its own story.

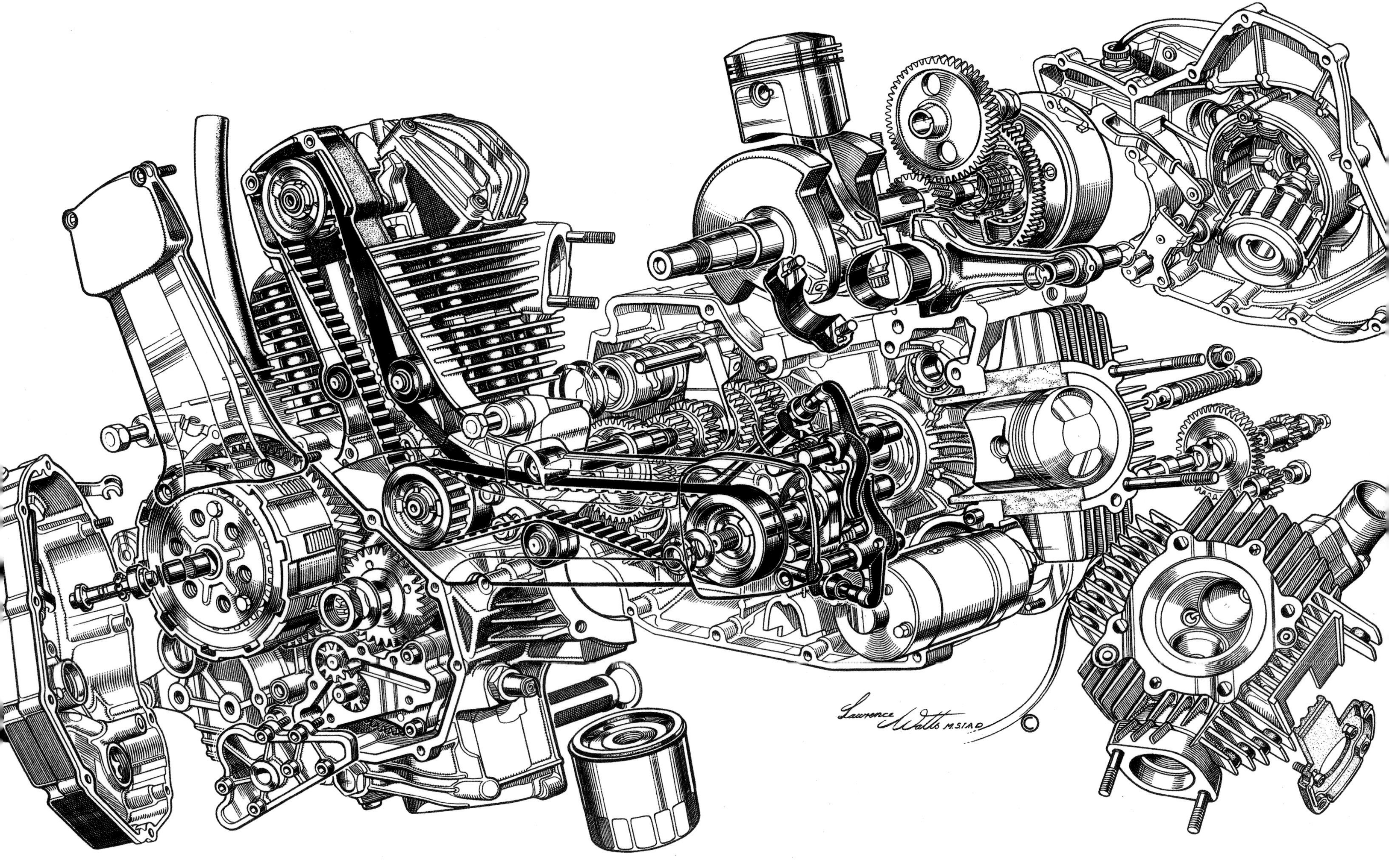

One of Lawrie's favourite drawings, the 600cc Ducati Pantah.

Motorcycles

Although the Japanese are today pre-eminent in motor cycle design, the pioneers of two stroke technology were the Germans, DKW pre-War and the East Germany's MZ post war, through their brilliant designer and engineer, the late Walter Kaaden. It was he, with limited resources, who developed MZ into a World Championship contender through the efforts of not only the brilliant young East German, Ernst Degner, but also British stars Mike Hailwood, Alan Shepherd and Derek Woodman. Lawrie found in that 1961 MZ engine a purity and simplicity of design which he greatly admired. As he recalled; 'Beautifully compact, it had a full-disc flywheel and, revolutionary for the time, the inlet mixture from the carburettor was controlled by a disc valve. This was mounted on the crankshaft, but outside the crankcase, and interposed between the carburettor and the inlet port in the crankcase. A precisely calculated cutaway in the disc enabled the engine designer to control accurately the amount of mixture into the engine, thereby enhancing its performance. It provided such an enormous performance advantage over the piston-ported two strokes that it was only a question of time before the Japanese cottoned on to the idea. What particularly impressed about the engine, apart from its simplicity, was its strength. At the point of maximum stress, the big end, MZ had a large caged needle roller bearing to cope with high loads. In comparison, the H-section connecting rod appeared quite ordinary. Attached to it was a full-skirted piston which had its crown machine finished with a chamfer around the edge; it had only one piston ring. Instead of drawing the cylinder head from above, or the side, as was the norm, I drew it from underneath to show the combustion chamber, with the squish area clearly shown. This meant something to engineers and knowledgeable types. As Walter Kaaden said many years later, anyone could have copied his design from this, and subsequent drawings of MZ's racers which I did, without resorting to 'stealing', a charge levelled at the unfortunate Degner when he defected to the West and worked for the Japanese.'

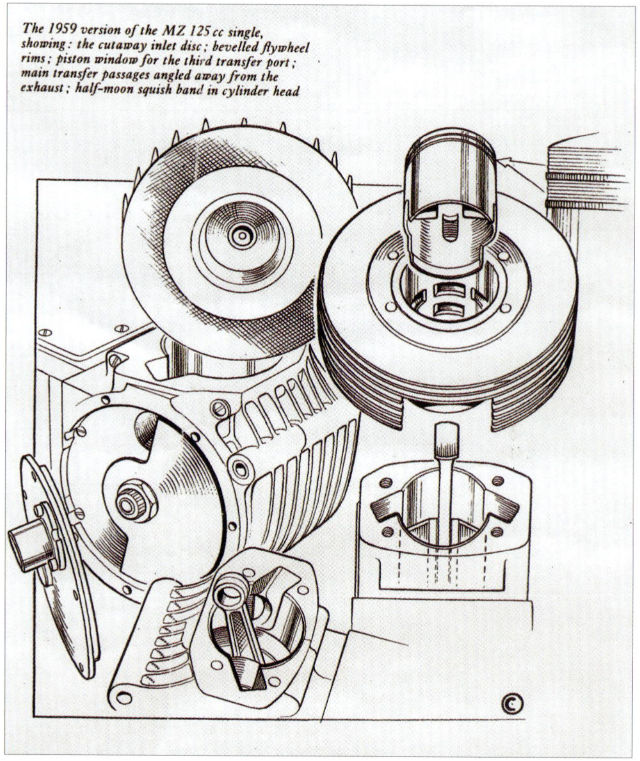

Simplicity of design of MZ's little 125cc racing two stroke inspired Lawrie to show the engine's features in this drawing. Note the rotary disc, connecting rod H section, underside of the piston, its two rings, the combustion chamber squish band, porting in the cylinder, transfer ports from the crankcase - a lot of detail in one small drawing!

Tempus Fugit

Time was ever a factor in getting essential information to enable Lawrie to do his drawings. For the 125cc MZ drawing, he had just one morning! He explained, ' We were in the boardroom at the MZ factory, Vic (Willoughby) was seated at the head of the table getting the story from Walter, I was at the other end, covered in oil and grease, dismantling the engine with a couple of spanners which Walter had given me. They had no mechanics to spare. As I took each part off, I sketched it. I was so absorbed in what I was doing that I was oblivious to what was happening at the other end of the room, interested only in the smallest detail of the engine in front of me. This happens all the time. Soon as I get started, I seem to go into some kind of trance, oblivious to everything and everyone around me... focussed entirely on what I'm drawing... or planning to draw.'

His own bike

Lawrie was 18 before he got his first 'proper' bike, a 500cc Royal Enfield attached to a Watsonian single seat sidecar painted a dashing silver. 'A gorgeous outfit,' he enthuses, all these years later. 'I'd never realised till now the significance of that because it was a Royal Enfield which I chose as a model to develop my ideas on enclosure on motorcycles, which became the Dreamliner project in the early '50s. It was a human reason which made me choose the sidecar outfit, impressing my girl friend Eileen. Not easy courting in a sidecar, you had to be careful what you were doing - I'll clarify that! You had to be careful where you went because petrol was still in short supply. One time we set off to explore the Malvern Hills, near Worcester, with Brian, an old friend of mine. On the way home, near Warwick, the rear tyre punctured. We were in a lane, no street lighting, in the dark, very dark, so I couldn't see to get the wheel out and mend the puncture. So I said to Brian, you and I will go and get help and leave Eileen to mind the outfit. Wasn't that a brave decision? There was no way I was going to leave him and Eileen alone! Afterwards, she claimed that I was too scared to go off in the dark on my own, so it must have showed! Anyway, we reached a farm house, the farmer kindly let us use his phone and I called my parents to explain what had happened. Father answered the phone, I could hear Mum in the background shouting 'He's dead, I know he's dead...' before Dad shushed her up. He made a great offer; if we could get ourselves home, he'd drive us back in the morning to fix the bike. So we got the bike into the farmer's barn, thanked the farmer, and walked the two miles into Warwick. Then we did the unthinkable, hailed a taxi. I'd got no money, Eileen had none, so Brian drew the short straw. Luckily, he had enough to pay the fare. Which reminds me, I still owe him... Anyway, next morning Dad drove us back to sort out the bike. Only then did I remember that Enfields had patented their quickly detachable rear mudguard and wheel. So all you needed to do was remove the mudguard, unscrew the wheel spindle without disturbing the chain or anything else. Simple! (in daylight...). That experience made a bigger impression on me than I realised at the time because in later life, especially when I became involved in the Dreamliner project, when I came to design anything I always looked to make it simple as possible to operate. Customer happiness is the key to commercial success.'

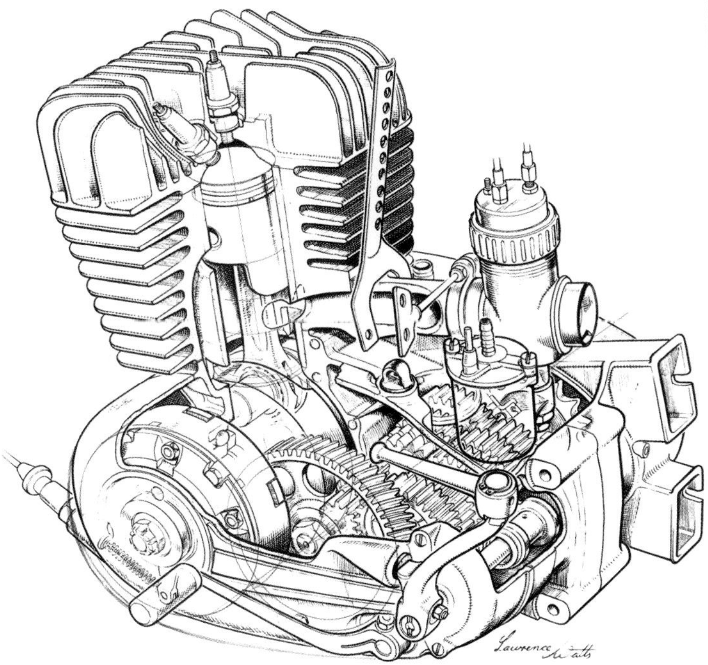

This time a roadster MZ, a five-speed 250cc air cooled two-stroke single showing the characteristic cast-alloy extensions for the moulded rubberised chain total enclosure on the right-hand side of the crankcase. Note the twin-sparking plugs for improved gas burn, an unusual refinement in that era.

From the sublime to the ridiculous

Comparisons are not invidious when they involve engines which provide performance benchmarks in their respective classes, from the tiny 125cc MZ to the complex four cylinder Suzuki two stroke, designed even before Degner joined Suzuki and which Lawrie drew when it was the quickest machine in its class. As Lawrie put it '…from the sublime to the ridiculous, that's what's made my job so challenging. Faced with four separate cylinder barrels, four cylinder heads, and two crankshafts, I thought; why not draw each of them in a different position, but relative to its correct place so that one could see the porting ? That would tell the story of the ports, imparting as much technical information as possible. With this in mind, I drew each crankshaft with only one connecting rod attached. This illustrated other aspects of the pistons which would have been impossible had they been left attached to the crankshafts. I also drew one crankshaft in mid air, outside the crankcase, and the other in the crankcase so I could illustrate the bearings. Not the easiest job, but so much more satisfying! I added little arrows to indicate the direction in which the cranks rotated, because one turned in a forward direction and the other ran backwards. Confusing? That's only the half of it! It was so unusual at that time - we're talking about 40 years ago - I did secondary drawings focusing on the layout of the crankshafts and showing how they were located and supported in that little gem of an engine. A big undertaking but it took only a couple of weeks to complete, otherwise we would have lost the news value. This time the mechanics stripped the engine, though some components I stripped further in order to see how they worked. I had lots of freedom to draw these engines because Suzuki trusted me. Everything was done on site at Suzuki importer's then HQ at Crawley, Sussex. The drawings were finished back at base.'

Shredding masterpieces!

'I went into London office (in Dorset House, Stamford Street) one day and discovered that between the desks were cardboard partitions acting as draught excluders. On closer inspection I found that the other side of each piece of cardboard contained a drawing. Then I found some art board with pictures done by the legendary Gordon Crosby. All of his masterpieces, done over a great many years, were there. An historic art collection was being used as draught excluders! I was so incensed that I stormed off to see Arthur Bourne (then publishing and main board director). I told him that it was a sacrilege to abuse such precious art works. Arthur went straight up to see for himself, while I went up the back stairs. At the top I found a young girl shredding documents. Among the stuff given her to shred were original drawings, some by Max Miller, some of mine, dozens! I told her to stop until Mr Bourne carried out his inspection. I was flabbergasted that anyone could be so heartless in destroying what were in effect historical documents, especially when I found that my original drawing of Malcolm Campbell's Bluebird was about to be shredded! Also going in the shredder was the original drawing of the first ever Mini engine and transverse gear box. And this was not even an old drawing! We never did find out who was responsible but it marked the establishment of our art archives. Amazingly, until that moment we never had any system for storing our drawings. Originals of Gordon Crosby's unique work are virtually priceless today, and are no longer being used as draught excluders! After saving them, I took photographs of Crosby's works going right back to the '20s, including some superb stuff from Pendine Sands, in south Wales, the scene of many pre-war, and early post war, world record attempts. Crosby's unique talent instigated the new archive and some time later, I helped stage an exhibition of his work in a Coventry art gallery.'

Opposite: All sorts of problems were posed, and solved, in this superbly detailed drawing of Suzuki's remarkable RG500 square four, as raced by Barry Sheene.

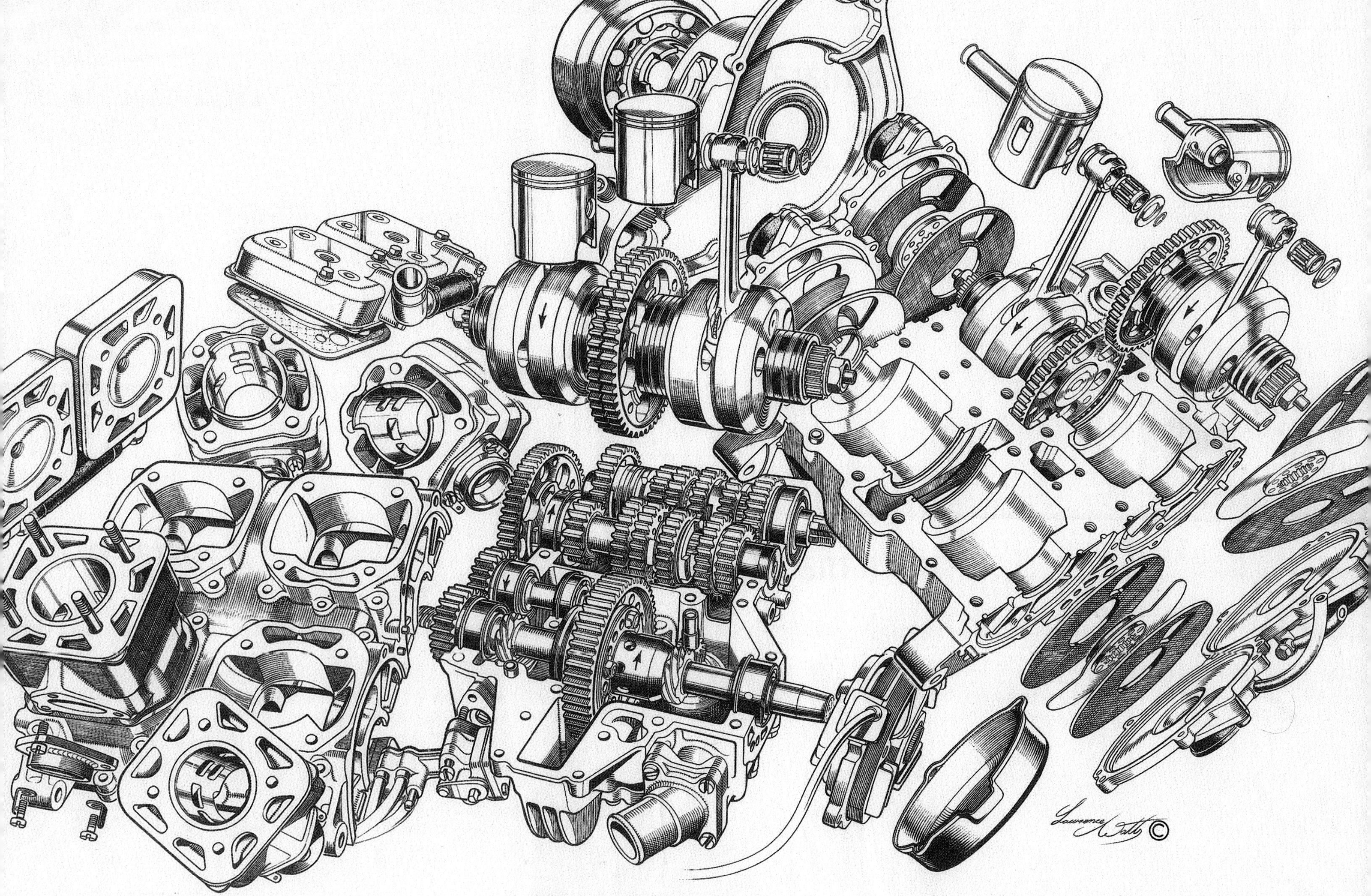
Lawrence Watts ©

Triumph Bandit

Among Loll's favourite drawings is one he made of Triumph's 350cc Bandit, launched just before the collapse of the factory in 1978. 'I can't think why Triumph wanted to introduce the smaller model first,' he remarked, examining the Bandit drawing. 'I greatly enjoyed doing this, but why they made it a three fifty instead of, say, a five hundred, is beyond me. On the face of it, it's an illogical decision.' The truth was later supplied by Norman Hyde, who worked in the Triumph development department from 1966, and when Triumph eventually collapsed, created his own business marketing most things related to Triumph. According to Norman, the Bandit had been designed for the American market, which had a capacity limit for learner machines of 350cc. So the Bandit would have had to compete head on with similar sized models from the world's leading manufacturers. To ensure that they got it right, the factory tried three different layouts. The first, a 350cc ohc twin, was designed by Edward Turner to a brief that manufacturing costs be no more than Triumph's single cylinder 200cc Cub. Cutting so many corners to meet this wildly unrealistic target, the resultant ultra-light model was timed at a highly impressive 115mph. However, it was so spindly and unsubstantial there was no chance of it being roadworthy or marketable.

However, Triumph's Doug Hele and Brian Jones developed a third model which was signed off for production. Sadly, only a handful of pre-production models were made before Barclays Bank pulled the plug and Triumph died. Norman Hyde later unearthed an incomplete machine, found the missing bits and made himself one of the very few Bandits in captivity. 'Not too difficult, really,' remarked Norman, 'I had a choice of nearly 26,000 pistons and 20,000 connecting rods! These had been delivered before Barclays put up the shutters. Other stuff found its way into the resourceful spares department of Birmingham's Len Vale Onslow, Britain's oldest working man when he died aged 104 in 2003. Ever the entrepreneur, Len snapped up Bandit bits on the off chance that the bike might one day go into production.'

Aimed at the American entry-level market, the 350cc Bandit was an advanced design for the 1970s. Compact and light, the dohc engine had the potential to succeed but only a handful were built before the factory closed.

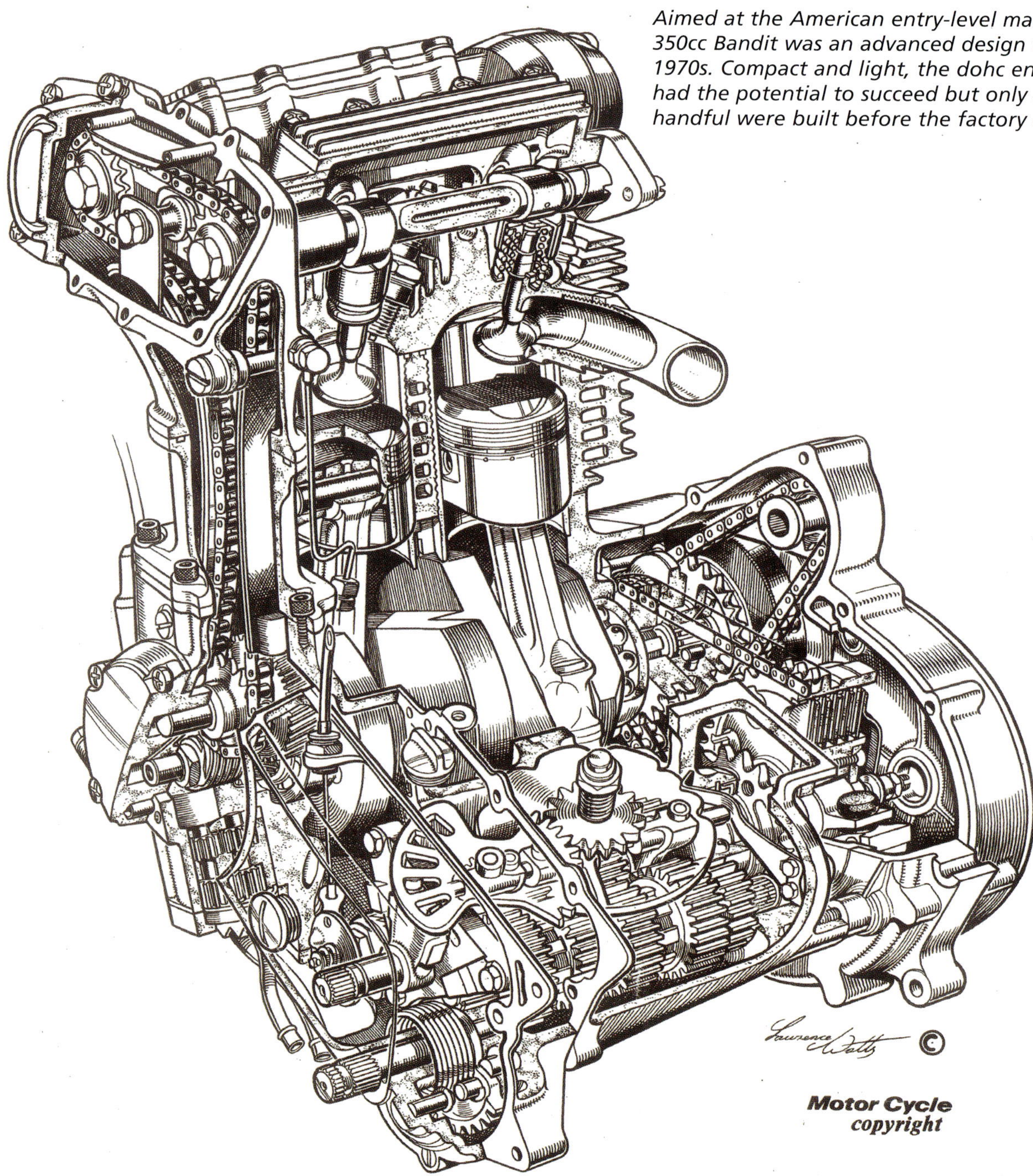

The KRM was unique in being the first racing engine to have cooling ducts forming part of the engine. They were designed to mate with ducts in the fairing which fed cooling air to the engine. Note the unusual overhead camshaft drive, similar to that on 1960s Jawa 350 twins.

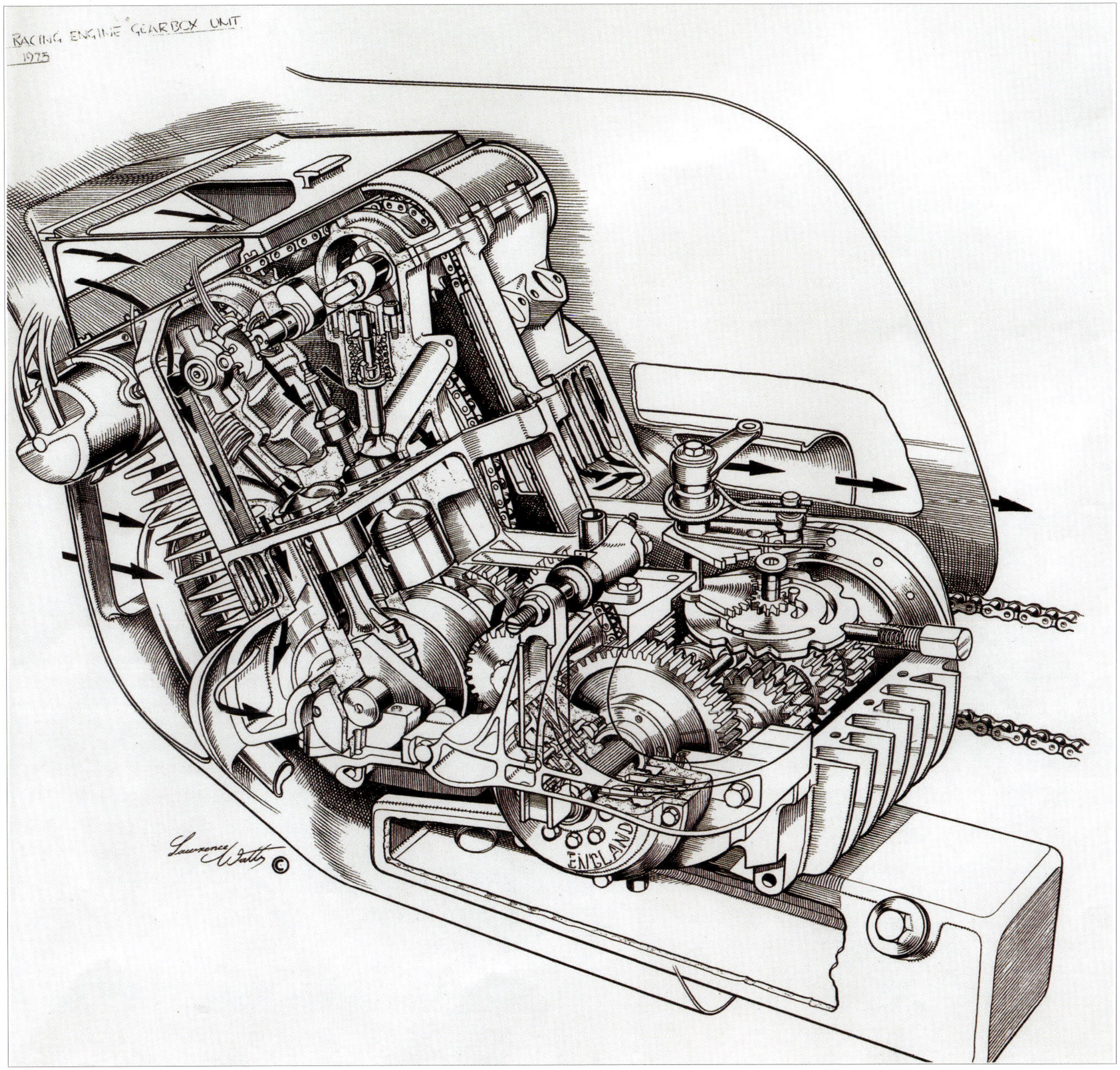

Kingston Racing Motor

'Remember the KRM?' asked Lawrie as he pulled out a sectioned drawing of a four stroke motorcycle engine. 'Designed by Jack Williams for Kingstons of Hull, the Kingston Racing Motors engine was an air cooled twin cylinder four stroke with a 360 degree crankshaft. Technically interesting, it never got into production. It featured a fairing which was attached directly to the engine, the purpose of which was to duct additional cooling air through passages in the engine casting around the sides and to the rear of the cylinder. High domed pistons with long skirts and three rings were against the norm in racing where slipper pistons with two rings reigned. At the time, Jack did not seem convinced of the effectiveness of four valves per cylinder so he opted for two valves each cylinder. They were operated by twin camshafts, one of which was driven off the end of the other, similar to the 350cc Jawa of the early sixties. So far as I'm aware, this was the first attempt to use a fairing ducting directly into and around the engine - as distinct from adding a fairing after manufacturing the engine. I often wondered if it might have started a new trend had it been produced in numbers.'

Time out in the Isle of Man

The adrenalin generated by the Isle of Man TT was such that everybody was up for morning practice at 5am, maybe going from a party directly to the circuit, then back to the hotel for breakfast. Afterwards, the post-breakfast tour of workshops dotted around the back streets of Douglas allowed us to follow up what we'd discovered earlier. Which is how we had a world 'scoop' in 1960.

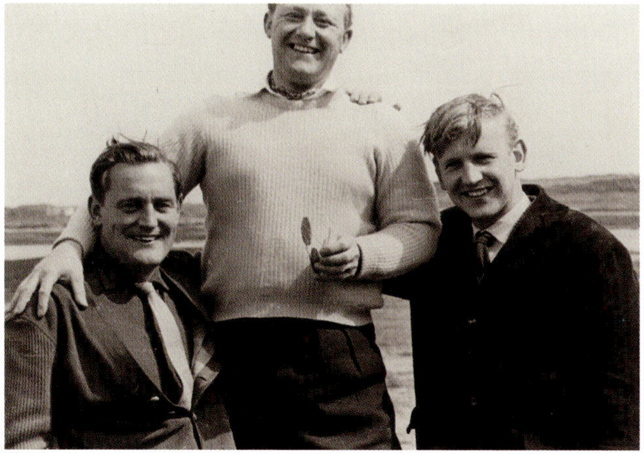

The late Geoff Riden, Lawrie and the author on a rare afternoon off during practice week, 1962 TT.

Bribery with Opal Fruits

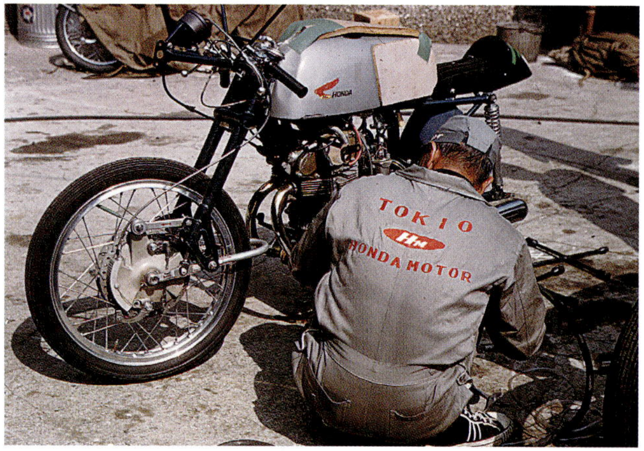

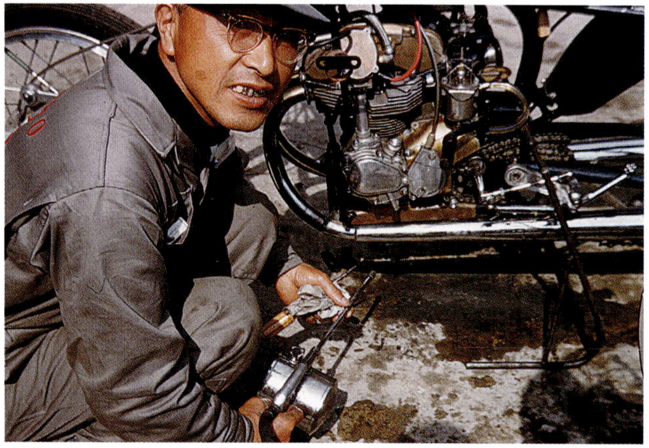

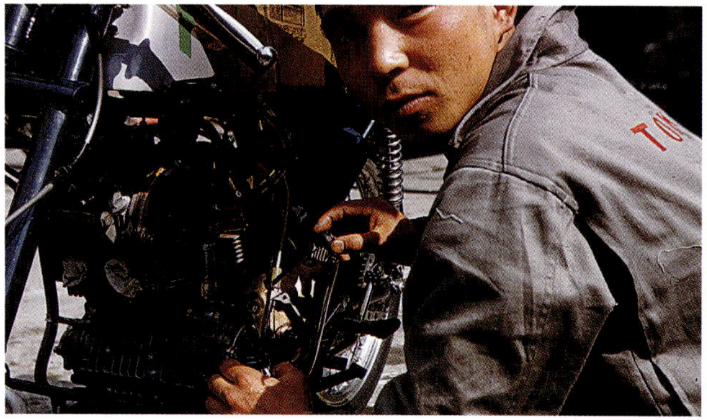

Japanese mechanics working on their 125cc twin cylinder machines at the Nursery Hotel, Douglas, Isle of Man during their first TT trip in 1959.

We had an inkling that Honda, on only their second visit to the TT, had things which they were not keen for us to see. So we paid them a visit at their race headquarters, the modest Nursery Hotel in Onchan, off the northern end of the Douglas promenade. From a non-intrusive distance, we watched as the mechanics removed the beautifully fabricated aluminium alloy streamlined fairings from their new 250cc four cylinder racers. Beyond public view, the mechanics were confident in the knowledge that a view of the naked machines would be a sufficient 'fix' for our information-hungry sortie. Little did they guess. As Lawrie ostentatiously pocketed his sketch pad and photographer Geoff Riden packed his new Pentax 35mm single lens reflex camera, the Japanese ambled away for their morning green tea break. For want of anything better to do, I poked a finger up an exhaust port, and felt something odd. Instead of being circular, the inner end of the port was divided in two! 'Lawrie, what do you make of this?' I asked. The expression on his face confirmed my suspicions. Geoff's reaction proved that our curiosity was not misplaced. This engine had two exhaust valves per cylinder, that's why the port was bifurcated! With the carburettors still in place we couldn't check if matching inlet valves were fitted. However, the exterior shape of the cam box casting indicated that this was almost certainly the case. So Lawrie drew some layouts of the cylinder head to test the reactions of the secretive, non-English speaking, Japanese. At the first two sketches, they smiled politely. But their astonishment at the third sketch, a schematic four valve layout, spoke volumes. Now Lawrie needed to know what type of layout, parallel, vertical or radial valves? So he produced a packet of Opal fruits and offered them around. Politely, the mechanics accepted. Between munching happily on the sweets, they helped to fill in the essential details of the valve configuration. As a bonus, they indicated that four valves per cylinder had been used in their 125cc twin cylinder engines the previous year (1959), though nobody outside of the factory knew. Back at the London office, our 'scoop' got a frosty reception from Technical Editor, Vic Willoughby,.. 'That's old hat, tell them Rudge and Excelsior gave up on the idea 30 years ago.....'

Vic Willoughby, technical editor of 'The Motor Cycle' from 1959 until his retirement in 1982, was a life-long friend of Lawrie. Vic was acknowledged by his peers as one of the finest writers of his era.

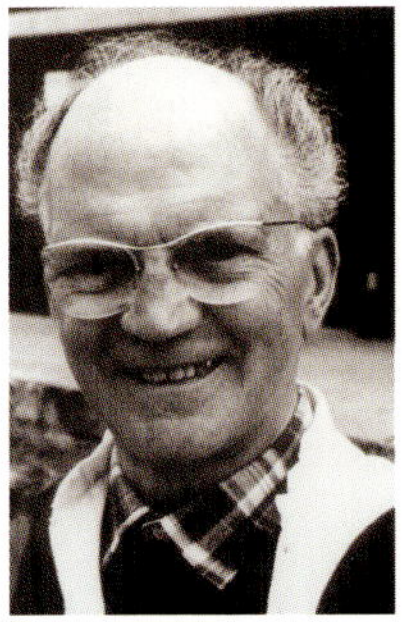

Flying without wings

Making a come back to racing after five years' retirement, ex-World Sidecar champion Eric Oliver and passenger Stan Dibben crashed during one Wednesday morning's wet and miserably cold practice. Lawrie needed to find the cause of the accident, because he had designed most of Oliver's Norton outfits. The accident had happened on the flat-out left hand bend on one of the most picturesque parts of the TT circuit overlooking Ramsey Bay above Guthrie's Memorial. The outfit left the road, flew down the mountainside for about 100 yards, leaving the front forks close to the road. Lawrie went straight to the steering head of the wrecked frame. He found part of the retaining bolt still in position, but sheared off flush with the underside of the head. He unscrewed it, and after painstakingly searching the crash site, the remainder of the bolt was found. From the marks on the thread, it appeared that the bolt had stretched, probably due to being over-tightened. It had obviously snapped under stress...'I did recommend that the head bolt be of a much bigger diameter to cope with the unusual stresses imposed around the Mountain Course,' Lawrie said to me, 'and I hoped that they had followed my advice. But this bolt is the same as used on the frames for the short circuit racing outfit....'

Now in intensive care in Nobles Hospital, Eric regained consciousness to find Lawrie beside the bed with the broken bolt. It was important to reassure Oliver that it was a mechanical failure which caused the crash, not driver error. Just able to speak, Oliver recalled that as he flew through the air he prayed, 'Please God, let it not hurt... and then I felt a great hand reaching out to catch me.' The marshals had found him, unconscious, caught up in one of the very few small trees on that otherwise bare, rocky, mountainside....

High jinks at the Sefton

Some of Lawrie Watts' happiest hours were in the Isle of Man in the 1950s and '60s. 'We worked all hours, but occasionally managed to get in a few hours' relaxation when we enjoyed ourselves'. Which probably explains how, at the 'Blue 'Un's' HQ at the Sefton Hotel, saucy pictures appeared on the woodwork between floors in the lift shaft, conduct which the management tolerated, along with piano 'recitals' by the likes of Dunlop's legendary race manager, Dickie Davis, or impromptu parties in whoever's bedroom happened to be vacant at the time. Occasionally, the timing was less than perfect, such as when Amal carburettor representative Barry Johnson returned unexpectedly for an early night only to find his bed parked upright in the corridor and his bedroom overflowing with 'guests'...

Be kind to Fairies...

Lawrie and Vic Willoughby were great pals. They seemed to spark the devil in one another, consequently were seldom far from depredations of one sort or another. One such occasion was a trip from Ronaldsway Airport back to Douglas with Lawrie as Vic's pillion passenger. Instead of saying 'Thank you little fairies' as tradition has it when passing over Fairy Bridge on the Ballasalla Road, Vic shouted, 'Bxxxxocks to the fairies!' Within 200 yards, a front wheel puncture had deposited them in a heap in the road!

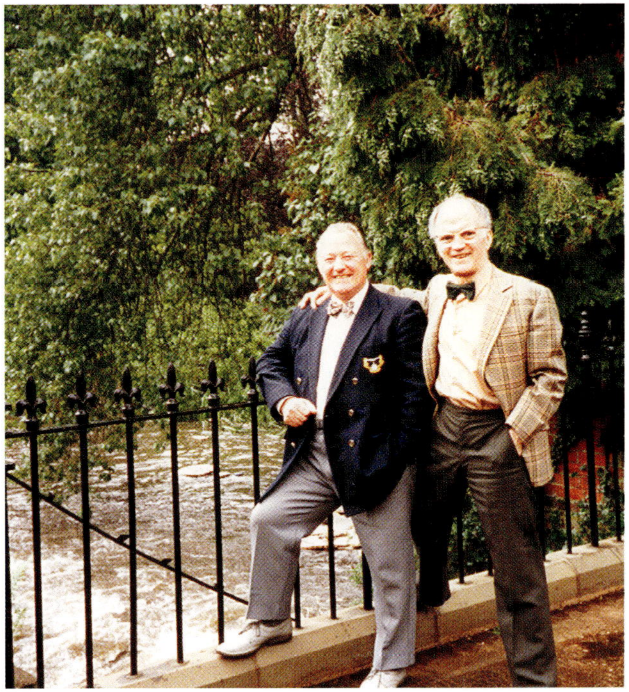

Buddies! Lawrie and Vic at The Old Mill, Coventry.

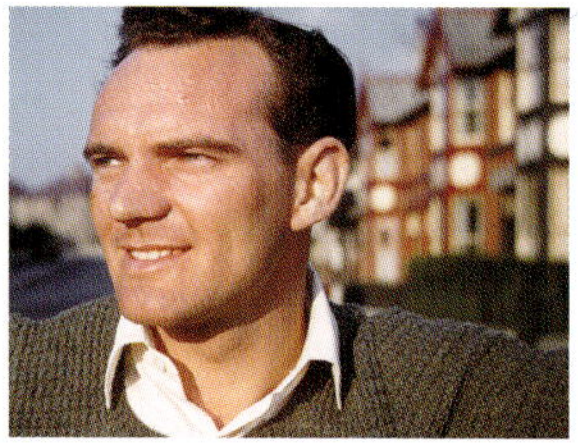

Bob McIntyre, first man to lap the TT circuit at 100mph, in 1957 on a 500cc Gilera four.

If lost, don't ask a witch

A glutton for punishment, Lawrie was not easily dissuaded from accompanying Vic on another excursion, a mystery tour over South Barrule in the deserted mountains in the south of the Island. In other words, Vic was lost. They stopped at a lonely cottage to ask the way back to civilisation. Bad move. The woman they asked turned out to be a witch! Determined to make the most of an unexpected meeting, our heroes were getting on fine with her ladyship - until she noticed the camera. Behind it, the grinning Lawrie, inspired by her, erm, exotic clothing (or lack of it) was taking photographs like there was no tomorrow. Instantly, her mood changed. Black looks, imprecations and body language suggested that Mr Watts should leave pronto, if not sooner. Not wishing to push their luck, the pair did a rapid retreat on Vic's Dominator 99. Thereafter, unaccountable things happened to Lawrie. His luck changed. Nothing went right. Only when, in desperation, he destroyed all photos taken of She Who Should be Obeyed (or not photographed) did life gradually get back to normal.

Remote control crash

Another ghostly happening occurred during a race when Lawrie had taken ace photographer Bill Banks to Whitegates Corner, on the exit from Ramsey. Lawrie recalled years later, 'We were safely behind the wall on the outside of the corner and I heard Banks yell as he hurtled over the wall and landed beside me. There followed an almighty bang, a cloud of dust, and then silence... Popping our heads above the wall we found a rider had hit it so hard that he'd triggered the old VN plate camera which Bill had left on top of the wall. In the photographic dark room, we later found a spectacular image of a rider who had photographed his very own crash! That he survived was a miracle but his photo didn't. Editor Harry Louis scrapped it under his 'no accidents' house rule.'

Riding the winners on open roads

As a schoolboy, I had looked forward eagerly to the post-TT article in the 'Blue 'Un' titled 'Riding the Winners'. This ritual was a throw back to pre-war days, the long tradition being preserved by Willoughby when he took over from Harry Louis and George Wilson. Loll got in on the act too. 'One of the perks of our job during the TT included riding TT winners on Snaefell Mountain, usually at dawn,' Lawrie recalled. 'In fact I had my first works ride in 1955 when Vic gave me a 350cc factory Moto Guzzi to warm up from Guthrie's Memorial to Windy Corner before he tested it. I could hardly believe my luck!' Although drawing was his main job in the Island, he also filled in doing other duties, one of which was chauffeuring famous riders. One such was driving the immortal Bob McIntyre in his wheel tracks around the circuit the day after the Scot's historic 100mph lap on his Gilera in the 1957 Senior TT. As Loll drove, Bob re-lived the lap for back-seat scribbler Vic. 'The thing which most impressed was Bob's composure about the whole thing,' remembered Lawrie. 'He recounted the occasion in perfectly matter of fact terms in that distinctive soft Scottish burr of his. There was no drama, no close shaves, merely a precise comment on his racing line to the nearest inch. Remarkably, his immense concentration on that lap itself did not preclude him from remarking on where he'd seen me waving to him on the exit from Quarter Bridge! A truly remarkable rider and one of the nicest people ever to grace a racing machine.'

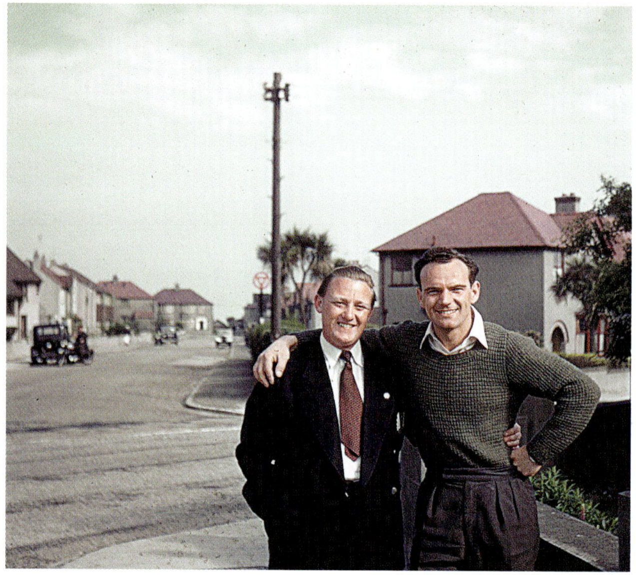

Mates! Bob Mac takes time out to greet Lawrie outside the 'Flying Scotsman's' TT workshop during practice week.

Lawrie spreading Vic's ashes around the Guthrie Memorial on the TT course in the Isle of Man.

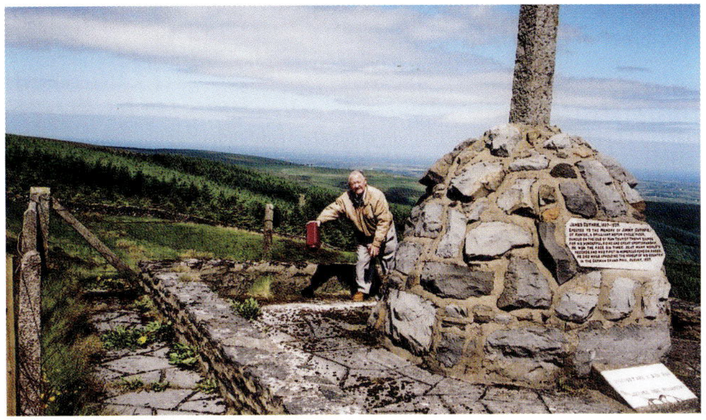

Ashes to seagulls?

Six months after our old mate, Vic Willoughby died, Lawrie took Vic's ashes to the Isle of Man as promised, with the intention of scattering them on the TT course. Dispersal would be more accurately achieved on foot than from the air. So much for the theory..... 'Everything went smoothly, we chose the Guthrie Memorial because Jimmy Guthrie was one of Vic's heroes, and the site alongside the TT course on the climb out of Ramsey up Snaefell mountain affords spectacular views over the northern part of the island,' explained Lawrie.. 'On this lovely clear day, we could see Scotland and Ireland. Solitary, peaceful, and just the two of us. Eileen set up the camera, I posed with the casket. Then, just as I opened the casket, a sudden gust of wind came out of nowhere and blew ashes all over the place. Caught unawares, I closed the casket and the wind dropped. Dusting myself off, I noticed a solitary seagull perched quite close on the bank, watching us.

'In calmer conditions, I carefully dispersed the remainder of Vic's ashes and we said our goodbyes to our old mate.

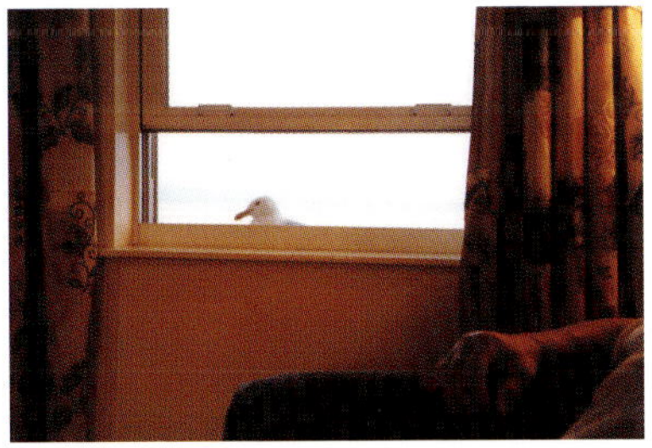

Seagull 'Vic' keeps an eye on Lawrie and Eileen at their bedroom window in the Sefton Hotel, Isle of Man.

Many's the time Vic stayed with us, especially in his later years when illness came more frequently, or he wanted his car straightened. Neither age nor illness dulled his intellect, his ability to voice opinions forcefully and with absolute belief in his view made him just as argumentative to the end, regardless of your point of view. Or anyone else's for that matter. In spite of this, he was a character who endeared himself to Eileen, myself and our family. We loved him, in spite of himself! And we missed him dearly. Now we had left his mortal remains in peace. Or so we thought....

'As we loaded the camera gear into the car, my attention was again caught by the seagull I'd noticed earlier. It had moved to within a few feet of the car, invading our personal space. It seemed to be taking a close interest in our activities, sitting quietly, not bothering us for food - as seagulls do - but seemingly at ease in our company. Eileen also remarked on its behaviour, how tame it was. We finished our packing and spent the rest of the day exploring this lovely island, my privilege was to show Eileen our old haunts during the previous 30 years of covering the TT races. At dinner that evening in the Sefton Hotel, 'The Motor Cycle's' former HQ, we were enjoying the view over Douglas Bay when up popped a seagull on the window ledge right beside us. Just to make sure that we noticed, it started tapping on the glass window! We waved, it stopped tapping, did not go away, but continued watching us eating our meal. Eventually, I got up, walked out of the dining room to the loo. Doing my business there, I heard a tap tapping on the frosted glass of the loo window. Startled, I could just see the outline of a seagull, the seagull! I assumed it to be the same one, because when I told Eileen she confirmed that as I left the table the gull had walked along the window ledges and then disappeared from view. By now I was becoming uneasy; what was going on? As if in answer, the gull reappeared on the window ledge alongside our dining table. We left, went for a walk on the prom before turning in for the night, and then retired to our bedroom. As I went to draw the curtains, yes! You've guessed it; a seagull - the seagull - was sitting outside on the window ledge! By now it was dark, a time when all self-respecting seagulls have long since found a resting place for the night. This one had not only found a resting place, but also our bedroom. Spooky?

'Next morning, forgetting about the events of the night before, I pulled back the curtains and... the seagull was sitting there. We hurried down to breakfast, and by the time we'd reached our table, same one by the window, so our winged companion had taken up station on the window ledge. By now we'd nicknamed him 'Vic' and, having done that, we somehow felt less spooked and more at ease. So 'Vic' followed our every movement right up to the time we boarded the SeaCat and sailed from the island... So, did we imagine all of this? No way! Is there a rational explanation? We'd like to think so. Until then, we can only assume there's a seagull probably still flying around the Isle of Man, a gull that we called 'Vic.'

Autographed beech dish made by Walter Kaaden and given to Lawrie, Christmas 1988.

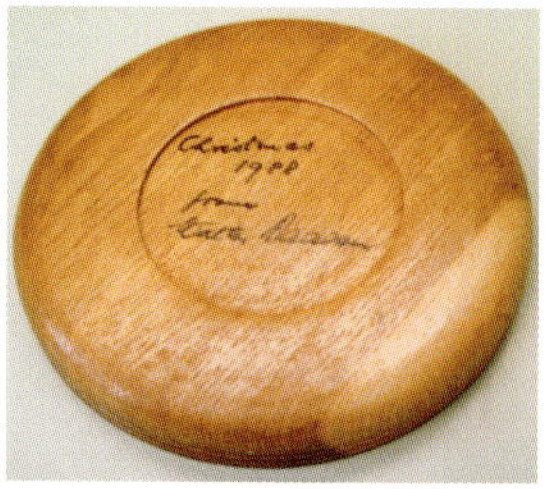

Lost and found by a living legend

What's the ultimate disaster for any artist? A lost sketch pad! That's exactly what happened to Lawrie one morning at TT practice. It was only back at the Sefton that he realised the pad was missing. Panic-stricken, he raced back to the paddock, now deserted but for a lone figure standing by the Cadbury cocoa tent. The figure was obviously waiting for him. 'Is this what you're looking for?' grinned the legendary Graham Walker, editor of the 'Green 'Un', the 'Blue 'Un's' great rival, and a TT winner in his own day. 'Honestly, I've not looked inside,' he grinned as he handed over the precious pad to an extremely relieved editorial artist.... Incidentally, Graham was father of the equally famous Murray Walker who, school permitting, accompanied his father and mother to the TT. But they stayed in the Castle Mona Hotel, well away from 'the opposition' at the Sefton.

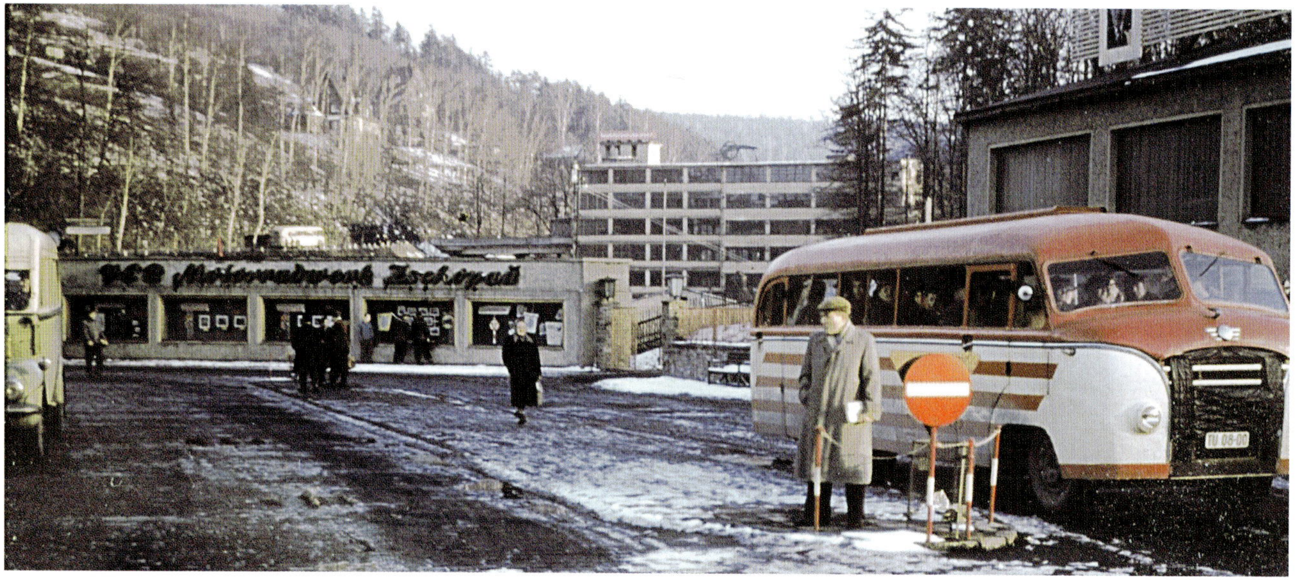

The factory bus outside the MZ factory at Zschopau.

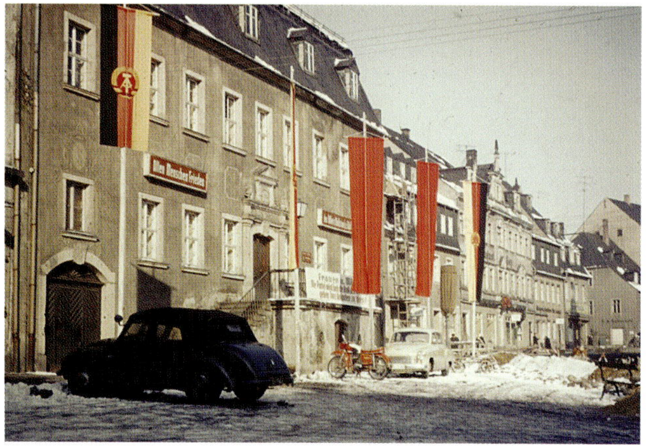

Lawrie and Vic had to report to the police at the Rathoff and give an account of their movements.

From Sefton to armoured carrier and guard

A carving of a woodsman made by a friend of Kaaden's and gifted to Lawrie.

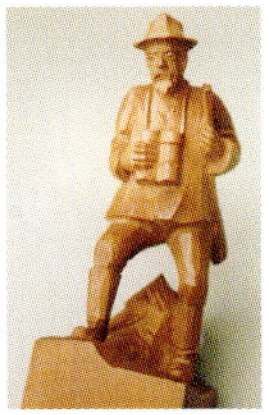

Kaaden and Willoughby outside the Zschopau Castle.

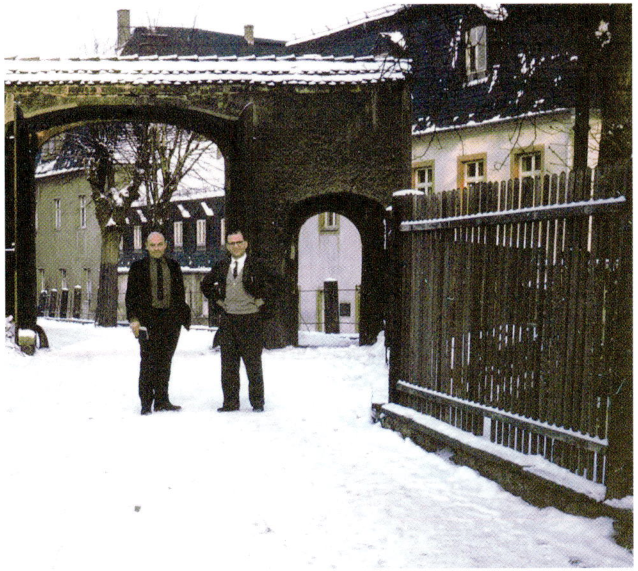

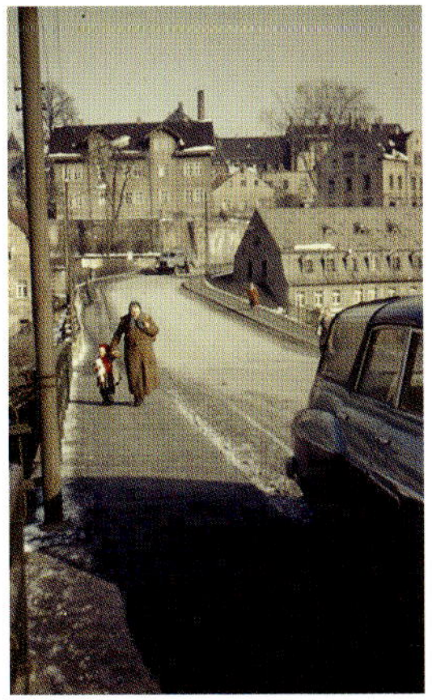

Looking back at the castle en route to the MZ race shop.

Regardless of where they were staying, many trade personalities made the back bar of the Sefton Hotel their watering hole of choice. It was there that Walter Kaaden, the MZ race chief, became good friends with both Vic and Lawrie, spending many a happy evening talking about racing, and two strokes in particular. Which is how Vic and Lawrie found themselves in an armoured personnel carrier as guests of the East German army during the Cold War. Lawrie recalled, 'Walter invited us to visit Zschopau but Vic assumed that Walter's written invitation was sufficient to get us through the border. It never occurred to Vic that visas might be necessary, as they were. Inevitably, we were in deep, deep ordure from the moment we landed. In no man's land, close to the newly built Berlin Wall, as we were walking to No 60 Friedrichstrasse, the border guards opened fire on us, thinking we were escaping East Germans. That we were trying to break into the wretched country never occurred to them! Dirt thrown up by the machine gun bullets was getting closer and closer and they would certainly have hit us had it not been for a Russian soldier, from whom we had earlier enquired the way, seeing our plight and rescuing us just in time. I still have nightmares about it. And Vic? I don't think he appreciated what a close call we'd had…

'Eventually we met Kaaden, who had been waiting five hours for us in his three cylinder DKW car, unaware that we had been under arrest in a cellar. Without the necessary visas (thanks Vic!), Walter had had to call in all sorts of favours to get a special vehicle for us to get out of Berlin. This materialised as an armoured personnel carrier. For added value, it came complete with an armed guard! In this we were driven across the city in the middle of the night to Chuneval airfield, an East German version of Templehof, where we had landed. At Chuneval, we finally got visas, plus a policeman who was to accompany us everywhere, including the trip to the factory in Kaaden's DKW. Wherever we stopped, for petrol or a pee, police appeared to check our papers. Uncannily, they knew our precise movements all the time.'

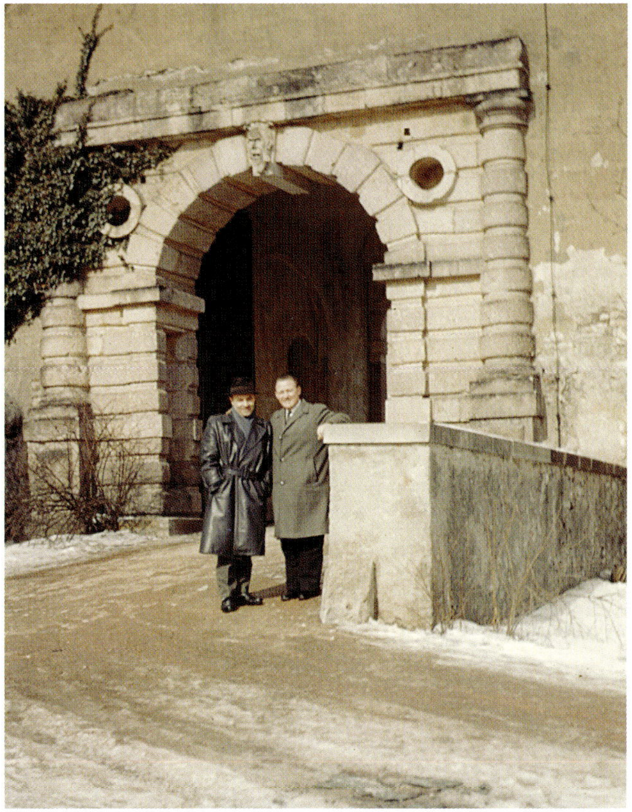

Walter Kaaden (left) with Lawrie at the imposing gates of Zschopau Castle.

Backwards with Walter

Lawrie and Vic were being driven by Kaaden back to Berlin following another visit in 1962 when the DKW's forward gears went missing. With only reverse still working, the intrepid Kaaden drove in reverse into Dresden, where Lawrie and Vic were to catch the Prague to Berlin express. Except it didn't stop at Dresden! By now hungry, exasperated and in a typical Eastern German queue which seemed to be going nowhere, Lawrie cried out aloud, 'Oh, to be in Coventry!' This provoked an immediate reaction from the man behind. 'Coventry? You from Coventry?' he yelled in English at a startled Lawrie. 'Yes, I live there,' answered Lawrie warily. The man stormed off, police arrived, took Vic and Lawrie to the stationmaster's office where they were told to stay. Eventually a train stopped, our bewildered heroes were ushered on to it, given tickets to Berlin and told, 'Normally this train does not stop here, but tonight is an exception. This is your carriage, the two female guards will ensure that you are not disturbed. Auf wiedersehen.' (Coventry had been destroyed by German bombing during the War, and Dresden had been destroyed by the RAF in a firestorm....).

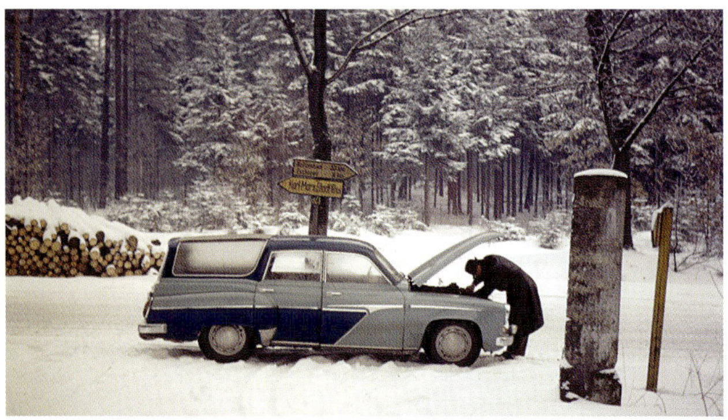

Walter Kaaden searching for his missing forward gears. He completed the journey in reverse!

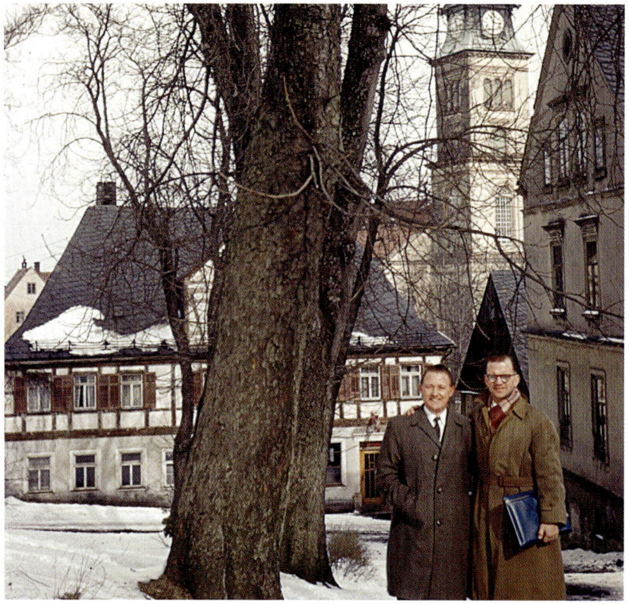

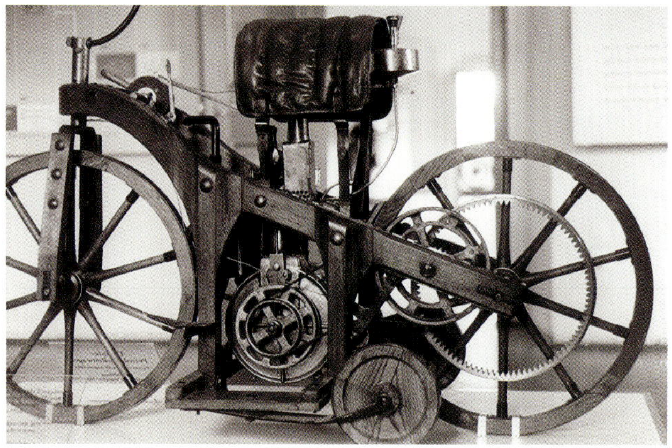

Lawrie (left) and Vic outside the engineering museum at Augustusburg. Among its exhibits is the world's first motorcycle, a wooden structure far removed from anything we know today!

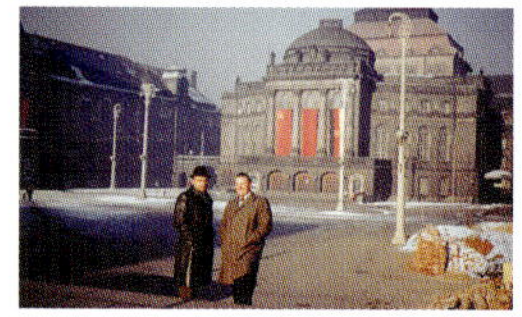

Walter Kaaden and Lawrie on a cultural tour of the Opera House at Karl Mark-Stadt.

Ernst Degner

The previous year, Ernst Degner had defected to the west. At the time, uninformed sources claimed that Degner had sold MZ's racing engine secrets to the Japanese Suzuki concern, then trying to establish itself on the world racing scene. Such claims Lawrie emphatically demolishes. 'Walter Kaaden was extremely generous with information, welcoming anyone to its use. He used to say that any secrets he had were still in his head! Technical information on MZs was already available from published sources, much of it written by Vic and illustrated by me. One example: for accurate dimensions of the ports, I inserted a rolled sheet of paper into the cylinder, made pressings of the port shapes, and then drew them full scale. Competent engineers could have replicated the insides of a racing MZ, or its rotary disc valve, which anyway had been pioneered by Zimmermann. When Walter joined MZ he further developed and refined Zimmermann's original design.'

I had known Degner for years and support Lawrie's view. When Ernst fled East Germany in August 1961, it was not to join Suzuki, but to a modest little job in the car and motor cycle trade in West Germany. But in October that year, Ernst received an offer from Suzuki to help them develop their new race engines, which were already built. New Zealander Hugh Anderson had signed for 1962, Degner was next to be approached, primarily for his riding ability (he'd been leading the 1961 125cc World Championship when he defected to the West). Yes, Suzuki's development accelerated after Degner joined but to perpetuate the myth that 'he sold his soul to the Japanese' does no justice to an honourable man and a brilliant rider whose death in a car accident years later remains a mystery. It would be no surprise if the cause of his death is locked in Stasi files.

Unlikely father for the Dreamliner

One of the most visually striking motorcycles of the 1950s was Lawrie's Royal Enfield-powered Dreamliner. More than 50 years later, the design is still an eye-stopper. Too far ahead of its time? Or, like others before it, just too much for notoriously conservative motorcyclists? On balance, probably a bit of both. Practical in use, and visually stunning, the Dreamliner seemed set for commercial success after its sensational debut. But let's hear the inside story of this revolutionary design.

The Dreamliner derived its heritage from the race track, not as a solo but with the Norton-Watsonian sidecar outfit of Eric Oliver as the unlikely progenitor. Lawrie had been summoned to Watsonian Sidecars by owner Ron Watson to draw his latest creation, a semi-streamlined sidecar for sidecar world champion Eric Oliver. Lawrie swiftly nicknamed it the 'duck-billed platypus' because of the ungainly shape. Only the sidecar was streamlined, the bike was naked and, as Lawrie recalled, 'The nearest thing to streamlining was a flat plate with a number on. Frankly, it was an aerodynamic mess! To lessen Eric's disappointment I offered to make the thing go quicker with a properly designed aerodynamic fairing. Never one to miss a trick, Eric's ears pricked up. When I drew a rough sketch to show how I envisaged the streamlining would look, Eric was hooked. But first off, I made a model, showed it to Eric who liked it, and then to his ex-passenger, Denis Jenkinson, who also gave it the thumbs up. ▶

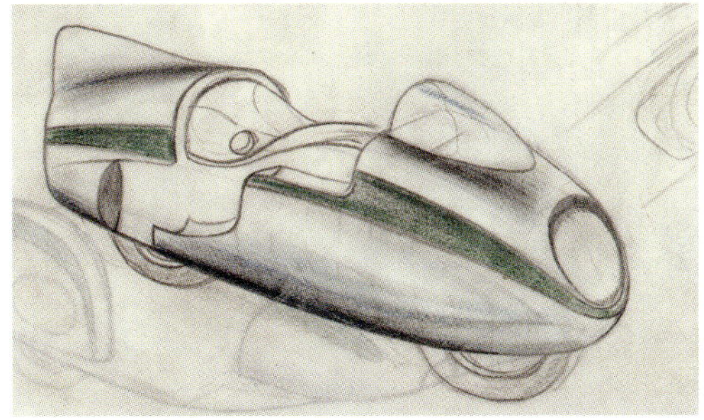

Kneeler Norton-Watsonian sidecar built for Eric Oliver. Note the air duct over the front number plate, created to spill air into an area behind the rider which would otherwise be a vacuum, net result a more efficient aerodynamic shape. Note the petrol tank alongside the rear wheel, later the fuel was carried in the sidecar mudguard.

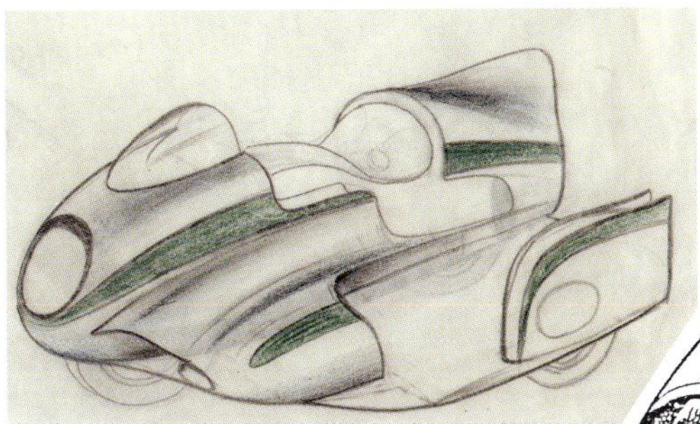

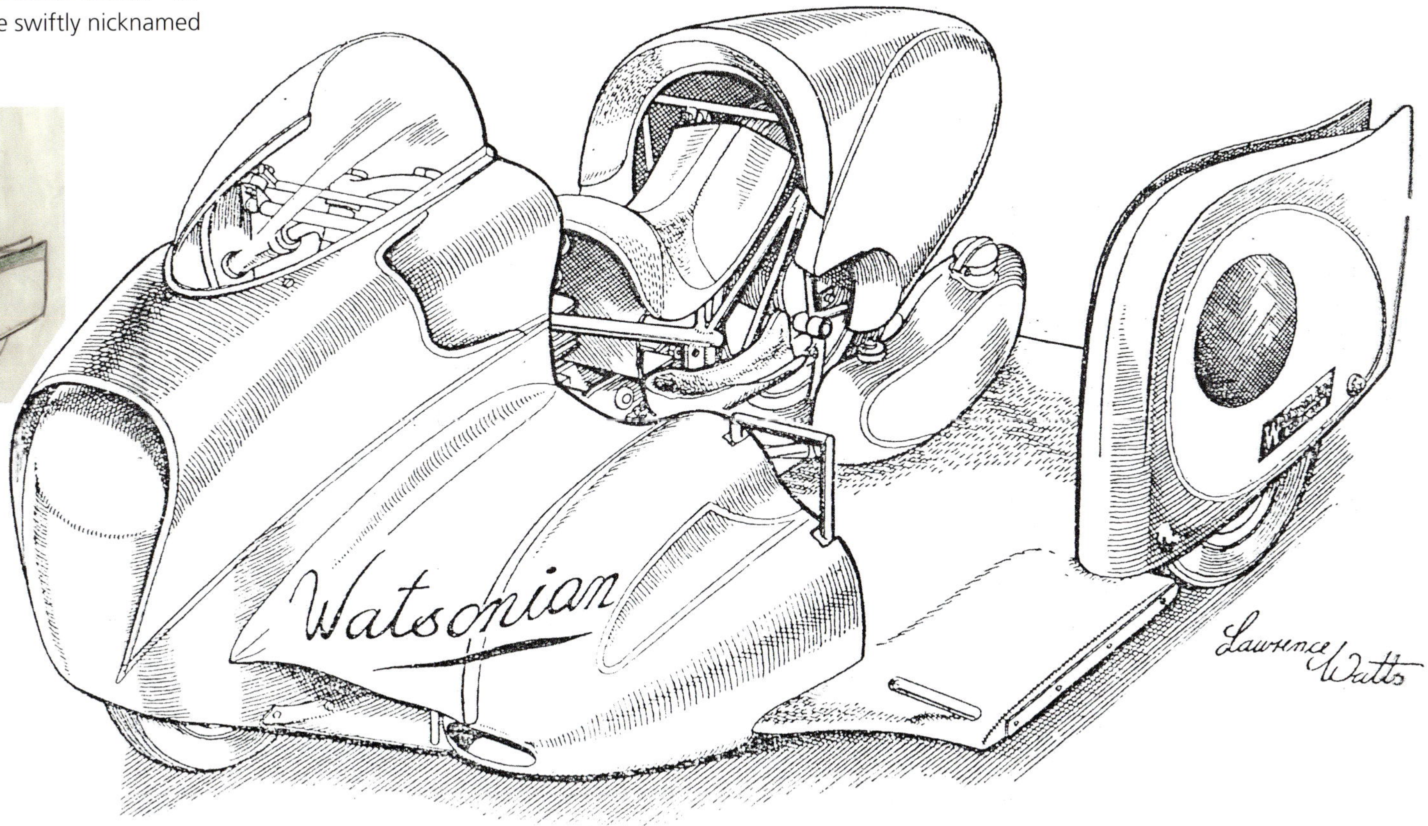

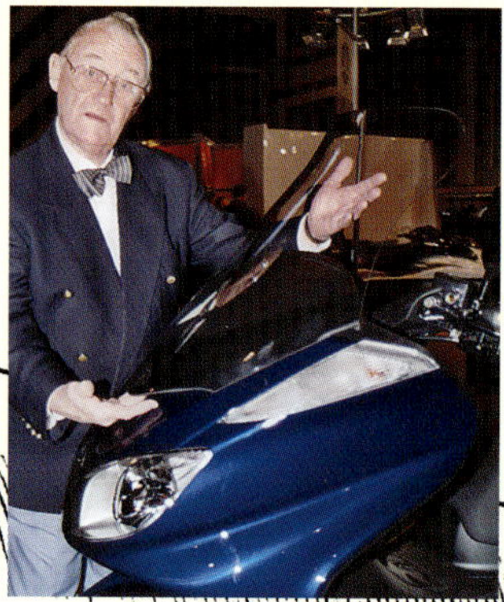

Lawrie finds on Yamaha's 2004 scooter screen a device which he first used on Eric Oliver's streamlined sidecar, breaking the airflow with a second 'skin' over the windscreen.

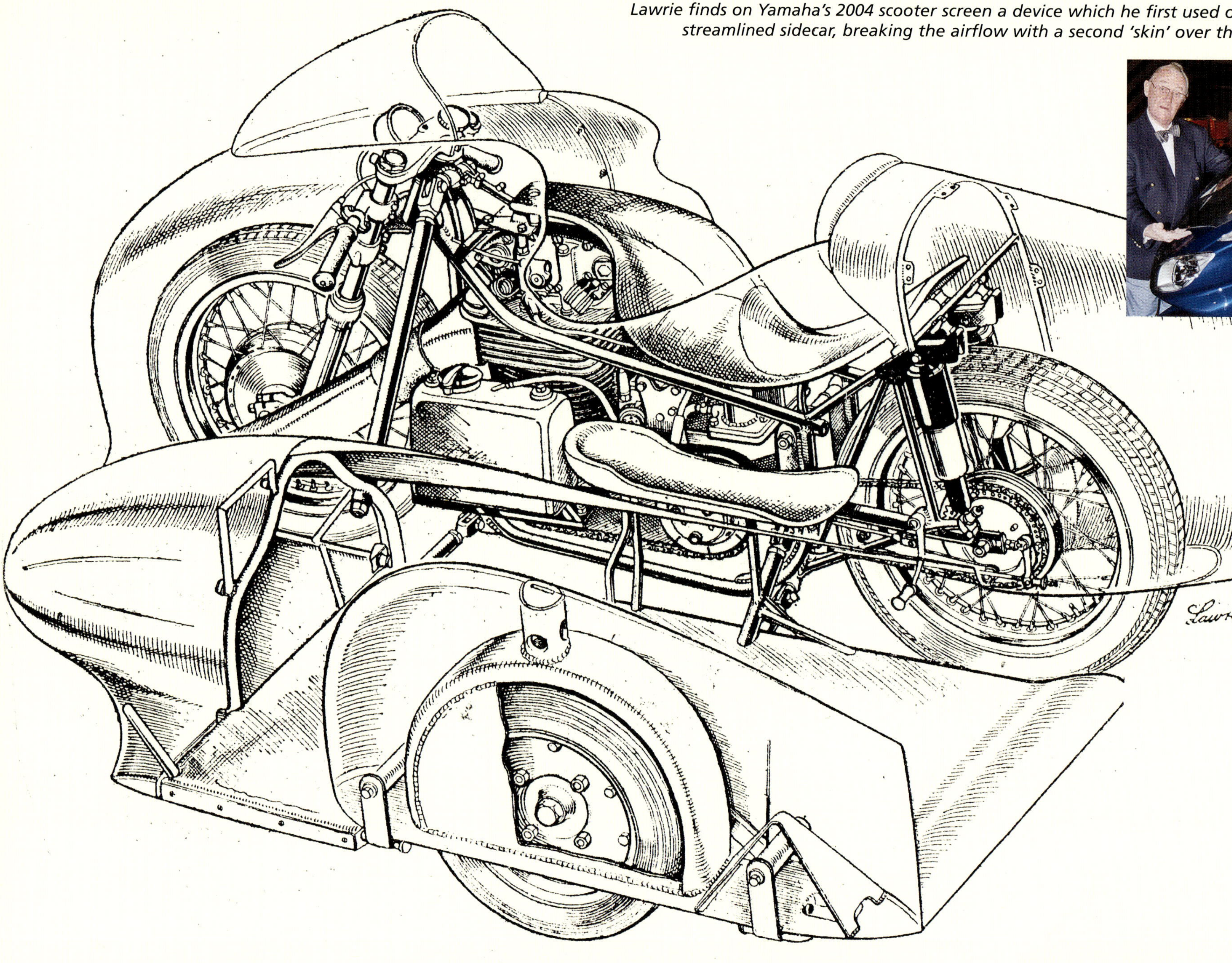

Earlier version of the Kneeler with forward-mounted fuel tank and bulkier tail. Note the large-diameter duct directing air from the sidecar nose to the engine.

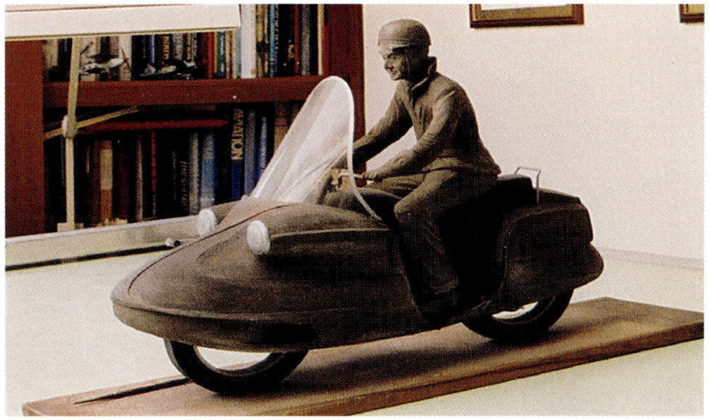

Plasticine model of the Dreamliner from 1952.

Jenks had raced very successfully with Oliver for 18 months during 1950 and 1952, and afterwards remained in regular contact. I also got Denis' successor, Stan Dibben, on side after he saw the proposed design. Ben Willetts, Watsonian's skilled in-house metal basher, fabricated the streamlining's nose in aluminium around an intricately-constructed wooden former. When Eric tested the outfit with the new streamlined nose, he reported it to be noticeably quicker than the unfaired bike. Ron Watson liked it too, but for a different reason; there was loads more display space for advertising Watsonian! By now, Eric was seriously excited. So he took up my offer to design a complete fairing for the machine, behind which Oliver could crouch, and which would also blend into and encompass the sidecar, thereby making a streamlined entity. That's how it all came about, there were no other streamlined outfits around, Eric's was the first.'

Although this gave Oliver a marked advantage over other, more powerful outfits, it was merely the first step in an evolutionary process. Oliver's odd physique, short legs and long arms, meant that he sat too high, creating unnecessary wind resistance. This prompted Lawrie to remark one day 'It's a pity you can't get lower on the machine.' Jesting, Lawrie suggested, 'Ever tried kneeling?' When Eric thought seriously about this he could see the logic and agreed to try it. Ben mocked up some knee rests which enabled Eric to adopt a lower crouch, but it still seemed to Lawrie that Oliver could crouch lower if the petrol tank was moved elsewhere. Pursuing this line of thought, Lawrie suggested that Eric re-assess the role of the passenger; was it strictly necessary to have him climbing all over the sidecar? This absorbed power, through increased wind resistance, slowing the outfit when the aim was to make it go faster. Better to have the passenger prone behind the streamlining and, with the petrol stored in a tank in the fairing over the sidecar wheel, the extra weight would ensure that the passenger did not have to climb out on left handers. The new layout allowed Eric to lie over the Norton's cam box. Additionally, there was now room to transfer a significant amount of his weight inboard on left hand bends and this, with the added weight of the fuel now directly over the sidecar wheel, obviated the need for the passenger to lean overboard on left handers. If you are still wondering how the petrol reached the engine with the new layout, it was pumped from the sidecar mudguard to a small header tank above the engine.

Lawrie's expertise in aerodynamics extended beyond drawing a pretty shape; he incorporated into the design an aerodynamic trick which managed the air to flow more efficiently over the machine. He explained, 'The first skin, the streamlined part of the nose, treats the air kindly. But the air does not react kindly when allowed to spill over the back of the fairing into a vacuum. This causes probably a fifty percent loss of the aerodynamic efficiency already achieved by the smooth passage at the front end. To remedy this, I put a second skin on top of the first in order to catch the boundary layer. This is the first compressed outer layer of air flowing over the shape which I delivered into the vacuum, causing that air to flow over the denser layer which was rushing across the top of the fairing. That it really did work was evidenced by the fact that they had to go up on the overall gear ratio to accommodate the extra °revs. Proof, if needed, came in the Belgian Grand Prix on the ultra-fast Spa-Francorchamps circuit in the Ardennes that year when Oliver cleared off from all the orthodox sidecar outfits. Soon everyone was copying, or trying to copy, Oliver's outfit. Chief among the copiers was Bill Boddice, who was unaware that I was behind Oliver's new design. Bill's version was nick named the 'Flying Saucer'. George Wilson, then 'The Motor Cycle's' Midlands editor and a good friend of Boddice, asked me to draw Boddice's new outfit. I explained to George that this might be difficult, bearing in mind that I'd designed Oliver's outfit, and Oliver and Boddice were deadly rivals. Unconcerned, George urged me to do the job. It didn't help that Boddice had no inkling that I had worked on Oliver's outfit and was still developing it behind the scenes. So it was that I tip toed into the enemy camp to draw Boddice's outfit for 'The Blue 'Un'. When the story eventually unravelled, Boddice was highly amused claiming - tongue firmly in cheek - that I'd been pinching his secrets. Truth was, without sounding pompous, I'd little to learn from Bill's design. He was trying to replicate the outfit which I'd already designed for Oliver! One day, as Oliver was testing his outfit at MIRA, I noticed a small group of people watching our activities from a remote vantage point on the testing ground. I became suspicious because MIRA was supposedly a secret testing ground for the automotive industry and strangers were not allowed in, nor were they allowed to watch from outside. My curiosity was well-founded because, as I neared the group, I recognised an unmistakable figure. Bill Boddice! By now he'd rumbled that I was working with Eric so he tracked me to MIRA, found a chink in the security and took his mates, and stop watches, along with him. Not just idly watching, but timing Eric's every lap of the high-speed circuit! Bill eventually saw the funny side of what he was doing and we became great friends after that. I still miss him.'

Very pleased with the way the new outfit was going, Oliver became even more ambitious and told Lawrie that he proposed ▶

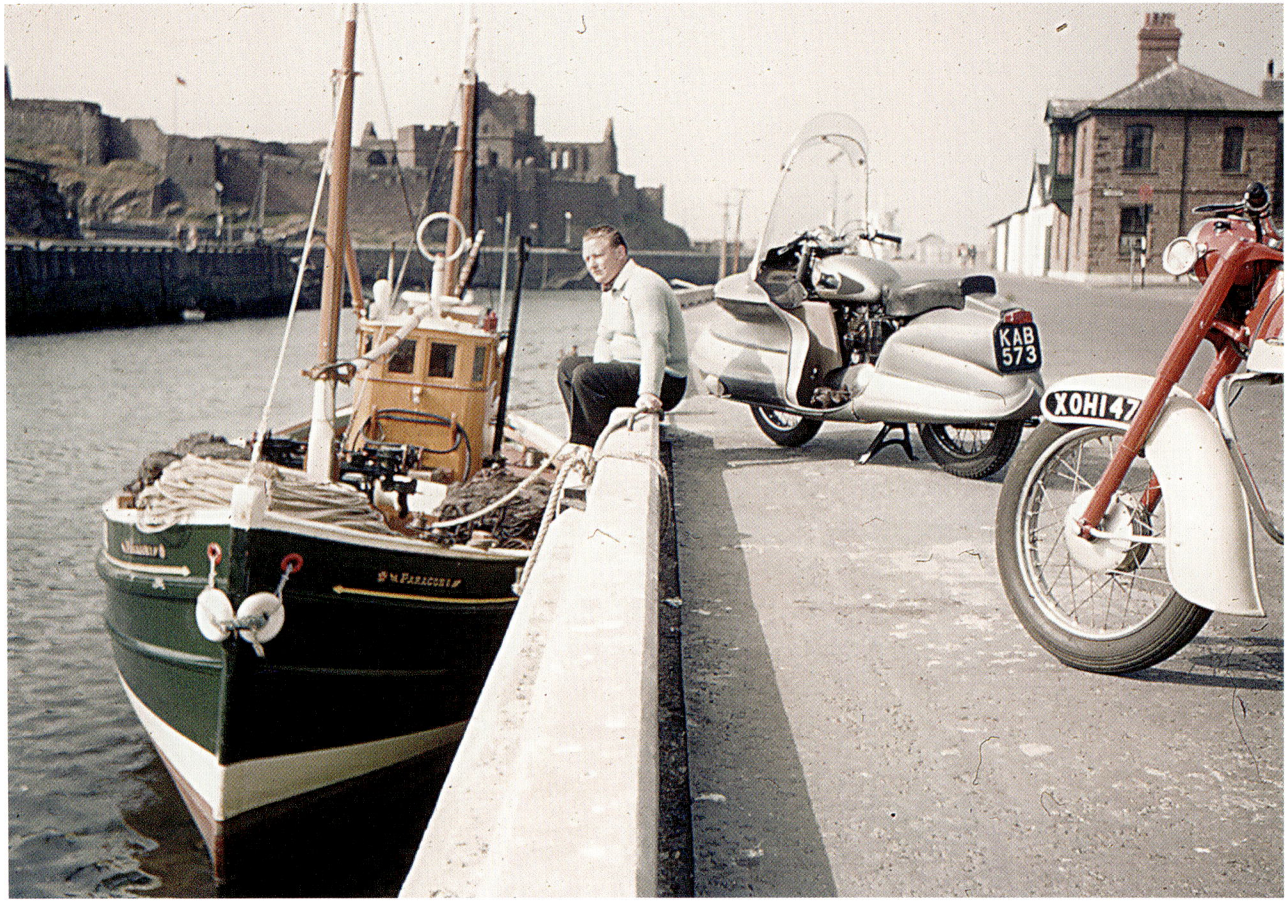

Dreamliner on the quayside at Peel, in the Isle of Man during TT week.

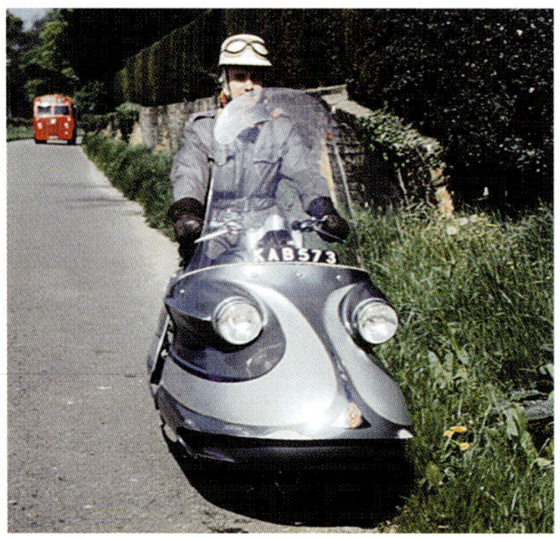

On the way home from Royal Enfield when the Dreamliner was at last completed. A memorable day!

fitting a smaller sidecar wheel to lower the outfit still further to improve its aerodynamics. But he was stumped where to re-locate the fuel tank. So Lawrie re-designed the sidecar to have a hollow floor which doubled up as a fuel tank. This combined to keep weight low as possible while providing a smaller frontal area to aid aerodynamics, which was now Oliver's priority. As a result, Eric Oliver won four successive sidecar championships and in the process created a new style of sidecar, and driving technique, which lasted for almost 20 years.

Throughout the development of Oliver's streamlined outfit 'The Motor Cycle's' editor, Harry Louis, had been an interested onlooker. One day he called Lawrie and asked if he would design some streamlining suitable for a road bike. Harry set up a connection with Tony Wilson-Jones, Royal Enfield development engineer, who was very keen to be involved, as was head man, Major Smith. The donor bike was a 350cc Bullet which, Lawrie felt, would benefit most from any improvements which the proposed streamlining would provide. 'The three fifty used the same chassis as the five hundred, weighed almost the same, but as it produced less power the power-to-weight ratio was markedly inferior. I hoped that the streamlining would rectify that,' explained Lawrie. With the full consent of his employers at Iliffe Press, and the enthusiastic support of the experimental shop at Enfield's, Lawrie started work transferring the one third size clay model of his streamliner into wooden patterns which fitted around the Bullet. 'I almost felt like an employee, so much time did I spend at Enfield's and such was the intensity of the work to complete the machine. My employer though was not Enfield but Iliffe Press, as represented by Editor in Chief Arthur Bourne. He was aware of everything, was kept up to date on progress, and was very enthusiastic about the project. It helped that he was also very friendly with Wilson-Jones. I was part of a very friendly team.'

As a development from the race track, where the primary requirement was better aerodynamics to enhance acceleration and top speed, the Dreamliner's priorities were more complicated, not least the ability to be manufactured in quantity at an affordable price, and to be easy to attach, and detach, to a road going motorcycle. A tall order, especially for an unproven product! However, Lawrie had these requirements firmly in mind as he developed the concept. Years before, with his first Royal Enfield sidecar outfit, Enfield's patented quickly-detachable rear mudguard and wheel had impressed him mightily when he punctured at night in a dark Warwickshire lane. That moment decided him if he was ever called upon to design anything, especially to do with motor bikes, he would make it user friendly. Now that he had his chance, the first priority with the Dreamliner would be practicality, then performance. Using full enclosure over the front wheel to optimise penetration, the design allowed the rider to adopt the standard riding position behind a windscreen extended upward from the top of the fairing. This provided upper body protection for the rider. With the standard handlebar width reduced from 27in to 22in, full lock turns were possible without fouling the windscreen. That was when Lawrie, metaphorically, ran into a brick wall. 'I wanted to substitute 16in diameter wheels for the standard 19in ones so that on full lock the smaller wheels would facilitate a comparatively narrow fairing which would provided excellent penetration. Ironically, the only person at Enfield's who was less than enthusiastic about the Dreamliner project was Sales Director, Major Mountford. He insisted that the standard size wheels be retained! No pun

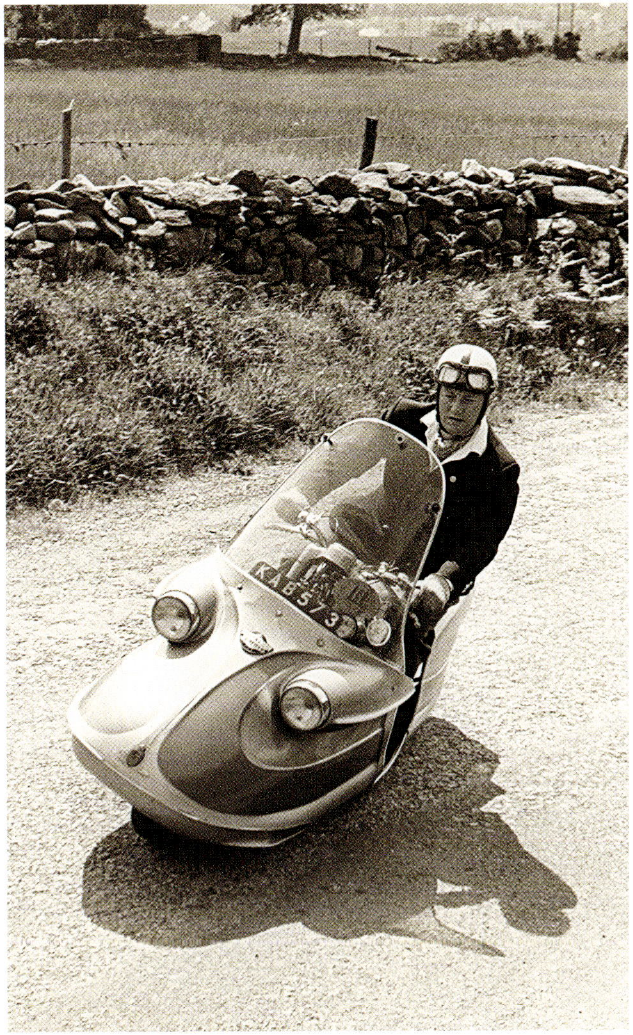

Lawrie tries the Dreamliner around the short Clypse course in the Isle of Man. This circuit was used for 125, 250cc and sidecar categories in the 1950s.

intended, this provided a major setback which required a new front end with a greater width than I'd planned. Mountford got his way and the bike could be turned on full lock when parking, the only time it was needed. That extra width penalised us in order to satisfy a manoeuvrability requirement at zero mph!'

To manufacture the Dreamliner economically was only possible in glass fibre, any other method would have been prohibitively expensive. Still in its infancy in the mid 50s, glass fibre was familiar to Lawrie. 'I'd used it for some years, mainly in model making, and I'd been involved when glass fibre was first used in the manufacture of radar domes on Lancaster bombers during the last war. So I was quite familiar with its properties. However, there was no question of me making the fairings, that was work for a glass fibre specialist. My task, mainly done in my own time, was to make the design compatible with manufacturing in the new medium and in this I was lucky to be associated with

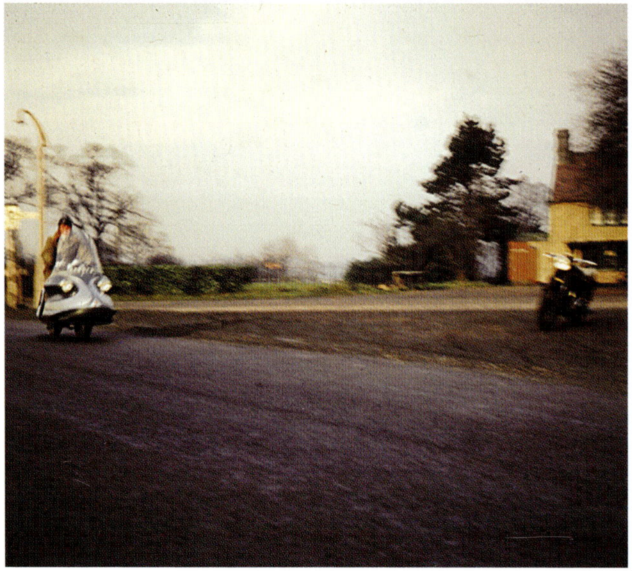

The Dreamliner being put through its paces.

Richard Wood, then one of the leading exponents of the day and editor of 'British Plastics', an associate Iliffe journal. Richard enthusiastically took on the task of doing the prototype at his home in Kent. A period of intense activity involved me trundling up and down between Coventry and Richard's place in Kent, usually with the wooden patterns around which the female plaster section of the mould would be formed to mate with the male section. The completed work was then returned to me for assembly. Richard's workmanship was magnificent, as evidenced when the mould was completed and the first glass fibre casts emerged. When the fairing was completed, the various metal bits which formed the mounting points, were bonded into the glass fibre either at my home in Coventry or in the Enfield factory. To ensure that the fairing would meet its criteria of being fitted, or detached, in three minutes or less the fixings had to align exactly with the allotted mounting points on the motor cycle chassis. No spanners were required; large wing nuts, tightened by finger, secured the fittings in place. Incidentally, steel hook ties, which dropped into a lug on the steering head, ensured a rigid locating point. Lower fixings on extensions from the engine provided mounting points for the bottom of the fairing. With a helper on the opposite side of the bike, removal of the fairing took about 60 seconds, although the fairing could be fitted, or removed single handedly. It just took a bit longer. At the rear, we adapted Enfield's quickly detachable mudguard assembly to lock into bonded-in brackets on the glass fibre bodywork. Again, undoing a finger-operated fastener released the bodywork which could be lifted clear in only 30 seconds! One waterproof multi-pin plug, from Joseph Lucas Industries, connected the twin front headlights to the main wiring

The upside down spark

'It never fails to amaze me how friendly and generous guys in the motorcycle business are towards one another. I had to rush over to Triumph to draw something one year and on the way my six cylinder Vauxhall was not purring along in its usual silky manner. This was strange, as I'd just had it serviced. Then I thought, maybe not so strange, somebody's disturbed something. So I mentioned this during the morning to Les Williams, more for conversation than anything. He asked for the car keys, then disappeared. Some time later, he returned the keys with a grin. 'It's fixed.' What was wrong, I asked. 'Upside down spark,' he replied. Les made a diagram, with a thin red line, showing peaks. That's a normal spark. Then he did another diagram below, also with a red line showing peaks, which Les called upside down sparks. So I asked what's the difference? Les replied, 'One's a negative, and one's a positive spark. One will ignite the fuel, the other will do it intermittently because the power is not in the spark.' Les spotted that the problem had been caused by someone who had re-connected the leads to the ignition coil the wrong way round, positive to negative, negative to positive. The difference between sucking and blowing! To me, that was typical of a motorcycle factory, take in your sick car and they fix it for you. But take your sick bike into a car factory and they tell you to clear orf!'

Eric was a decent bloke

Lawrie was at pains to refute one of the commonly held impressions of Oliver. 'Stories about Eric being tight-fisted were not true. The opposite was the case. Once he came back from the Swiss Grand Prix with a model Swiss Alpine hut as a present for me. Another time he gave me a super riding outfit and helmet. When with him, he always stumped up for drinks, meals, etc. So OK, I was working with him and it was in his interest to keep me happy, but the same thing applied to others.'

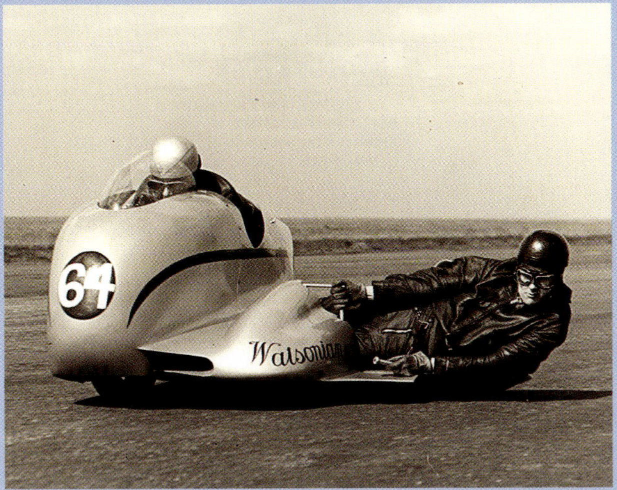

Eric Oliver and 'The Motor Cycle' Midlands editor, Kevin Gover, test a late version of Lawrie's design at Silverstone. The improved aerodynamics meant that Oliver could pull a higher gear on all the faster Grand Prix circuits. He won four successive world championships and changed the face of sidecar racing forever.

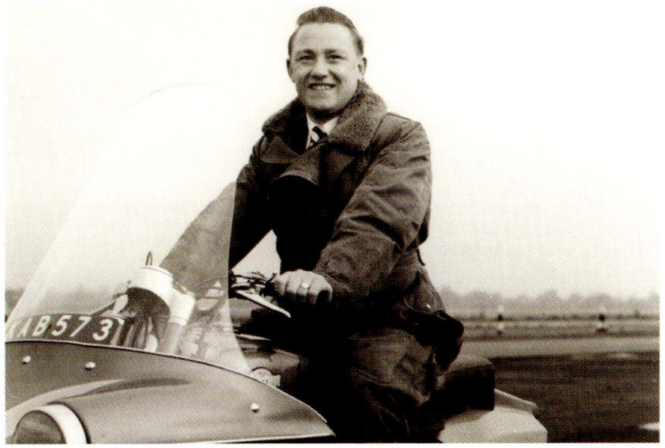

Lawrie looking pleased with himself on his Dreamliner.

loom; the rear lights were done similarly. It was thanks to Geoff Johnson at Lucas for making the additional looms, essential for joining to the standard wiring and facilitating rapid fitting and dismantling.

'We took advantage of the quickly detachable mounting system to get multiple sets of performance figures at MIRA for comparison with and without the fairing. I didn't expect much of a performance advantage under 40mph, because of the extra 45lbs weight, but there was a marked improvement not only in fuel consumption but also in acceleration! At the slowest speed, 30mph, the improvement in fuel consumption was 23%; at 40mph this improved to 26% but by 50mph, the improvement had galloped to 35%! Almost a second was saved in second gear acceleration (15-40mph) and 2 seconds saved in top gear acceleration from 40-55mph. Almost a second was saved on the standing start quarter mile. Maximum speed improved by no less than 7mph! Translated into on-road use, these figures became even more meaningful as the converted Bullet showed its willingness to climb main road hills in top gear for longer without the need for changing down. And re-fuelling stops became less frequent as the fuel economy kicked in on long journeys. All round, it was an eye-opening experience, and that's before taking into account the remarkable protection from the elements! I soon got used to making journeys in all weathers clad only in a coat and slacks, without the need for dressing up in my customary over boots, muffler, etc. Bliss!

'What happened to the Dreamliner after such astonishing exposure? Unbeknown to us, Sod's Law was about to strike when the FIM (Federation Internationale Motocycliste) intervened and banned the full frontal fairings, the so-called slab-sided 'dustbins', in 1957 on the grounds of lack of safety in side winds. This pleased nobody, not the riders, the manufacturers, the entrants, nor the organisers. Instead of consulting with aerodynamycists on alternative designs which ▶

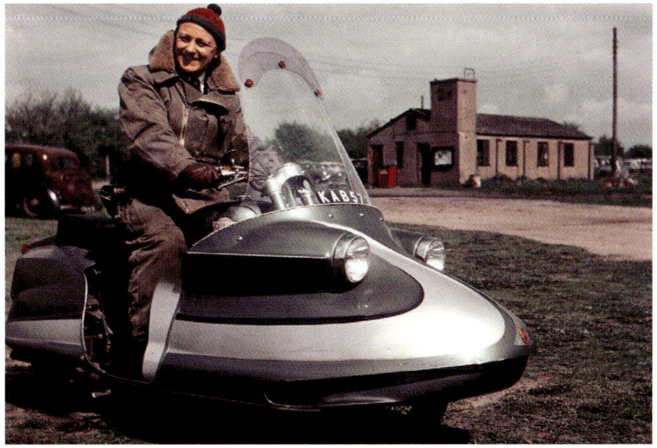

Lawrie in flying suit on the Dreamliner at RAF Bicester. Note the extension on top of the screen - to deflect air over the rider's head.

Gentleman Joe

'Eric was getting support from Norton, which is how he got into MIRA. Some of their plodding big singles made ideal sidecar machines and even though there was no link between their ancient push rod engines and the overhead camshaft Manx engines which powered Oliver's outfits Norton could see the value in success advertising. One day while I was at work drawing something at Norton's HQ at Bracebridge Street, Birmingham, I'd retired into a quiet corner at lunchtime to enjoy Eileen's sandwiches. Gradually, I became aware of a low moaning somewhere close by. On investigation, I discovered a black man next door who was crying his little heart out. Asking him what the matter was, he eventually told me that he was a 'nobody, worthless'. I reassured him that this was not so, but he said that Mr Craig had said so. He said that Mr Craig had popped his head around the corner, obviously looking for someone, and had gone away muttering, 'Nobody's here.' On the basis of that remark my new black friend had assumed that he was the 'nobody' to which Mr Craig had referred! Leaving my sandwiches aside, I went downstairs to race chief Joe Craig's office, and told him what had happened. Without a word, he leapt to his feet and rushed upstairs. Following, I was just in time to hear him apologising profusely to the distraught black man for the misunderstanding, and then putting him at his ease. That simple action was the mark of a gentleman, and another reason I always held Joe in the highest esteem. Those who claimed that he was cold and calculating could not have known the man beneath that exterior; he treated me with courtesy and generosity from the first moment I met him to the last.'

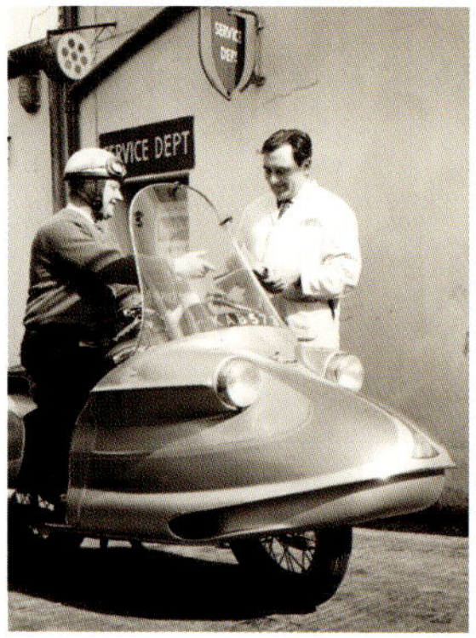

The Dreamliner passes its first MoT.

would not be unduly affected by side winds, as demonstrated by the Dreamliner, the FIM over-reacted by imposing a blanket ban. As the classic full-nose racing fairing had been the inspiration for the Dreamliner, ordinary motorcyclists seemed to lose interest after the ban was imposed on racing frontal enclosure.' An honourable exception was Doug Mitchenall, one of the Mitchenall Brothers, makers of Avon Fairings, Bulford, Wiltshire. An imaginative engineer, Doug was pioneering glass fibre in his handlebar fairings, saw the potential in Lawrie's Dreamliner and borrowed extensively from the design. His version, with well-engineered fitting kits for most British big twins, was successful to the point that once during a party in the Sefton Hotel he hailed Lawrie across the room with the remark 'Here comes the man who made me a millionaire!' Less elaborate than the Dreamliner, Avon versions were successful for some years but then fell victim to traditional British biker conservatism which fell back on partial streamlined designs based on contemporary race fairings of the '60s. With that, the concept of full frontal enclosure died, not even the newly arrived Japanese dared to risk it except many years later, and then, only on scooters!

Back at Enfield's, host to the Dreamliner, the management adopted the enclosure concept for their big twins, but without the very thing which made the Dreamliner such an eye-catcher, that beautifully sculpted nose cone. Enfield's version, called the 'Airflow' was also in high-quality glass fibre made by the Bristol Aircraft Company, and embodied weather protection but without that beautifully sculpted full frontal enclosure over the front wheel. 'Son of Dreamliner' it was not. Even the best efforts of Enfield's sales department failed to sell the numbers of Airflow models for which it had planned.

Velocette, the old established Goodman family marque with a rich racing heritage, seemed an unlikely source for a Dreamliner-evolved design but that is what they surprised the motorcycling world with in the middle of 1963. Called the Vogue, the bike used the well-proven water-cooled 192cc horizontally-opposed side valve twin from the ubiquitous LE 'Noddy' model, used by UK police forces for many years. The brainchild of Doug Mitchenall and Bertie Goodman, the Vogue used many of the Dreamliner's features, not least being well-engineered and practical use of all-enveloping glass fibre bodywork which concealed the power unit and completely protected the rider. I know this because I used a prototype loaded with camping gear for a race-reportage trip to the 1963 Dutch and Belgian Grands Prix. Even fully laden, the little bike cruised all day between 55 and 60mph, showing 70+mph on downhill stretches. It averaged 86mpg for the entire trip.

Those were the days when streamlined weather protection was virtually unknown so to enjoy protection of the same order as on the Dreamliner was sheer bliss and enabled long distances to be covered in comfort, if sedately. I remember leaving London in a torrential downpour which eased up only marginally for the journey to Dover, yet I remained dry throughout and the willing little engine never missed a beat, which was in stark contrast to the dozens of waterlogged cars and bikes lining the roads out of London. Doug had also incorporated the Dreamliner's twin headlamps, blending them into the Vogue's front nacelle. But, in deference to perceived public taste, he played safe and left the front wheel exposed. Tastefully finished in two tone cream and tan, the little bike was an eye-stopper wherever it went, doing sterling taxi work on shopping trips from the camp in the race paddocks to the nearest town. Pillion passengers were complimentary about the comfort of the machine which, unusually, used footboards instead of footrestst, allowing the rider and passenger some freedom to move about, an aid to comfort on a long run. So what happened to the Vogue? It met a similar fate to the Dreamliner, being ahead of its time. Forty years on, the design is sufficiently smart to look modern and, if introduced to contemporary markets, could sell in numbers which it sadly failed to do when first marketed.

Ungentleman Edward

'In contrast to the friendly atmosphere at Enfield's, the opposite was the case at Triumphs when Edward Turner was in charge. He had to be called 'Mr' on every occasion, but he never returned the civility. He embarrassed me on numerous occasions by tearing strips off his own employees who, in some cases were friends of mine, in front of me. He did this repeatedly with Jack Wickes and John Nelson, loyal employees and key men in the development section. Jack never retaliated but John stood up to Turner, which I would have done too. Turner's attitude turned me off him for ever and I had as few dealings with him as necessary.'

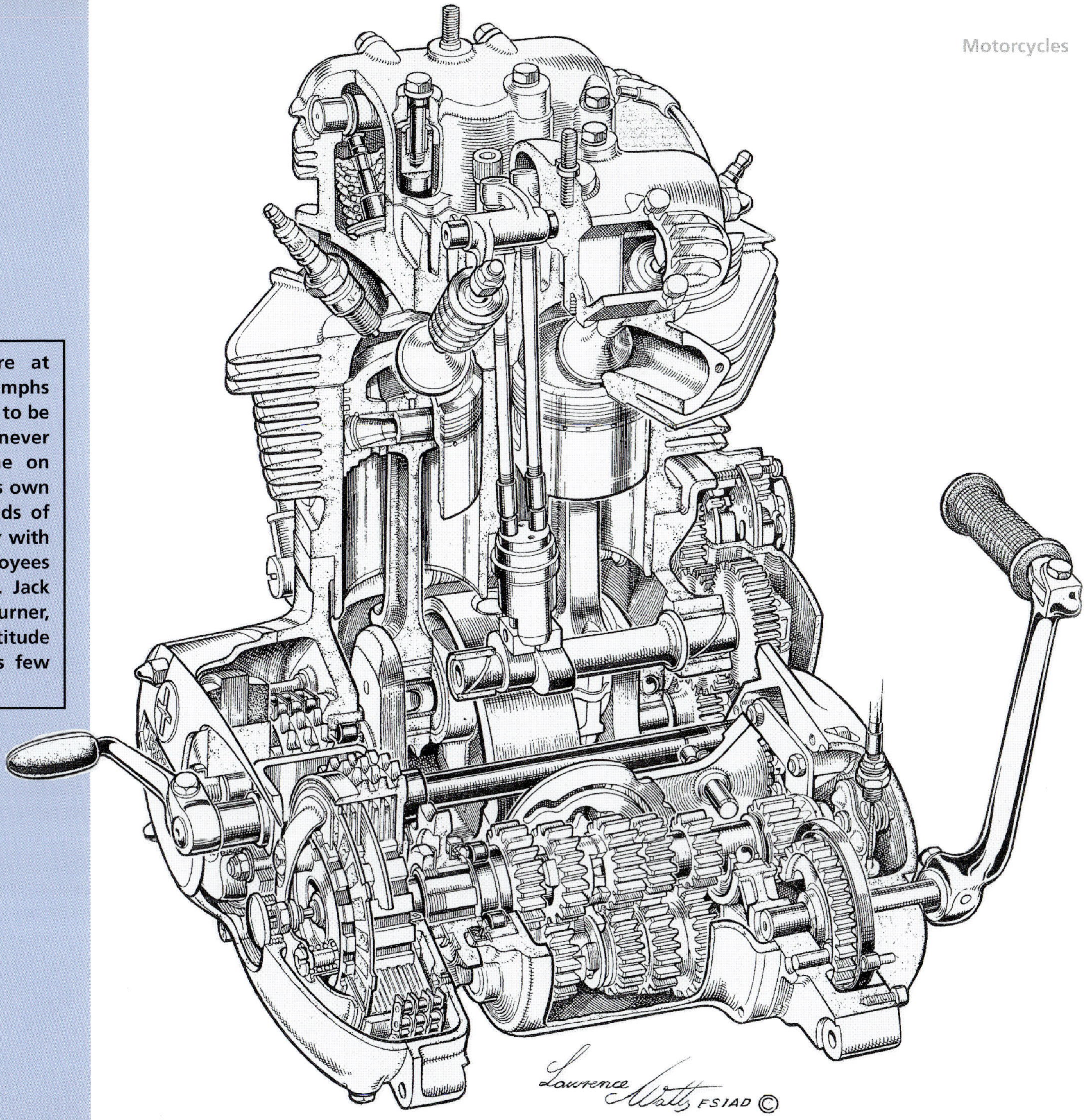

Triumph 750cc twin showing the new 5-speed gear box and left-hand gear change for the USA and Continental markets.

In at the deep end

'One day a phone call summoned me to hurry and join Oliver's crew who were testing over at MIRA. My query as to what was wrong evoked the enigmatic response, 'You'll see when you get here.' So Eileen and self piled into my sparkling new Standard Vanguard, to which I'd just fitted a tow bar, and we whistled over to MIRA at Nuneaton - a short hop for us. When we got there, Eric Oliver was sitting on top of his lorry's cabin. Just for fun they had driven into a water splash not realising that MIRA had two splashes, one of which was shallow and the other deep, very deep. The lorry was totally submerged. Ben had swum ashore and called me to the rescue while Eric had climbed onto the roof to stay dry. Worst of all the outfit, which was in the back of the low-loader lorry, was totally submerged! They clearly expected me to rescue them! All I had was my precious new Vanguard, with its untested tow bar and, most fortuitously of all, a tow rope in the boot. When Eric saw the rope, he immediately jumped into the water, grabbed the rope, swam underwater and, in a series of dives, secured it to the underside of the lorry. I tied the other end of the rope to the tow bar and slowly eased the lorry, sidecar, and Eric out of the water splash onto dry land. It took an age to drain the water out of the lorry's engine but we did and it ran again, just in time to set off home before dark.'

Car on two wheels

During one of his social visits Chez Watts, Bertie Goodman, an admirer of the Dreamliner, startled Lawrie by asking if he'd be interested in designing a car on two wheels! While Lawrie was recovering his wits, Bertie explained that he'd been having discussions with a Government department about solving inner city traffic congestion. That's right, congestion was a talking point 40 years ago! What Bertie had in mind certainly caught the imagination of the civil servants because it was no less than a car on two wheels! It would use a motorcycle engine for economy and performance, but would seat two people in tandem within a narrow body which would also provide them with a modicum of weather protection. The body would be so designed as to provide safety in a crash. After many discussions, and furious sketching by Lawrie to interpret Bertie's thoughts, a design slowly took shape. Like nothing else that Lawrie had ever drawn, the boxy shape took form in metal and went on test. For simplicity, the prototype was ▶

Lawrie at speed in Tom Kileen's K17 space-saver 'Car on 2 wheels'. The outrigger wheel each side is lowered just before the K17 stops. Retraction is automatic with the vehicle under way.

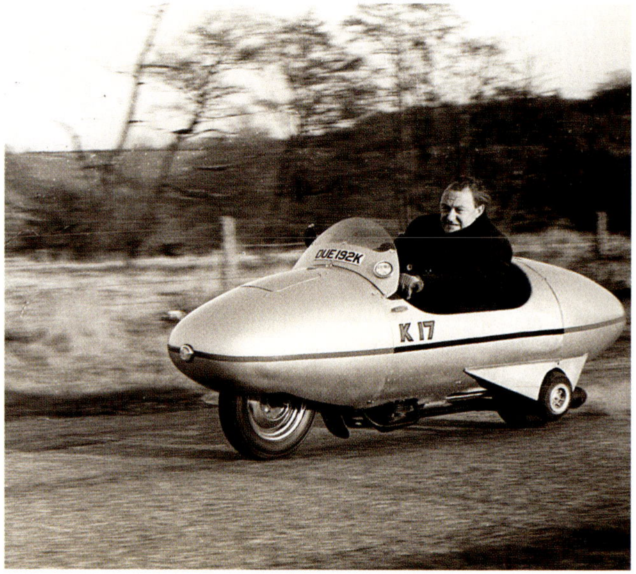

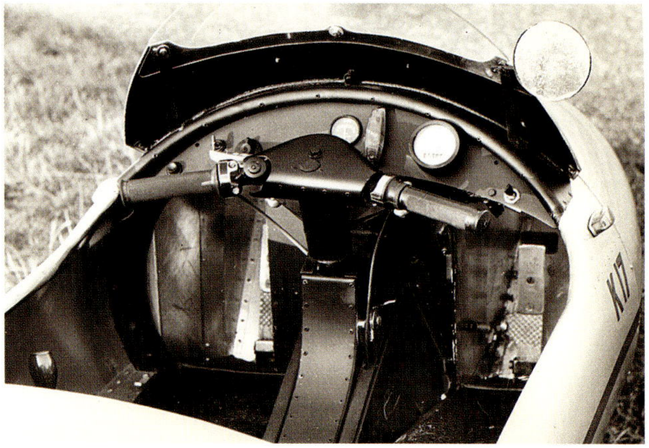

Motorcycle controls are fitted.

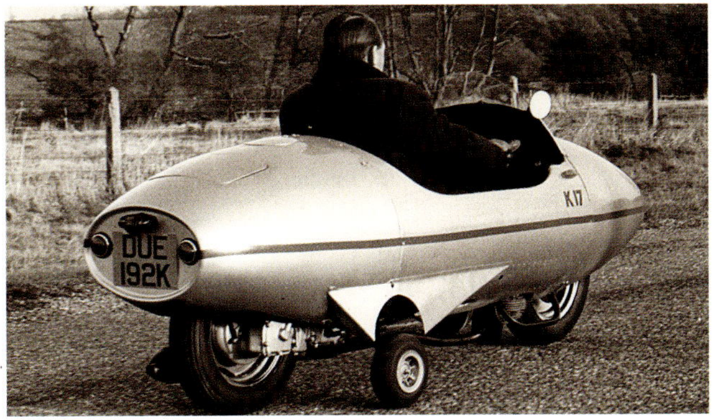

Left and below: Rear view showing the clean lines of the K17. Though the prototype was a single seater, room was available for a passenger behind the driver.

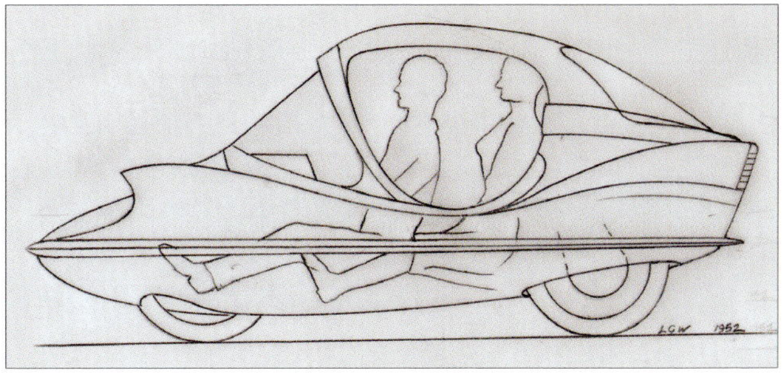

built for one rider only, though space was left to add a rear seat. Retractable wheels each side were a major design feature because once inside the vehicle, the rider could not reach out and touch the ground. They were designed to retract once the machine was moving and when the speed dropped, such as stopping at traffic lights, the outrigger wheels dropped down and supported the machine. Lawrie not only called on his aircraft design experience to engineer the retractable mechanism to ensure that it was entirely fail safe, but also did many experiments on the Dreamliner to see how slowly he could ride before starting to topple or getting into balance wobbles. As a result, he designed an undercart which would drop down at 5mph. Increase the engine revs and the wheels retracted. To be on the safe side, he spring-loaded the wheels so that if they did come down slightly early, or start to retract prematurely, they would not interfere with the machine's steering. 'This was a safeguard because in practice I found that 2mph was about the slowest before I needed to put my feet down, but others might have felt safer doing it at a slightly higher speed. It all worked and Ken Sprayson made it for me at Reynolds Tube Co. Incidentally, this project was also taking place around the same time as the Dreamliner was on the go. Harry (Louis) was keen for me to do an article for The Motor Cycle but I have no idea when it appeared.'

Years later, a chap called Tom Kileen, who was a car body designer at Aston Martin, came on the scene, made a machine based on drawings he'd seen in The Motor Cycle, and invited Lawrie to see and to ride his machine. 'It was very stable, I could ride it down to almost stalling point, hands in the air! This was what my old mate Vic Willoughby did when he tested Gustav Baumm's NSU-powered 'Flying Hammock' some years earlier. But Vic didn't appreciate it when I replicated the gesture! My machine and, as far as I know, Kileen's, went to the Ministry for assessment but nothing ever came of it. That they fulfilled the criteria laid down by ▶

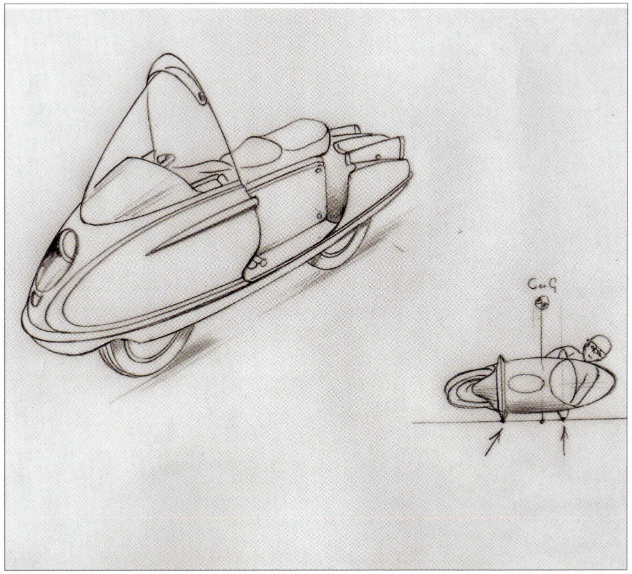

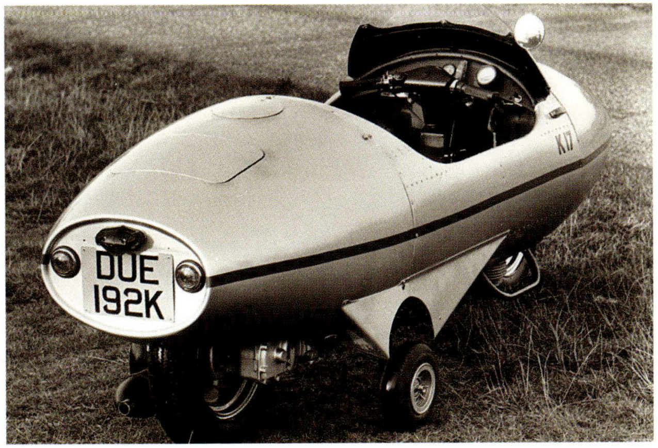

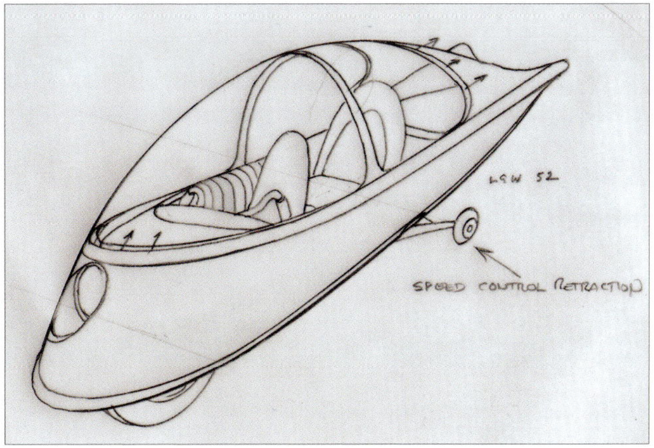

Using material based on his Dreamliner, Lawrie drew these concept 'two wheel cars'.

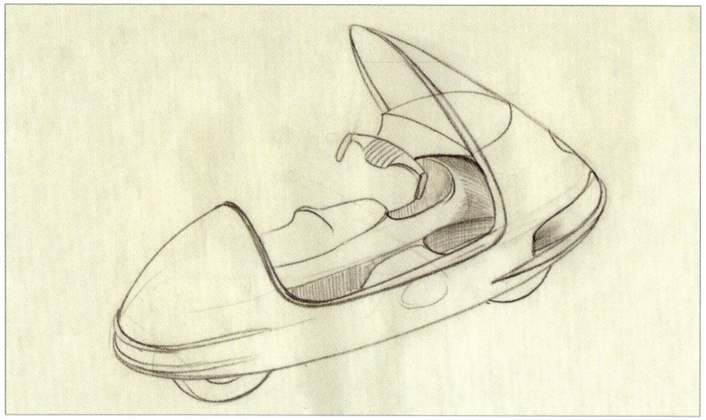

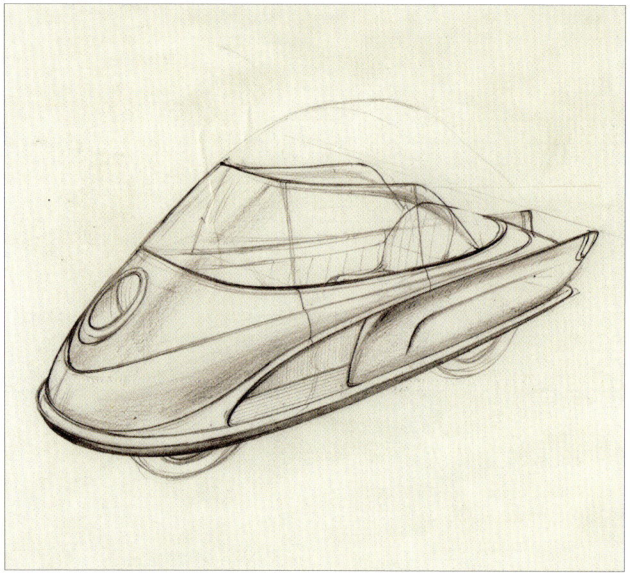

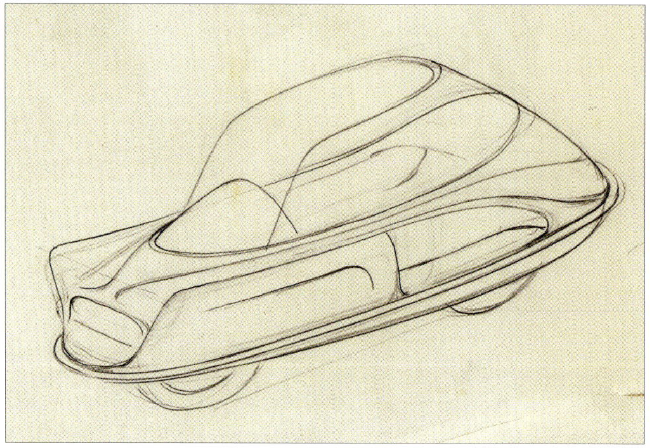

More 'Two Wheel Car' concepts, to meet the ministry brief. None were built.

the Ministry was beside the point; what the bureaucrats overlooked was that such a vehicle could never be as narrow or manoeuvrable as a solo motor cycle and it was questionable whether it would be as 'safe' as they required. It's extraordinary how civil servants seem to assume that accidents will be reduced to zero if you wrap enough metal and cotton wool around the drivers! It never occurs to them that accidents are almost invariably the result of one human doing something daft to another. The type of machine, whether it be a Tesco trolley or a Ferrari, is immaterial.'

Brave left hand

'After a few years in the office I was entrusted with an outside assignment in deepest Southport, of all places! Brockhouse Industries had been licensed to import and sell the Indian Brave, a modest little bike with a 250cc side valve engine. A really nice guy, whose name I forget - looked after me very nicely. Which was just as well, because I couldn't use my right hand to draw! Gout had afflicted the muscles and it was impossible for me to hold a pencil. That's when I discovered that, I could draw equally well with my left hand! So I did a full drawing of the Indian engine and it was published shortly afterwards. The office never knew the circumstances, which was just as well, because the affliction stayed with me for some months before disappearing.'

Clive the crankshaft tuner

'One day, when I was working on a drawing of the prototype 750cc 3-cylinder BSA in the comp shop, Clive (Bennett) suggested that we break for lunch. On the way out, a mechanic who was obviously having trouble balancing a crankshaft asked Clive if he could spare a moment. Clive replied that we would be back in an hour, so keep working and we'll lend you a hand when we get back. On our return, the mechanic was still struggling. Clive offered to help. He carefully set the crank up on the balancing rig, checked the run-out minutely with his dial gauges, took hold of a huge rubber hammer, then belted the crank an almighty wallop on a pre-determined spot. The gauges showed that the crank was now perfectly balanced. Not a word of a lie, saw it with my own eyes! Now that's what I call intuitive engineering skill. True to form, Clive called it a scientific pisser! I might add that this was NOT how the production engines were built! The balance factor would have been worked out by then and incorporated into the production cranks.'

Ariel Arrow

In conversations dealing with motorcycling, Lawrie frequently refers to Clive Bennett. 'I first met Clive when he was working at Ariel Motorcycles and I was sent there to draw the new Ariel Arrow. We subsequently became good friends, shared many experiences, and learnt from one another. He was interested in the Dreamliner and rode it a lot and I taught him to fly and he taught me things I might never have otherwise considered about motorcycles. Introduced as a sports version of the Leader, the Arrow had a more sporty engine and was raced successfully in the Thruxton 500 Miles and, later in race trim, by Peter Inchley. I understand that the engine was based on a post War 250cc Adler and I liked it a lot - except that you had to get the timing of your gear change precisely right when changing up or down to avoid clashing of gears. Well, Clive had a lot to do with the development of the Leader and the Arrow. I watched Clive make the original frame out of sheet metal, they didn't have a press then, and he was trying to prove to the Ariel management that a box-section frame was a good idea. Anyway, Ken Whistance, the managing director, was one of those comparatively rare captains of industry, he had an open mind. He was prepared to accept new ideas if he was convinced that they would work. So he accepted Clive's concept, gave it the go ahead, the Leader went into production and caused a sensation when it was introduced to the press in the summer of 1958. Because of my association with Clive, I had a long-term loan of a Leader and loved it. The weather shielding reminded me of my own Dreamliner but the Leader had delightfully light steering, was effortless through corners and the two stroke engine was a better performer in all respects but one; a gallon of fuel went a lot further in the Dreamliner, especially when cruising close to the respective maximum speeds.'

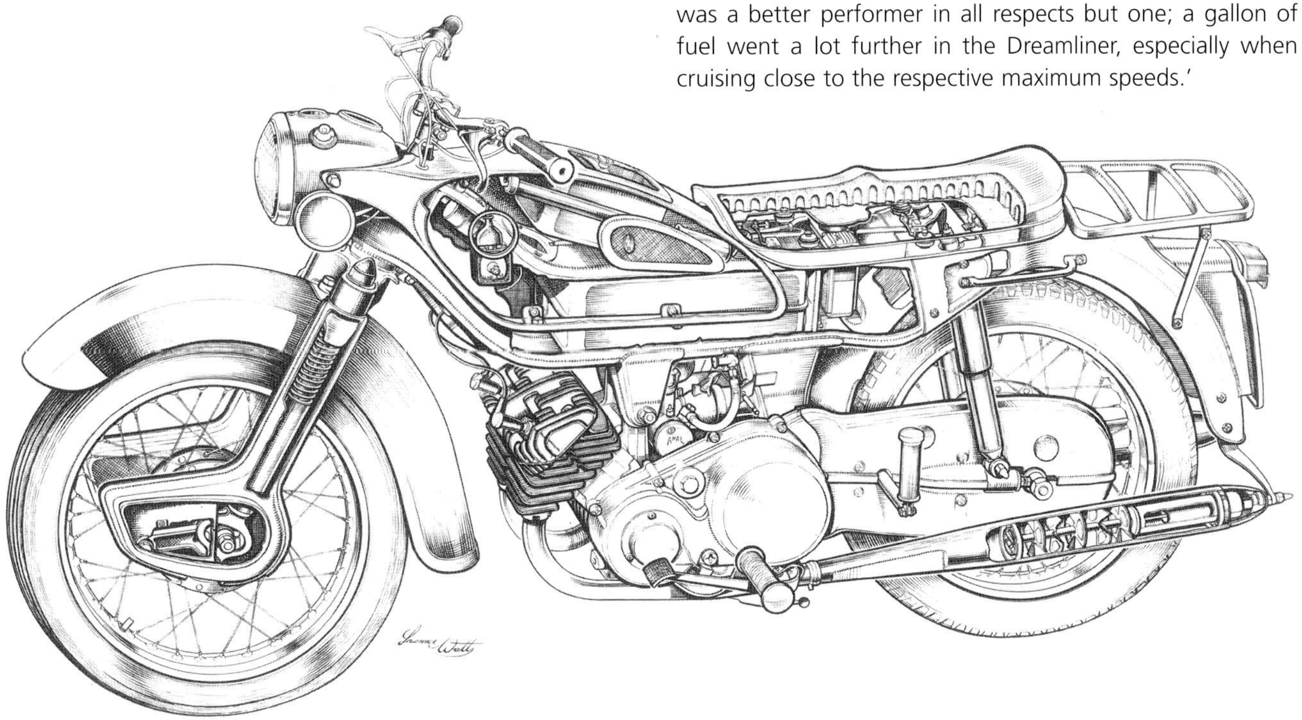

The Arrow, based on the Leader, was aimed at riders requiring a sportier ride, which it suppplied in bucket-loads.

Aimed at the sportier end of the market, the Arrow came two years after the Leader and two-stroke tuner Hermann Meier soon became involved in some indecently rapid Arrows. The most impressive was Mike O'Rourke's, which finished seventh behind the works MVs, Morini and Hondas in the 1960 250cc TT, at an average of 80.13mph, an average speed higher than the standard model's maximum! Proof that it was no flash in the pan came later when Cecil Sandford and Sammy Miller scored a fine second place in the 250cc class at the Thruxton 500 Mile production machine race. Lawrie claims the Arrow to be a 'milestone' machine and the Leader to be so far ahead of its time that it could have made Ariel a very prosperous company. 'Order books were full soon after the machine was launched, and Ken Whistance told me that the Leader was the best thing they had ever done. Triumph, the senior partner in the Group, were trying all they could to boost their name and profits and as there was no love lost between Triumph's Edward Turner and Ken Whistance, Turner ensured that Ariels would not threaten this product by obstructing Ariel's expansion. It may seem daft now, and it was daft then, but that's how politics were in the days when the British industry could see no opposition over the horizon. They assumed that the opposition could only come from within their own industry! That attitude, coupled with under-investment in new models, such as the Leader and Arrow, had to be a reason for the decline of the British motorcycle industry. Too few Ken Whistances, too many Edward Turners, in my opinion.'

Lawrie's drawing clearly shows that the designer intended to use the production pressed-steel frame for engines of greater capacity than the 250cc two stroke. 'Yes, I noticed that when drawing the layout,' confirmed Lawrie. 'The method of mounting the engine below the pressed beam frame member lent itself to bigger units, if necessary, and this was confirmed some time later, unfortunately after Ariel went out of business.' Under a bench in the experimental shop a transverse four cylinder overhead camshaft engine was found. Sammy Miller, then employed at Ariel, told me that it was a one-off, but under development with a view to

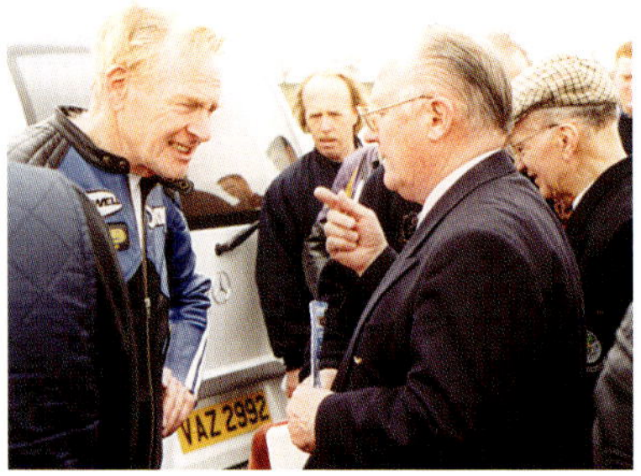

Lawrie forcibly making a point to Sammy Miller, who worked at Ariel when the Leader and Arrow were being developed. Later he raced one successfully.

production. It was constructed, and run, in great secrecy in case Turner became aware of its existence! Had it come to fruition, it would have been the flagship of the Ariel range - and a major thorn in Triumph's side. So no surprise, then, when Ariel was closed down, still with unfulfilled orders for the two stroke range… 'I think that this horrible story is one of the great tragedies of the motor cycle industry,' recalled Lawrie, sadly. 'It illustrates, as if that were needed, that the failure of the industry was as much, if not more, due to the incompetence of individuals rather than a mechanical disaster. Bob (Currie) and I used to discuss this aspect of the industry ad nauseam, but Bob could not bring himself to do anything about it, he was not that sort of guy.'

About that time, Ariel introduced a 50cc three-wheeled moped with a hinge in the middle, which was designed specifically for women. Lawrie drew it and as he progressed, he became increasingly intrigued. 'I didn't think of doing a three-wheeler when I set out to design the Dreamliner but this later, hinged concept, was different. The first thing I would have done was to take the wheels off and replace them with bigger ones for stability. Ivor Davies, of Triumph, learnt that I'd done a cutaway drawing and he arranged for me to have a loan of the little three wheeler, which was powered by an Anker 50cc two stroke automatic transmission manufactured in Holland. It was ideal for an experiment I wanted to do, but that's another story.'

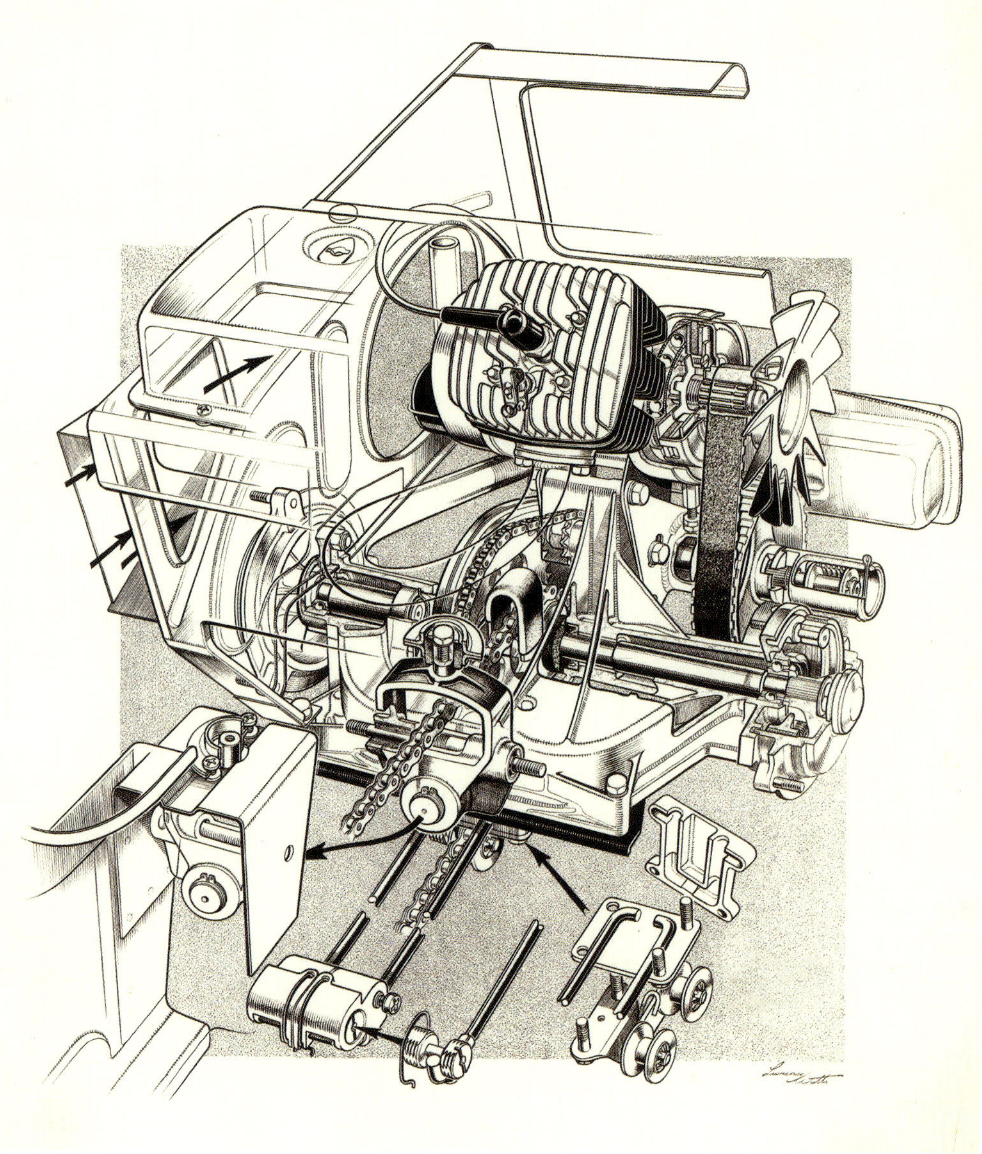

The ingenious 50cc Ankermatic engine which powered Ariel's 'Bendy' three wheel moped. It was not a commercial success. Honda later repeated the experiment with their Stream, and had marginally more success, especially with female customers who liked the 'safety' of the third wheel.

500cc Silk

Compact Silk 500cc water-cooled twin cylinder two stroke was an in-house design to break away from Scott heritage. In the drawing, Lawrie has cut in two the left hand cylinder to show the porting and the water passages around the cylinder. He's also sectioned the adjacent full-disc flywheels in order to show the crankshaft and big end layout. Unusually, the drive is taken off the centre of the crankshaft

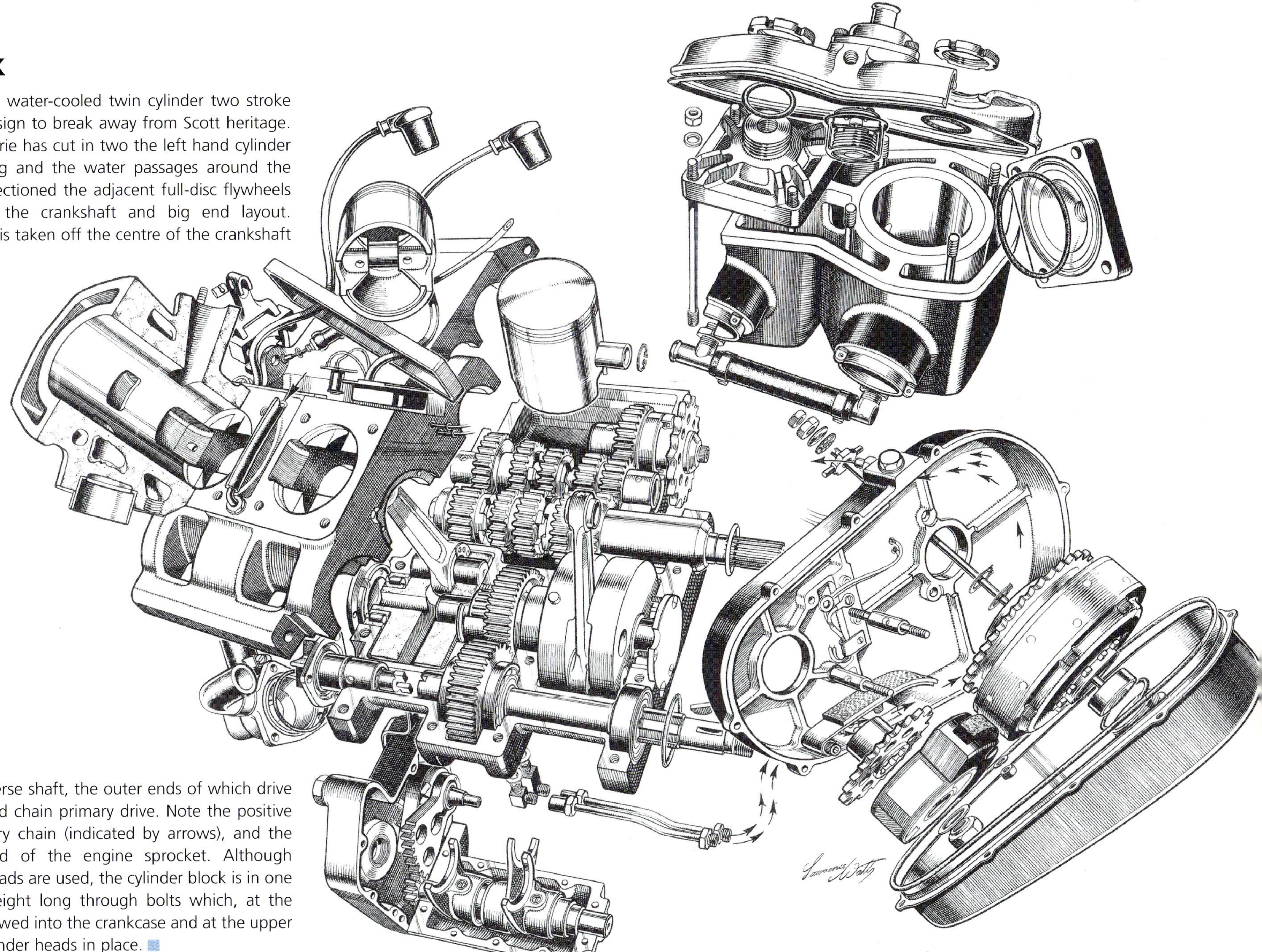

to a massive transverse shaft, the outer ends of which drive the water pump and chain primary drive. Note the positive oiling to the primary chain (indicated by arrows), and the alternator, outboard of the engine sprocket. Although separate cylinder heads are used, the cylinder block is in one piece, secured by eight long through bolts which, at the lower ends, are screwed into the crankcase and at the upper ends secure the cylinder heads in place.

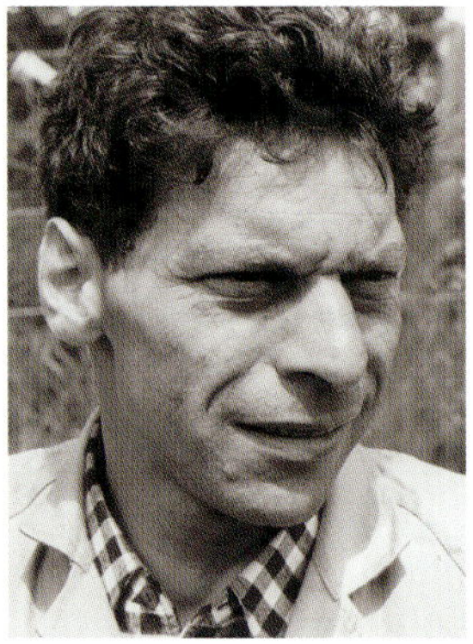

Helmut Fath.

Fath flat four

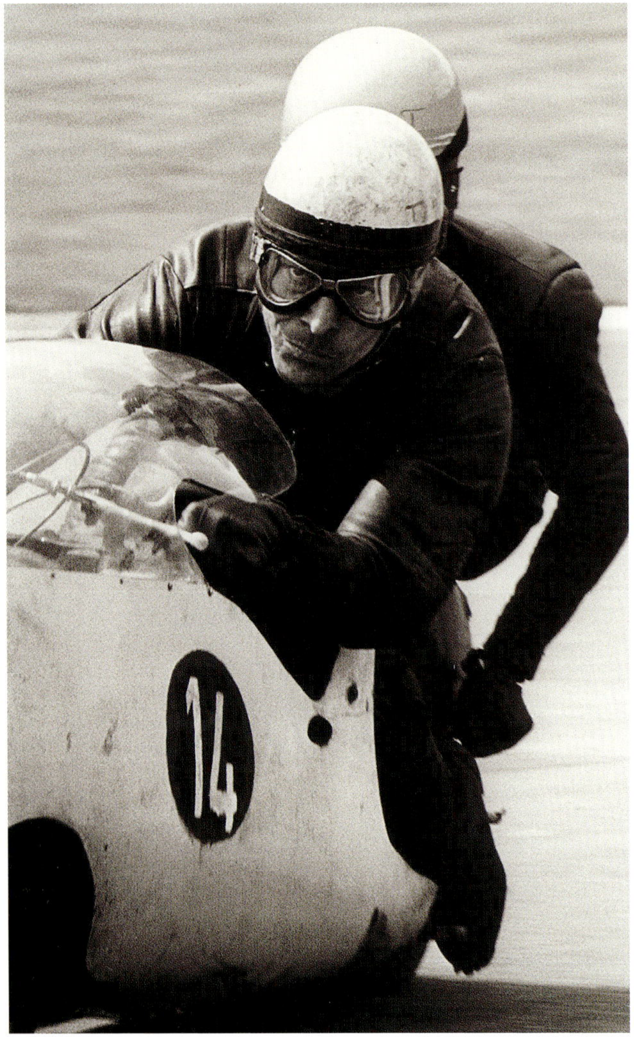

Determination! Helmut Fath in typically press-on style rounds a fast left-hand bend.

One of the most charismatic and talented racers ever was Germany's Helmut Fath. For more than 20 years he mesmerised all who saw him race either BMWs or his own URS and Aro Fath machines. Unusual for a top flight rider Fath was a brilliant engineer and, when BMW refused to give him the necessary parts to re-build his race engine after a long lay-off following a serious crash, he revealed his true talents as an engineer by designing and building his own engines. He had already won the 1960 sidecar World Championship for BMW, now he'd go and win it for himself and in so doing end BMW's 15 year domination of the sidecar championship. A tall order for a well funded factory but for a self-taught engineer, still on crutches after his big crash, with only a modestly-equipped home workshop, a seemingly impossible task. Not for Helmut Fath, life for him had always been a challenge, this one was just a bit bigger… Shutting himself away from the world in his workshop in Ursenbach, near Mannheim, Fath grafted from dawn till dusk for four years. By now he'd been virtually forgotten by race public and fellow competitors, except for an occasional whisper that 'Fath's up to something, nobody knows what…..'. Unannounced, Helmut pitched up at the 1966 West German Grand Prix at Hockenheim. The surprise caused by his appearance was surpassed only by the uncovering of his sensational new URS, the fruits of four years' graft; an across the frame double overhead-camshaft four cylinder with fuel injection and a six-speed gear box. From the start, the machine was quick, displaying great potential and Fath, in spite of difficulty walking, soon regained much of his typical robust style with the sidecar's wheel usually well clear of the road on right hand bends (in comparison with British and many other continental outfits which had their sidecars mounted on the left, Fath's was on the right). As a yardstick, the four cylinder MV Agusta which had won the 500cc World Championship for eight successive years revved to 10,500 rpm, but the first version of Helmut's four revved to 14,000rpm! (Through later development, he found another 2,000rpm). To achieve this while maintaining reliability, he paid special attention to metallurgy, with very slim titanium connecting rods, winding his own valve springs from a high quality Swedish steel, and using a crankshaft configuration in which no two pistons were together at the top or bottom of their strokes. This was achieved by spacing the four crank pins at 90 degrees to one another resulting in dead-centre and mid-stroke positions coinciding. Another unusual feature was that the crankshaft was made in two halves with pinions in the middle of each driving opposite ends of a half-speed countershaft. ▶

The crankshaft was carried in four bearings, had forged connecting rods and a central labyrinth gas seal. The pegs on the left hand end drive the clutch drum.

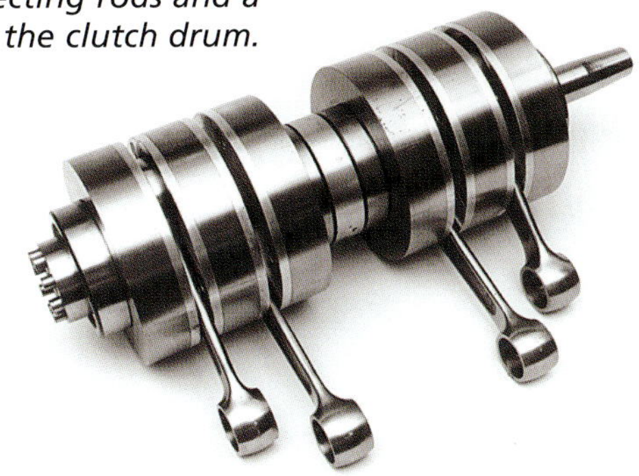

Fath two stroke 500cc flat four

Individual carburettors fed each of the four cylinders via disc valves. The external belt drives are clearly visible. To ensure the engine was fitted as low as possible in the frame, the exhaust pipes exited above the horizontal cylinders. Fath favoured separate pipes and expansion boxes.

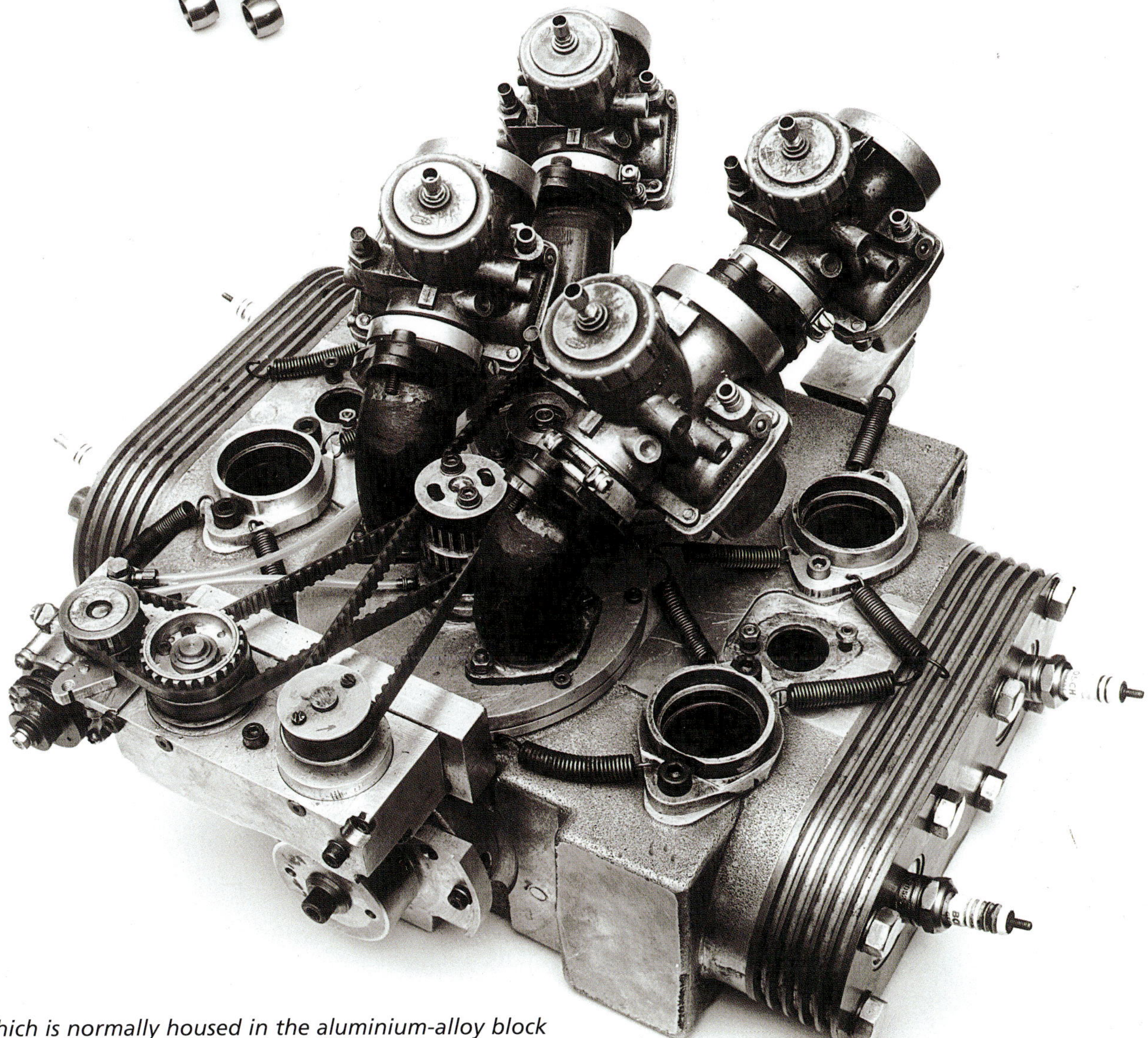

The uneven firing intervals of 90 degrees, 180 degrees, 270 degrees and 180 degrees created ignition problems. Chief among these was the 1,000th of a second for the ignition coil to re-energise within the shortest interval, but number 1 cylinder also suffered from weak sparks at comparatively low revs. Fath partially solved the problem by fitting one double ended coil per cylinder, so each of the twin plugs was fed from opposite ends of the same coil. The ultimate solution came when he made a contact breaker with four pairs of points spaced at 45, 90, 135 and 90 degrees.

Collapsed pistons were cured by re-designed forged items. When I asked Helmut what the power was he replied, 'I don't know! My home-made dyno only tells me if the power is more or less than the last time I tested. That's all that matters.' When I suggested that the obvious power advantage he had over everything else, whose bhp was known, he agreed that an output of 80-bhp was 'possible'. However much it produced, Fath took great pleasure in using ▶

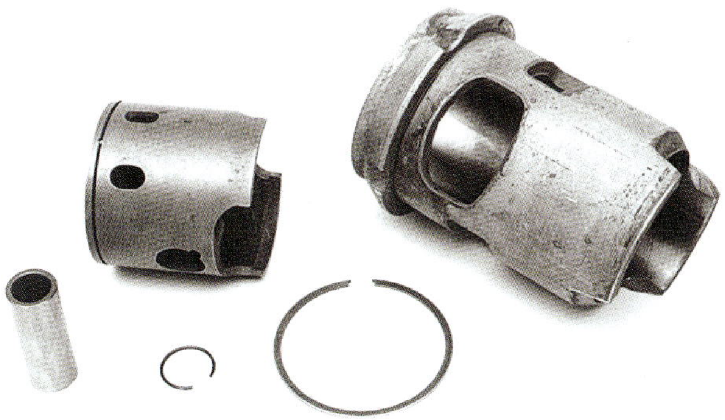

Each piston had only one ring, shown here with the liner which is normally housed in the aluminium-alloy block

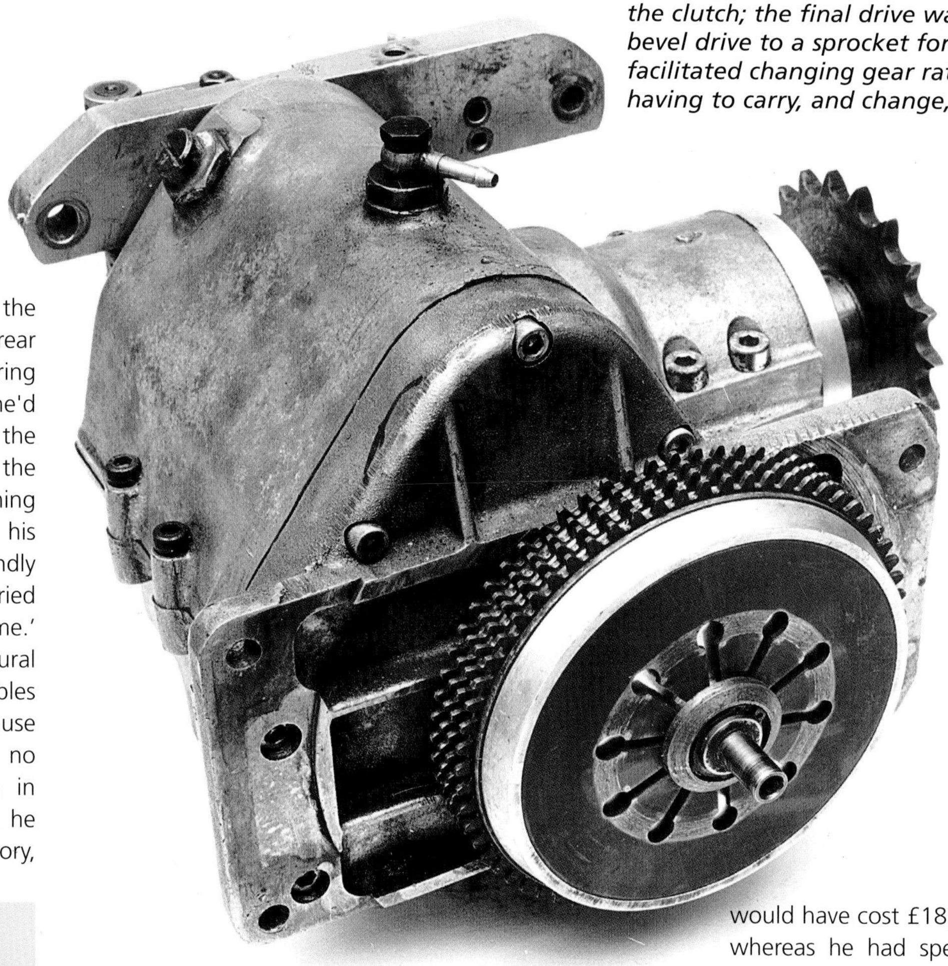

The longitudinal crankshaft drove the six speed gear box via the clutch; the final drive was taken through 90 degrees by a bevel drive to a sprocket for the chain final drive. This facilitated changing gear ratios, a more economical means than having to carry, and change, complete clusters as on BMWs.

it. I watched him the following year (1967) in practice for the French GP at Clermont Ferrand when he burnt out his rear tyre in enormous power slides when nobody was pressuring him. He did the same in the race, retiring just after he'd taken the lead, a pattern which was to be repeated for the rest of the season. Talking to him at the 1968 TT, I made the rather obvious remark that he had a great chance of winning that year's World Championship if only he restrained his right hand. A look of sadness spread across his friendly features as he nodded before replying, 'I know, I have tried but always I wish to go quickly. That is really racing for me.' At 39 years of age, he managed to curb his natural exuberance to finish fourth - after numerous troubles delayed him. He apparently drew solace from that because for the rest of the season he drove 'sensibly', though no doubt frustrated, to claim the title at Hockenheim in October, where he had started. In a season's end chat, he told me that his engine, had it been made by a factory, would have cost £18,000/£20,000 (a lot of money in 1968), whereas he had spent £5,000 on his. The only financial assistance he'd received had been from oil, fuel, and accessory companies. Too much of a gentleman to repeat the words of congratulations greetings from BMW.....if they did. Unhappily, there was to be no fairy tale ending for this Fath four; after an initial spell of success, when it challenged Giacomo Agostini's all-conquering 500cc MV Agusta in the solo class, ridden by Karl Hoppe and Englishman Bille Nelson, the end came when the team was purchased, with American backing, by Friedl Münch while Fath was recovering from

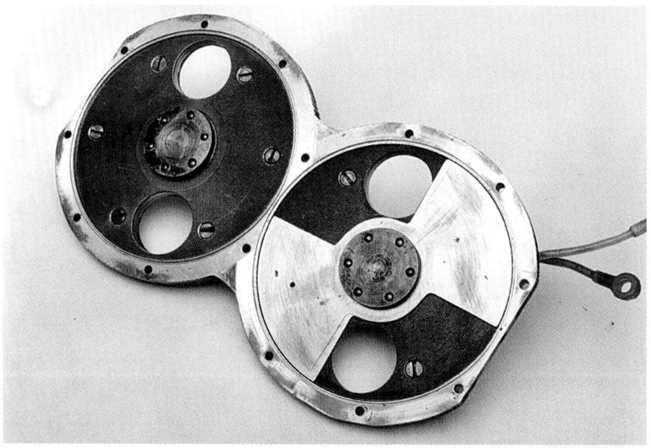

Each of the four cylinders had its own disc valve. The top cover is off, showing one of the single discs in position; each inlet port is opened and closed by alternate discs.

breaking a leg in Finland in August 1969. When the agreement took effect in May 1970, all Fath's machinery went to Münch. Did Helmut retire, or mooch about looking cheesed off? Not a chance! Instead, he resurrected his interest in racing two stroke engines and offered his services as a two stroke tuner, in a short time becoming one of the most successful in Europe. Inevitably, he thought about making a racing come-back himself and, typically, planned a new engine, not only for himself, but also for his great friend and collaborator, Billie Nelson, to ride in solo events. But this time the engine was to be a brand new two stroke flat four cylinder....How different can one get? But it was not so surprising to those who remembered that back in the early '60s he had planned a water cooled flat four cylinder four stroke to replace his BMW twin. Ten years later, the scene had changed dramatically with two strokes beginning to dominate racing in all categories, providing good work for a talented tuner such as Helmut Fath. With his experience of tuning two stroke racing engines, of which he was one of the leading private exponents in Europe, it seemed logical to make his own two stroke engine. Historically, Germany had been the home of two stroke design since Zimmerman; Walter Kaaden had kept the East German MZs in the frame at race tracks all over Europe and much of this knowledge was available to supplement Fath's own two stroke racing knowledge. So he dusted off his old plans and the original flat four became a water cooled, disc valve, two stroke laid across the frame BMW-style and driving through Fath's own six speed gear box integral with the engine.

Vic Willoughby, a long time admirer of Helmut Fath, became interested in writing a story about the new engine and, naturally, he asked Lawrie to draw it. 'It was a long time ago and I don't remember a lot about it,' said Lawrie, shaking his head sadly, 'but I do remember what a wonderful bloke Helmut was and I also shared Vic's enthusiasm for him. I'd not seen an engine like this before and I so admired Helmut's ingenuity that I really got stuck in and became totally absorbed in the drawing. Then something happened, I'm not sure what, and I was told to put the drawing on one side. Apparently, Vic did not complete his story so the drawing has never seen the light of day until now. Which is a great pity as it would have stood as a monument to an indomitable character who never gave in, no matter what the odds stacked against him. He was also one of the most unassuming ▶

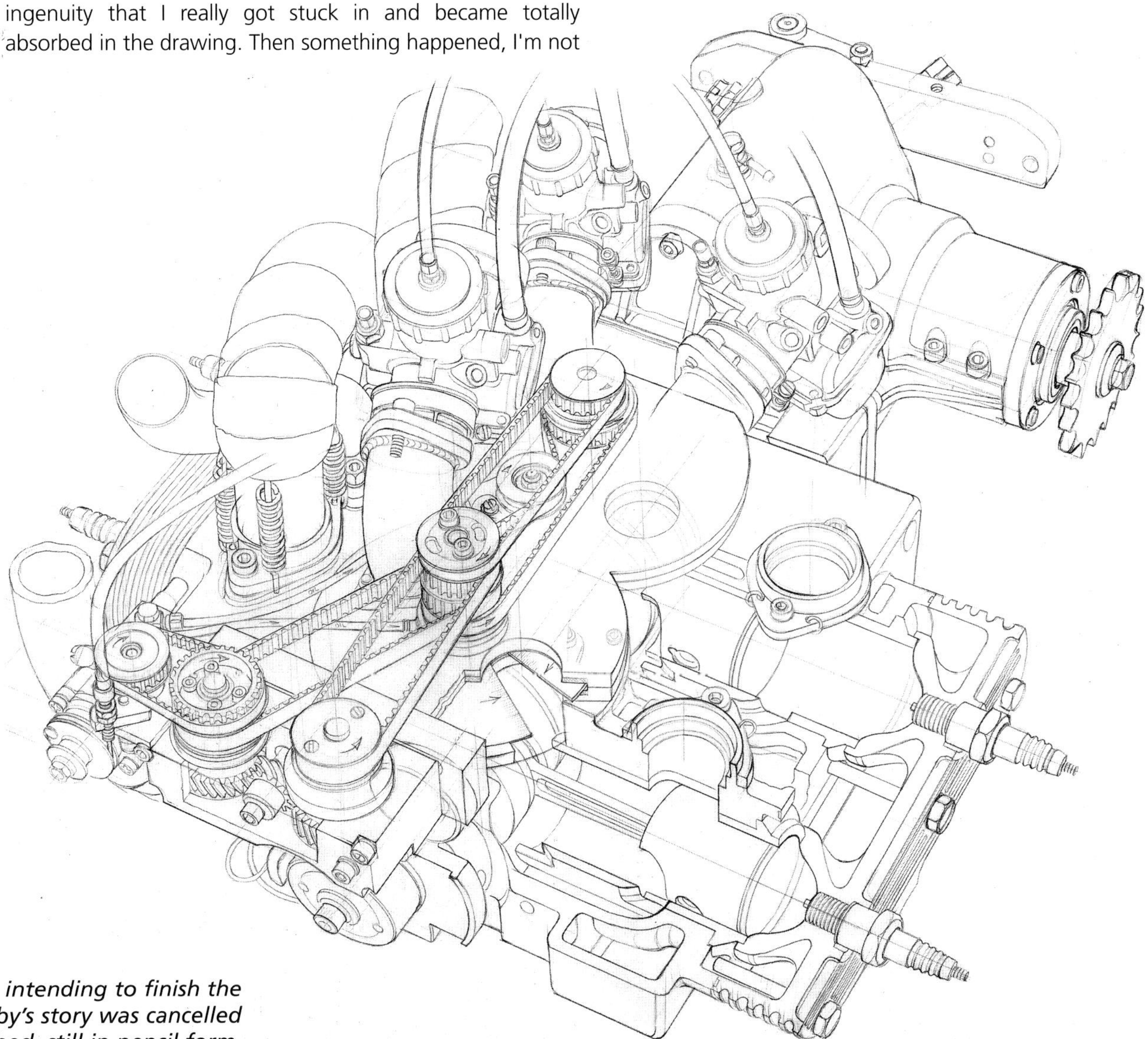

Lawrie did the preliminary sketches in the Isle of Man, intending to finish the drawing at home. Something intervened - Vic Willoughby's story was cancelled - and this is the first time the drawing has been published, still in pencil form.

and friendly characters in racing and deserved greater success than he achieved, if nothing else than as a reward for his incredible engineering achievements as for his spectacular driving. That's why I'd like this drawing, even in its unfinished pencil state, to be a tribute to a truly remarkable man, and an all round good egg, as we say in Coventry.'

Some people at the time thought that Fath had drawn inspiration from another German flat four, the Berlin-based Köenig. Apart from being horizontal flat four cylinder two strokes, the similarity stopped there. With a bore and stroke of 56x50mm, giving a capacity of 495cc, Fath's engine had a longitudinal crankshaft, an integral six-speed gear cluster with a clutch at the rear of the crankcase. BMW fashion bevel gears turned the drive through 90 degrees to a sprocket on the left hand side for a chain final drive. Another major difference from the Köenig was Fath's adoption of a belt-driven oil pump to force-feed the caged needle roller big ends, main bearings and disc bearings. The aluminium alloy cylinders had Nickasil chrome bores and, initially, Fath used Yamaha pistons pending delivery of his own design single-ring pistons. Uniquely, each cylinder casting incorporated half the crankcase which ensured that the crank chambers themselves were also water cooled. Fath did not use a water pump preferring a thermosiphon system with water entering the underside of the cylinders and exiting at the top. Externally, Lawrie has shown the four separate 34mm Japanese Mikuni carburettors which fed the engine via four disc valves (Köenig used one). Not shown are the four separate expansion chamber exhausts, which Fath preferred for extracting maximum power. This was 112-bhp at 12,200rpm. Ignition was initially by Bosch flywheel magneto but, to save weight, Fath later used a battery to power the pulse unit fitted to the front of the crankshaft.

In the hands of Billie Nelson, the new four made its debut as a solo in May 1974 at the German GP at Hockenheim. The power was exceptional but this accentuated deficiencies in the frame which caused Nelson's retirement. A new frame for the Dutch TT gave Fath the opportunity to show the bike's potential when, after changing a plug at the start, Nelson got going just after the leaders completed their first lap. In spite of not feeling totally confident with the machine's handling on the fast curves, Nelson had no bother catching the leaders, who just happened to be Agostini (Yamaha), Barry Sheene (Suzuki) and Phil Read (MV Agusta)! A less sporting rider might have been tempted to spoil the party but Nelson said that he himself hated people who started dicing with him when he was leading a race. He encouraged Fath by reporting that the new bike was not only exceedingly rapid but also easy to ride because of its wide spread of usable power from 8,000 to 13,000rpm. Development continued through the summer of 1974, with Nelson helping to develop through his testing and accurate feedback, and Fath translating that into mechanical improvements, gradually making the machine more competitive. ▶

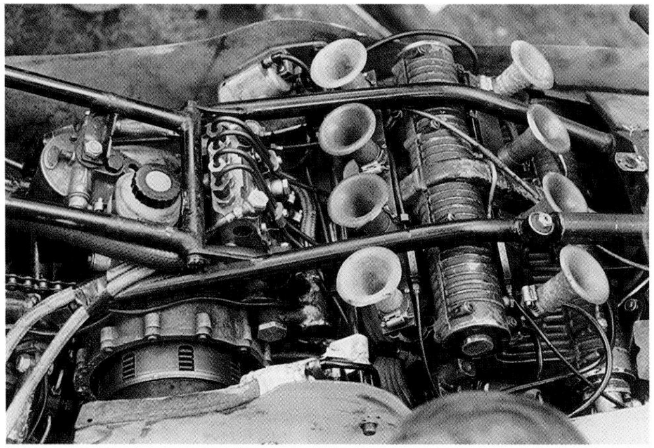

Unusual feature of the later Fath four stroke engine was an auxiliary inlet tract (parallel to the cylinder axis). This helped acceleration. The inlets are of glass fibre, the fuel injector is just behind the engine.

Karl Hoppe on the 500cc URS-Metisse, German Grand Prix 1969 Hockenheim.

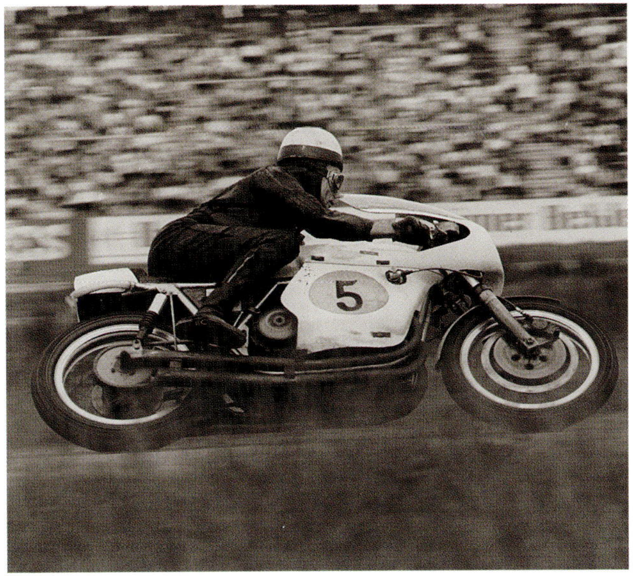

That all came to a tragic halt at the Yugoslav Grand Prix at Opatija on Sunday, 8th September. Billie Nelson crashed his Yamaha at 125mph in the 250cc race and died later in hospital. One of the most experienced, and safest, riders on the grand prix circuits, the 33 year-old from Eckington, Derbyshire, had been a successful sidecar passenger and solo rider for some years. Indeed, Fath's crash in Finland in 1969 which deprived him of retaining his world sidecar crown also deprived Nelson, who was passengering for Fath, a chance of second place behind Agostini in the 500cc World Championship on Bill Hannah's Paton twin. Billie Nelson's death was a massive blow to Fath because they were not just colleagues but close friends. Instead of finding another rider to replace Billie, Fath used the engine from Nelson's machine and fitted it into a sidecar outfit for Siggy Schauzu. But the outfit was not to Fath's liking, being too heavy and wide. By that time, Fath had completed a new, more powerful, engine in the hope of making the modified outfit more competitive. It was not to be. When the Aro-Fath, as it was now known, finished the 1975 season at Hockenheim, it had only once finished in the top ten. A mixture of drivers and solo riders over the following two seasons gradually gained respectable race placings in Grands Prix for Fath's four cylinder as it gained reliability to match its undoubted speed. After the 1977 season started, Werner Schwarzel, runner-up in the 1976 Sidecar World Championship on a König, was provided with an Aro-Fath outfit. It was he, the ex-König driver, who provided Fath with the pinnacle of achievement for his flat four with a dramatic win in the Belgian Grand Prix. Added to a superb third place in Schwartzel's first outing in the previous week's Dutch GP, a win at the British GP at Silverstone, was sufficient for Fath's bike to claim third place in the championship. Just as he was about to reap his well-deserved rewards in the 1977 season through wins and high placings by the dashing Schwarzel, Rolf Biland and his English passenger, Kenny Williams, claimed the Championship with their Yamaha-powered BEO, the first of the new breed of fully enclosed 'worms' modelled on F1 cars, but minus the fourth wheel. This was too much for Fath. He abandoned any idea of making a come-back into racing, although the FIM's B2A category for 'conventional' sidecars would have allowed him to continue racing his flat fours. After more than 20 years in racing, during which time he had won two World Championships, then lost his Urs four cylinder in the 1970 Münch deal, came back with his brilliant two stroke flat four, only to suffer the loss of his rider and close friend Billie Nelson, it was time to retire to a quieter life in Ursenbach. That Helmut left his mark on racing is der grossest unterstatement. Throughout, he had shown a strength of character and achievement rarely seen before, and certainly not since, a remarkable man. Helmut Fath died at the comparatively early age of 61 in 1994.

The mother of all rushes! Lawrie had only one night in a draughty deserted factory, to complete this drawing which was subsequently enlarged to 12 ft!

All done by mirrors

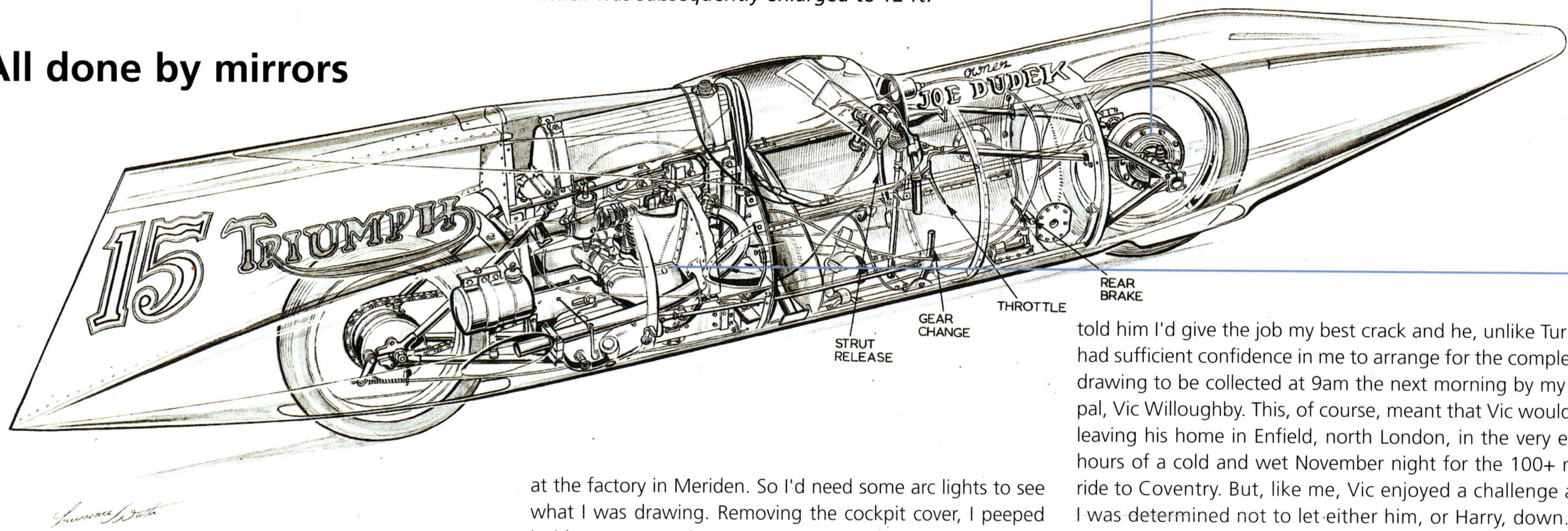

In his time, Lawrie has had some tight schedules to meet but none so tight as the drawing he had to do of the Bill Johnson 224.57mph world record holding Triumph for the 1962 Earl's Court London Show. 'Ivor Davies, the ever-helpful PR of Triumph, called to say that the bike had just arrived at the factory en route to London. It was available for me to draw provided I could do it before the bike was transported to London early next morning. Oh, one other thing, Lawrie you can remove the cockpit cover, but no more, all panels must stay in place. No time for dismantling or reassembling. The penny dropped; I'd not only have to work through the night, but I'd not be able to draw anything but the exterior unless I could get myself some x-ray eyes. Thanks a bunch, Ivor!

Hot footing it to the factory, I sussed out the situation; the streamliner was propped to one side of a gloomy store room at the factory in Meriden. So I'd need some arc lights to see what I was drawing. Removing the cockpit cover, I peeped inside; to see anything worth drawing, I'd need my aircraft inspection kit. This contained mirrors and extension rods with angled mountings which allowed me to look around corners. Just the job for prying behind the streamlining! A powerful inspection lamp I had specially designed for peering into dark recesses in aircraft would be worth its weight in gold. Just as I completed my check list, who should appear but the Big Man himself, Edward Turner. 'You can only draw that if you work all night, Watts. Can't be done. Not enough time. Good night!' With those encouraging words ringing in my ears, I asked Ivor for written permission to stay overnight in the factory, a very unusual occurrence, and for arc lights to be positioned around the bike. Rushing home to collect my gear, flask, and Eileen's sandwiches, my mind was working overtime. How could I do the job in the time available? I must admit that Turner's words spurred me on, I'd do it just to spite him! I called my boss, Harry Louis, told him I'd give the job my best crack and he, unlike Turner, had sufficient confidence in me to arrange for the completed drawing to be collected at 9am the next morning by my old pal, Vic Willoughby. This, of course, meant that Vic would be leaving his home in Enfield, north London, in the very early hours of a cold and wet November night for the 100+ mile ride to Coventry. But, like me, Vic enjoyed a challenge and I was determined not to let either him, or Harry, down. So that was my state of mind as I returned to Triumphs for my long night shift.

'It's an eerie feeling being totally alone in a factory at night, without the normal hustle and bustle of a busy production plant in full flow. The pool of light from the arc lamps defined my own personal space, which I found comforting. Sitting there contemplating the project, planning how to execute it, I marvelled at the ingenuity of the designer and constructor, Polish-American Joe Dudek. Like me, he was into aeroplanes having worked on the X-15 research aircraft, and he was also an inspector. So he would have appreciated how I was proposing to translate his superb craft onto paper. As he had absolutely no backing, Joe, an accomplished drag racer who had a record of building fast bikes which others rode at Bonneville record week, had to make do with whatever materials he could source. This also appealed to me. Inspiration for Joe's design started when he first saw

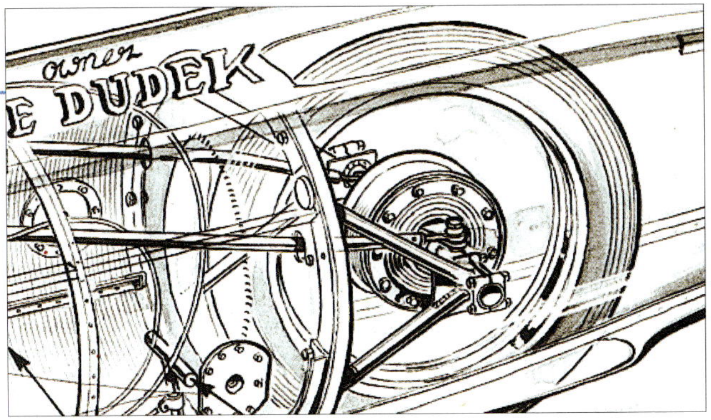

Detail showing the hub-centre streering, using a modified sprung hub from a Triumph 650 Thunderbird.

pictures of Gustav Baumm's NSU-powered 'flying hammock' on the Bonneville salt flats in Utah when Baumm was taking a bucketful of world records in 1955. Appreciating the significance of aerodynamics, Dudek used a discarded drop fuel tank as the body of his streamliner, grafting on a tapered nose cone at the front and, for stability at straight line high speeds, adding a long tapered tail fin which, at its highest point, rose to 37in. As rider Bill Johnson reclined in a 'hammock', the top of the cockpit canopy was only 34in high. The slim frontal area was helped by opting for hub-centre steering, Dudek adapting a sprung rear hub from his old Thunderbird to accept a steering king pin with a caster angle set at the optimum of 12 degrees. With the 649cc Triumph twin engine, which ran on a mixture of methanol and nitro-methane, installed behind the rider, the weight distribution was more or less evenly distributed within the 96in wheel base. The completed machine measured 17 feet and weighed in at 400lbs.

'To translate the nub of this information into a drawing in the time available was a tall order, so I opted for a pen and ink drawing with minimum lines to save time, and without any frills. Every drawing has to start with a reference point and, as I could see both wheels beneath the fairing, I chose them. I worked steadily through the night, gradually building up the picture until, flat on my back beneath the machine, I could hear something approaching me. I could see nothing in the blackness beyond my own pool of light from the arc lamps. As I froze, not knowing what to expect, the factory mouser strolled into view! Seeing me lying there, he wandered over and greeted me by rubbing his nose in my face and making a big fuss of me. Thereafter, we kept one another company during the long hours until around 6.30-7am, when I finished as much as I could. Bidding my furry friend adieu, I hurried off home, asking the watchman on the factory gates to ring Eileen and warn her that I was on my way, so she could start getting breakfast ready for Vic and myself. Back home, I wasted no time in finishing off the drawing, just as Vic arrived, ravenous from his night ride. On his way after a hearty breakfast, he made good time to the printing works at Watford, north London, where my drawing was seized and went through its processes to appear two days later in the Special Show Issue.

'Unknown to me, Ivor Davies had a second plan. Liaising with Harry Louis, Ivor had had my 20 inch long drawing blown up to about 12 feet long, mounted on board, and hung above the record breaker on the Triumph stand. I walked into Earl's Court to be confronted by my now gigantic drawing! Oh dear. That initial shock gave way to deep embarrassment as I realised that any little mistakes I'd made were now glaringly apparent - as I thought - due to the 12 times magnification of the original drawing and a similar magnification of my signature! I wanted to creep away and hide. As is sometimes the case, nobody seemed to notice and the Triumph people and their customers were more than happy to bask in the glow of Dudek and Johnson's wonderful achievement. After a time, so was I, especially when I remembered the almost impossible circumstances facing me at the outset.'

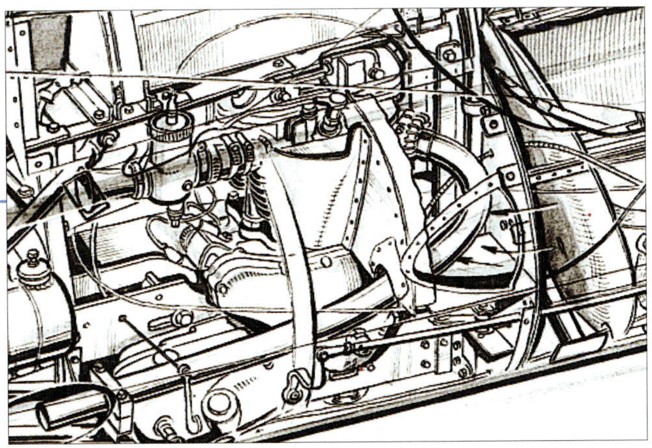

Installed behind the rider, the tuned 650cc Triumph ran on a mixture of methanol and nitro-methane. Note the ducting around the cylinder barrel - cooling air comes in via air vents each side marked by directional arrows.

Automobiles

Three dimensional

Lawrie evolved a skill in draughtsmanship which, applied to the automotive industry, produced a three dimensional drawing. This also showed the major internal features of the vehicle via 'cutaway' panels in the drawing, a trick he had learnt from his childhood experiments burying things in Plasticine, then revealing them through partially cutting away the top layer. No surprise that after 18 months invaluable experience he had graduated from Armstrong Siddeley Whitworth (as it had become) to the Daimler drawing office, close to his home in Radford, Coventry. Little could he have guessed that his new job, drawing body parts which had to mate with other parts, was work which today has been delegated to computers with the advent of computer aided design, CAD.

The Daimler spell

He had always wanted to work in a drawing office, Daimler had a vacancy, but even better, it was in the body styling section. Of course, the war was still on and from his desk he could see 1,000 aircraft making their way south over Coventry and Birmingham, preparatory to bombing Germany. As he recalled, 'It was quite easy to count them. They came in groups, of ten, fifty, hundreds, and then joined the main group until the sky was full of aircraft. And there was I, not yet 17, in the styling department of one of the oldest and most prestigious car firms in the world helping to draw cars for when the war was over. It seemed almost surreal at the time, because it was far from clear that we would win the war, but our bosses were absolutely confident that we could. You just had to admire their optimism! Settled in my new job, I was working on a new design of radiator. The Daimler radiator had for many years been an upright, lovely structure, with a trademark dimpled top. Messing about, I sketched a radiator which was sloping back and more streamlined. At the same time I sketched a windscreen which increased the rearward slope from 5 to 15 degrees. The management nearly had a fit! But big boss, Sir Bernard Docker, loved the new design and they used it for many years. In fact, you can still see the outlines today.'

Lady Docker's dogs

'I had a soft spot for Sir Bernard Docker, he was a smashing bloke, I loved him. Sadly, my opinion of Lady Nora Docker may have been coloured by the actions of her two black spaniels. On arrival in the Daimler drawing office, which they visited regularly, they made a beeline for me. Then they peed on my drawing board! Needless to say, this caused much merriment among my colleagues. However, I had to button it because the Great Lady Docker was right in front of me to see my latest drawings. In all that time, she never once attempted to stop her bloody dogs pissing on my handiwork!'

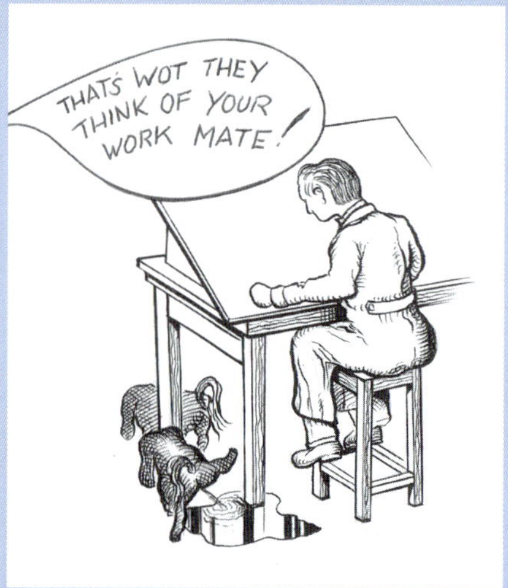

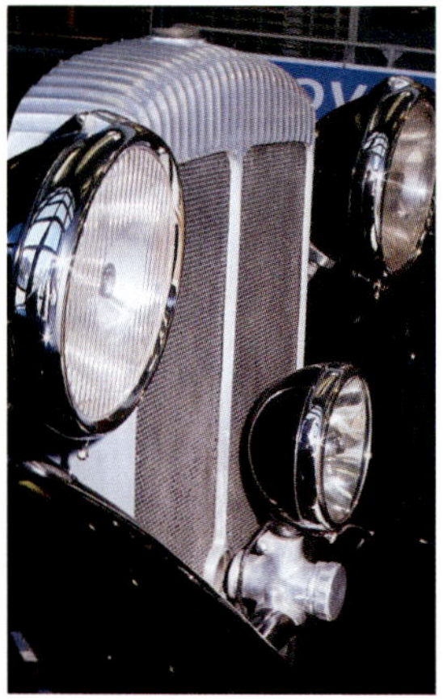

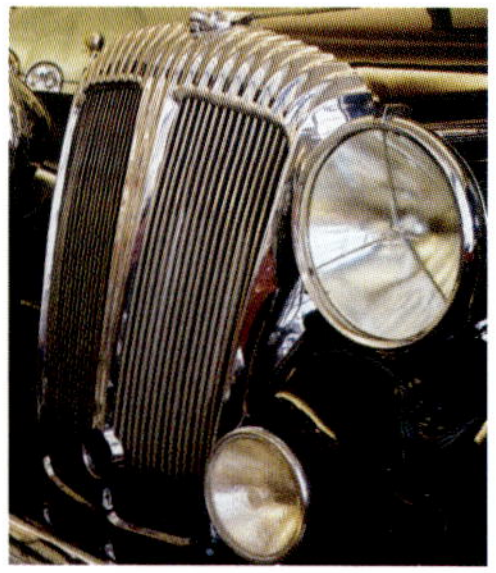

The flat, bolt upright traditional Daimler radiator as fitted to the King George VI car and the first of the new styled radiator shells.

Siamese prince inspired Docker's Daimler

Occasionally, the war had its lighter moments. Prince Birabongse, or Bira as he was known, was a very successful (and wealthy) Siamese racing driver who had made his name racing an ERA pre war. At the height of the war, he ordered a special-bodied sports Daimler finished in Siam's pea green racing colours. No, it certainly could not wait until peacetime. He wanted it now! In great secrecy, Lawrie and a few colleagues at Daimler designed a beautiful sports body to fit an existing Daimler chassis. To ensure secrecy was maintained, not from the Germans but from the British wartime government, the Bira car was to be constructed in Daimler's Browns Lane 'shadow factory', a factory out in the country which, it was hoped, that the Germans would not bother to bomb. Lawrie takes up the story. 'The head of the drawing office, Eric Scalture asked me to go with Bira to the shadow factory, taking along the model of my new radiator to see if they could take a casting of it for the new body which Bira had ordered. Which is how I found myself in a military tank, with a dummy of my new radiator, being driven across Coventry from the Daimler factory to Brown's Lane. Why use a military tank? Well my boss figured that we'd look as if we were on serious wartime business when, in fact, we were doing something quite naughty - building a one-off sports saloon for a Siamese prince - which had absolutely nothing to do with the war effort! So the first time I saw Brown's Lane was from inside a military tank. Daimler made the chassis, the specialist coach builder Hooper made the bodywork and, finished in pale pea green with matching pea green leather upholstery and a walnut trimmed interior, the completed car was stunning,' Loll related. 'With a low roof, long boot and spats covering the rear wheels, it was the most streamlined car Daimler had ever made, quite breath-taking. Bira hid it until the end of the war. Subsequently, some features were included in the famous gold-plated Docker Daimler, adding a welcome touch of glamour to those dreary post-war years. People were amazed how Daimler had produced a stunning shape in such a short time, especially during a period of great austerity. Had they been told the truth about its development, they would scarcely have believed it.'

His and Hers

'Sir Bernard was always complaining that the Daimler designs we were working on in the design office were too sporty, while his wife preferred something more regal. This got to such a stage that one day, just for a joke, I designed a car on a Daimler Straight 8 chassis which I called 'His and Hers'. The front was really sporty, and the rear was typically regal, with a glass panel separating the driver from the passengers, which was normal in limousines of that era. I also included panels of little gold stars attached to the paint work, which were a throwback to the gold plated Docker Daimler, which was the star of an early post-War Motor Show. The sporty lines and strapped down bonnet were borrowed from Morgan, as were the knock-on front hub caps. When the finished drawing was shown to Sir Bernard by my design office boss, Sir Bernard was so taken with the concept that for a time it looked as if we were going to have to make it! Luckily, common sense prevailed and my little joke concept car was quietly put to one side, no doubt to Sir Bernard's sorrow. I'm not too sure what her Ladyship thought... Anyway, this will be the first time the drawing has been published.'

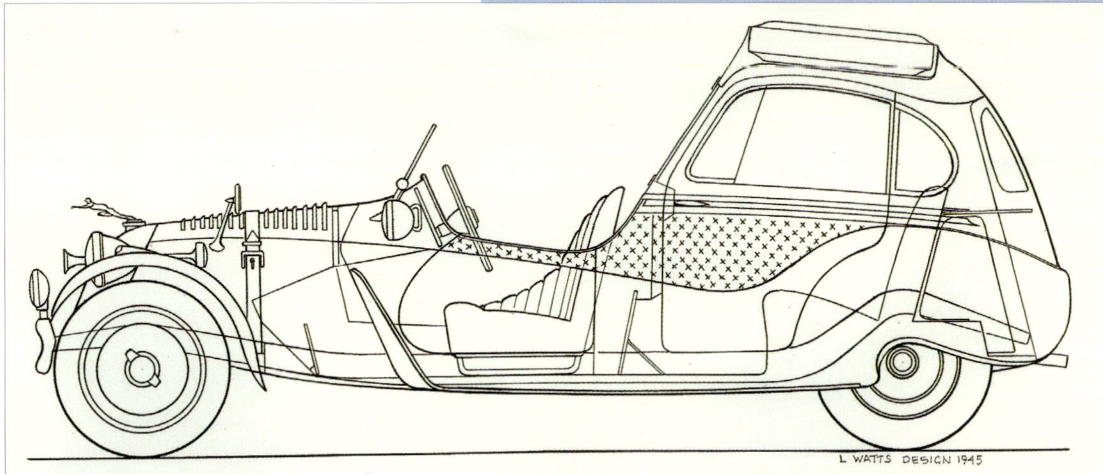

'His and Hers' concept car from a doodle by young Mr Watts (daydreaming dept) Daimler Cars Ltd.

Montego

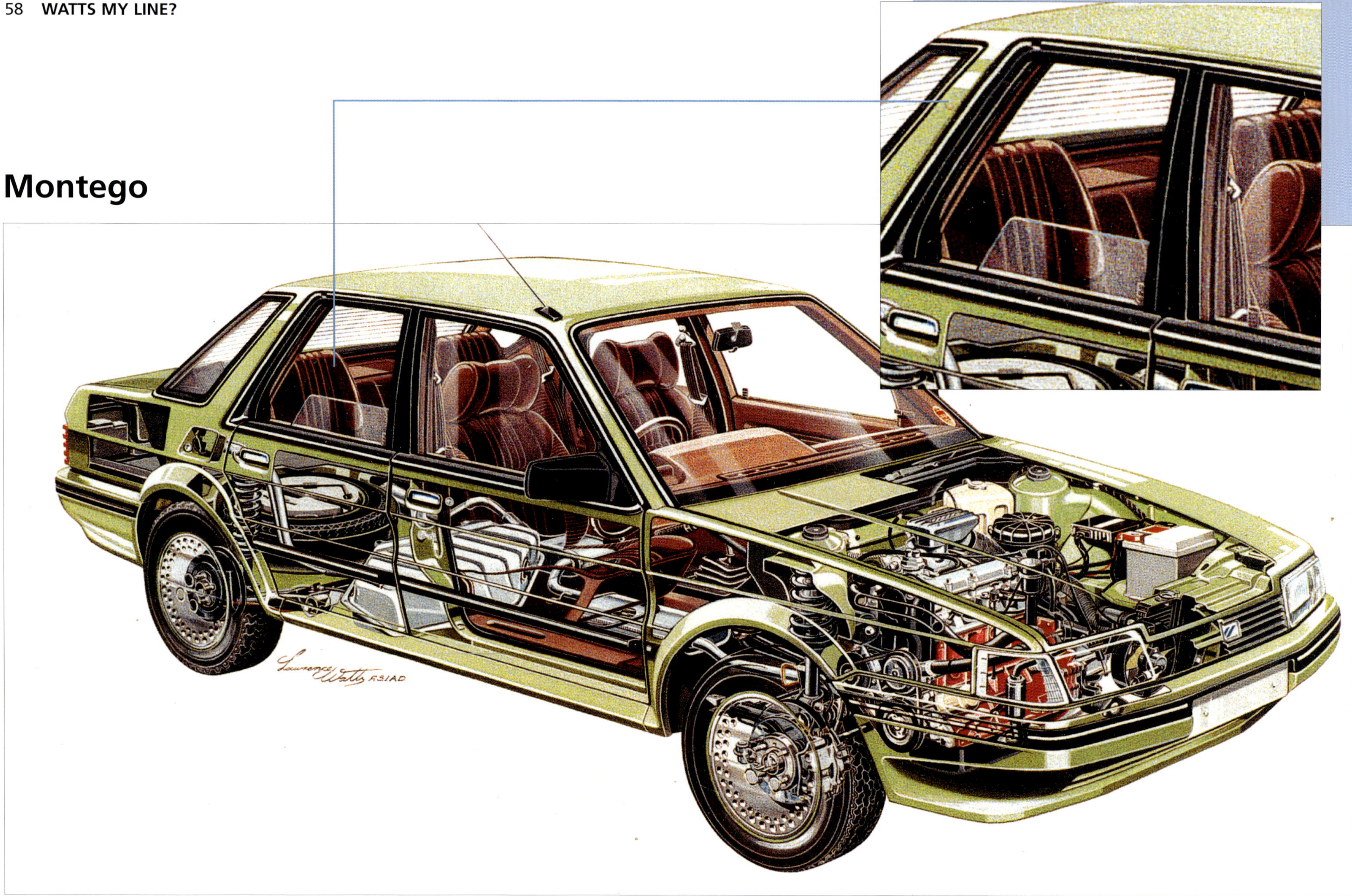

How to make a dull subject interesting, give it the 3-D treatment! Montego shows the effectiveness of the painted drawing, pioneered by Lawrie. New for that year was leather interior and heated rear window. By showing the interior through the half-opened rear window Lawrie has obtained a three-dimensional effect. Also new were the light-alloy wheels which the coloured drawing shows to good effect. End result, quite a smart car!

Making a silk purse...

'Some cars that make beautiful drawings may not be such good cars. Here are two coloured drawings of very different new cars, a Montego and a Rolls-Royce, both done prior to launch. The Rolls is a vastly superior car, but exactly the same amount of work went into each drawing, which is about a week. However, I could not get excited about the mundane Montego and I think this shows in the painting, when compared with the Rolls. So, to attract visual interest to the Montego, I partially wound down the rear window, shaded the glass and added heavier colour behind the glass inside the car to provide a three-dimensional effect. This also helped to show the new heated rear window. Finally, I highlighted the upholstery, which added some sparkle to an otherwise drab interior. There was no time to cock up this picture simply because that Montego was one of the first models to come down the production line and I was working furiously, trying not to get run over by trucks arriving with new parts.'

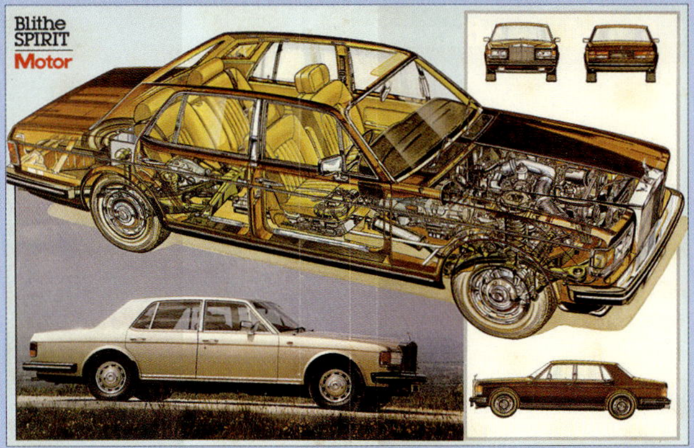

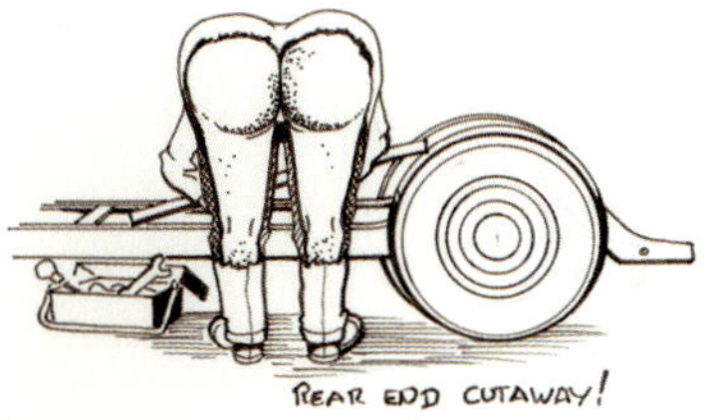

Body check

'In this job you have to see the funny side of things, even when the opposite is the case. Here's an example. There was this guy at Standard Motors who would always, but always, stand in front of me deliberately obscuring my view. Nobody knew why, but it became a ritual with him. One day, when working on the Triumph Mayflower, I got my own back. There, as usual, he was bending down apparently looking at something. I continued working, on the principle that you draw what's in front of you. So I drew his great big backside! He didn't know, of course, so when it was finished, I tapped him on the shoulder, handed him the drawing I'd just done, and said, 'Is this what you want?' He never got in my way again, especially as his fellow workers displayed the drawing in pride of place on the factory wall!'

Speedy paper

Those who have seen the large pull-out of Thrust in 'Motor' magazine remarked on the wonderful effect of speed. 'This was achieved by masking the finished machine, using a special brush and gouache colour which matched the desert, and repeatedly pulling the brush horizontally across the painting to build up layers of desert colour for a 'streaking' effect. Afterburner flames were added before the masking paper was - very carefully - peeled off. The gamble worked well only when the masking paper came off cleanly!'

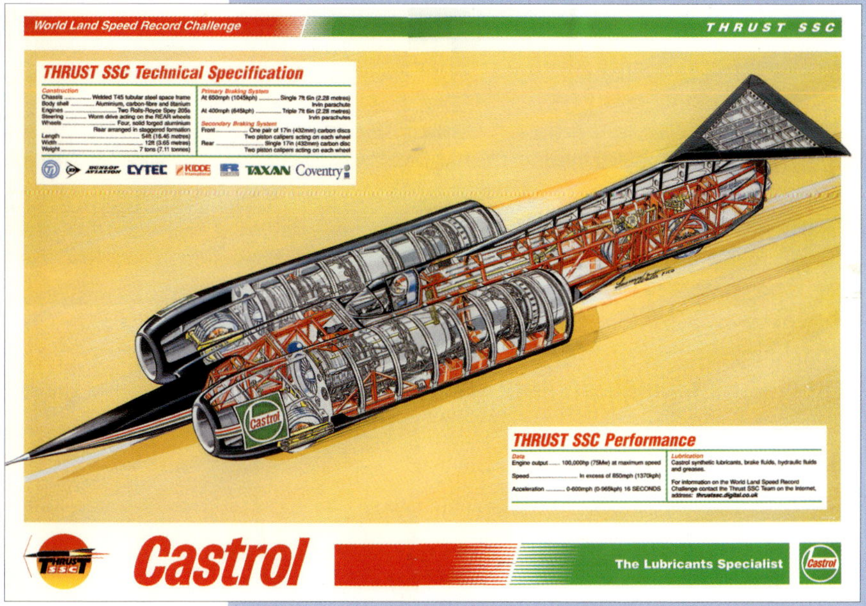

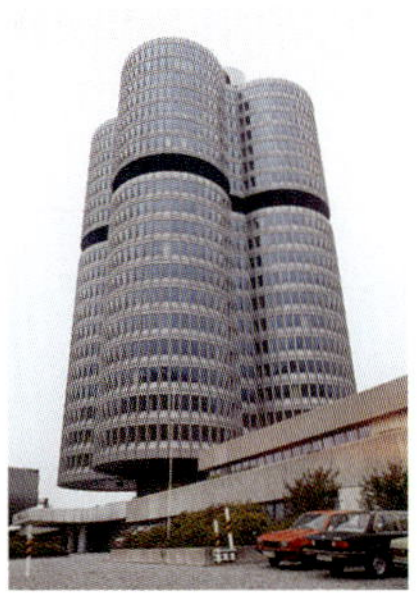

The striking BMW HQ in Munich.

BMW

This sectioned view of the 1970s Series 6 BMW was drawn following a one-day visit to the factory. Lawrie remembers the 100% co-operation of BMW's staff.

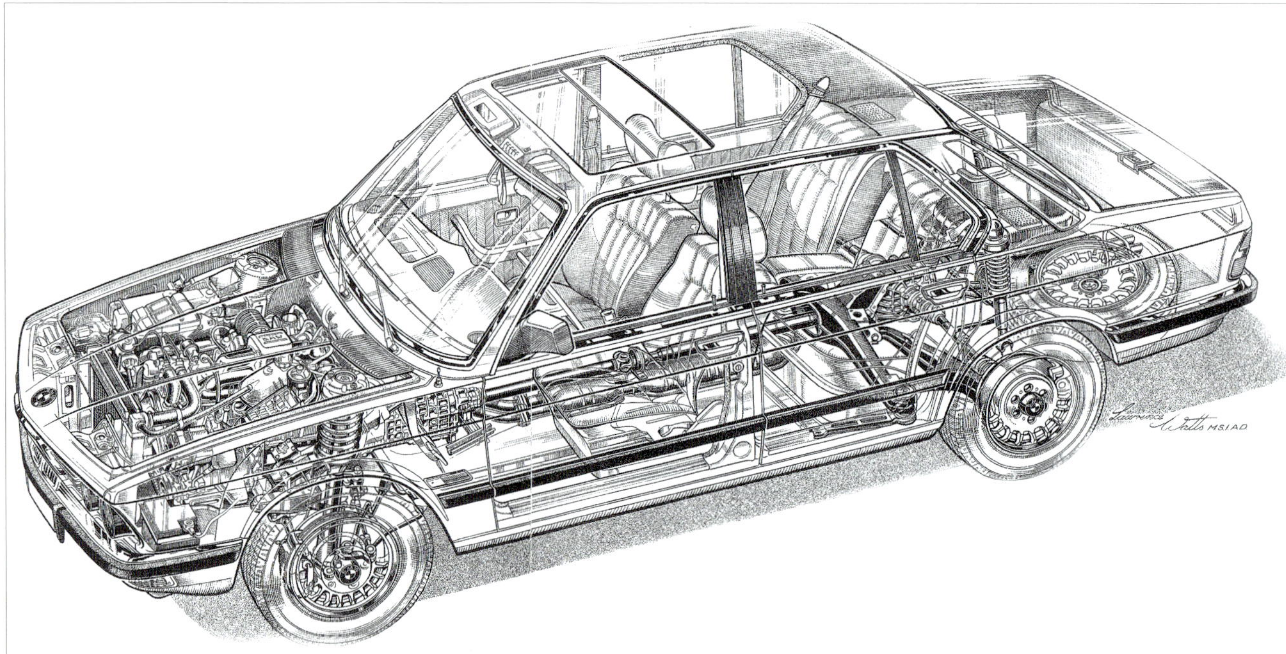

During the time I've known Loll, he's always gone out of his way to praise BMW, not just for their products, but because he especially enjoyed working with BMW people. That warm feeling is still there many years later. 'They were wonderful people to work with, I have lots of happy memories. This drawing of a 1970s 6 Series saloon is a classic example of how well we worked together. I went to Munich, got all the information I needed, including loads of detail, and was back in Coventry drawing the car the next morning! I could not have done that without their 100% co-operation. Sometimes this was the result of the personal interest BMW's legendary Press Officer, Karl Hoepner, took in what I was doing. One time he took me on to the roof of BMW's futuristic circular steel building in Munich. There he photographed me, after that I was taking panoramic views over Munich. So, far from being pressured, I actually had time to mess about.'

Typical of Loll's attention to detail in every drawing is the joint which he drew in the middle of that BMW 6 Series final drive shaft. As if anyone would be interested in such an unglamorous piece of metal! However, it reminded him of his own Volvo cars. 'They also had joints in the middle of their prop shafts and when the shafts wore out, as they did, it's right down by your seat. So if they break, they're likely to fly up into the cockpit. You can imagine the havoc that would cause!'

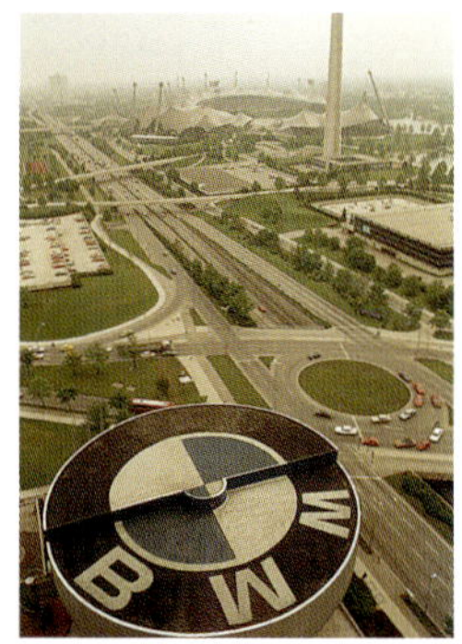

The top of the main tower, from here Lawrie had a panoramic view of Munich.

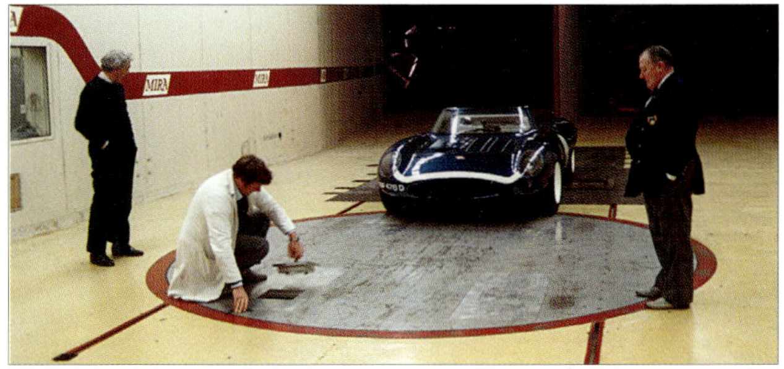

In the wind tunnel at MIRA with the XJ13 trying to cure front end lifting.

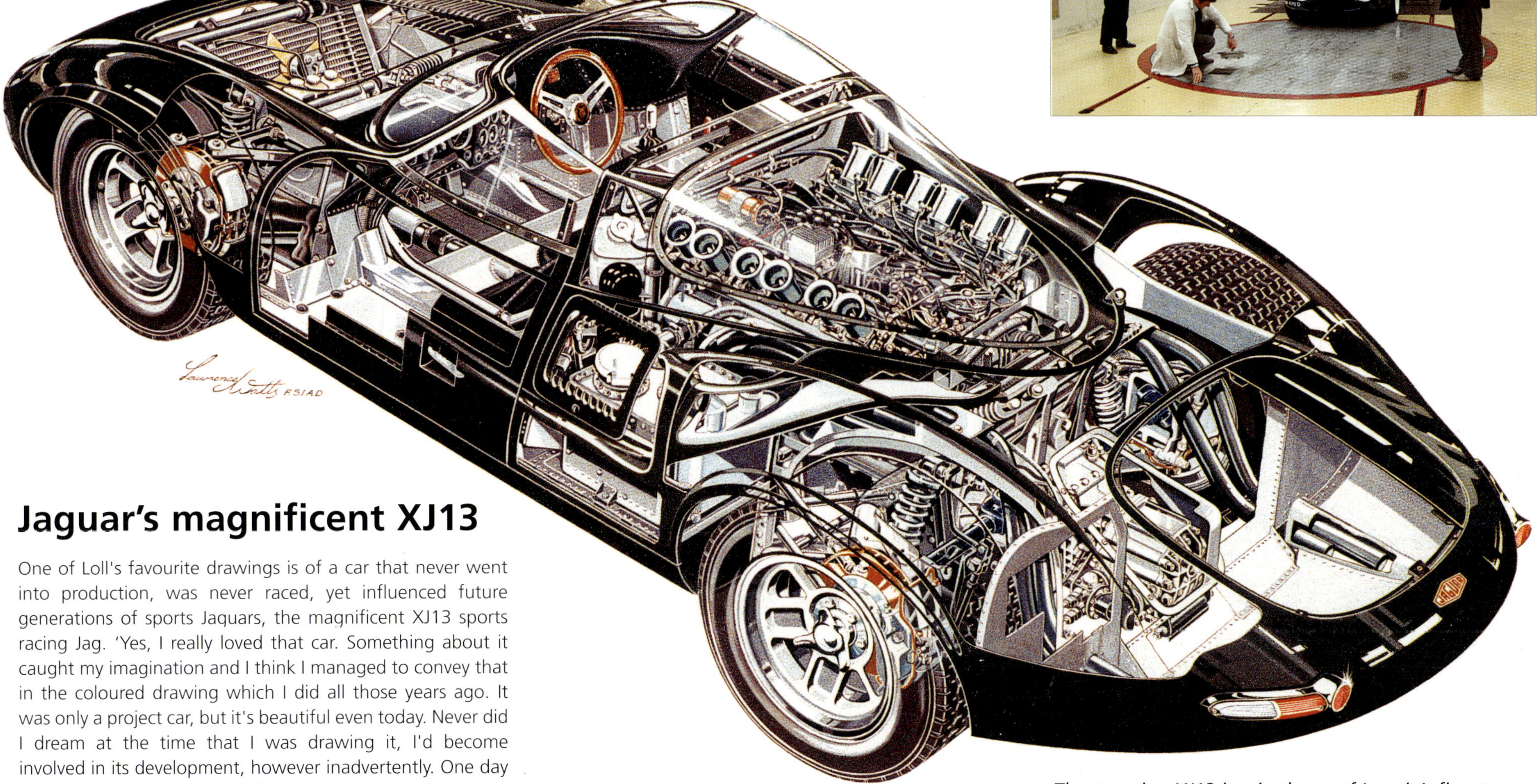

The stunning XJ13 inspired one of Lawrie's finest drawings. His attention to detail is remarkable - 'I got carried away by the sheer beauty of the car.'

Jaguar's magnificent XJ13

One of Loll's favourite drawings is of a car that never went into production, was never raced, yet influenced future generations of sports Jaquars, the magnificent XJ13 sports racing Jag. 'Yes, I really loved that car. Something about it caught my imagination and I think I managed to convey that in the coloured drawing which I did all those years ago. It was only a project car, but it's beautiful even today. Never did I dream at the time that I was drawing it, I'd become involved in its development, however inadvertently. One day Jaguar called me over to MIRA (the old Motor Industry Research Association testing grounds, near Hinckley, Leics) to do some aerodynamic work for them. Apparently, the front wheels were lifting at high speed and the car had gone through several hedges. We did some aerodynamic tests and found that the bonnet shape was encouraging the front of the car to lift above 130mph. We altered the bonnet and those tweaks kept the front wheels firmly on the road, even though the design was by then getting a bit out of date. Instead, Jaguar made the XJ220, which was a totally different concept, and the sole XJ13 went into Jaguar's museum. Many enthusiasts reckon it to be the quintessential Jaguar shape. I'd go along with that.'

A grasp of the basics

'Over the years, various designers came to trust me to the extent that they would sit around a table and tell me exactly what was in their minds when they embarked on the project under discussion. From this conversation I'd be interpreting into sketches what they told me so that after every such meeting I'd be able to leave sketches to evaluate. Then I'd go away and do proper sketches for them to see if I was on the right lines. That's always been the way, because for me to draw something that doesn't yet exist I need to be able to see inside the designer's mind before I can translate his ideas onto paper. An easier job was to turn up at the factory with the product already made. However, that was a luxury I seldom enjoyed. My job was to get the drawing done to coincide with the launch to the public.'

Getting the right perspective

'Assuming that I'd got the right message from the designer, the most important part is the wheel, because the wheel is a fixed dimension. Once you have that diameter, the next step is to consider from what viewpoint all the other parts are to be shown. If the engine is at the front, that's where I'd start. The engine size in relation to the wheel is important, looking down on the car from, say, a 30 degree angle to the ground, right onto the centre of the wheel. So I draw the centre line of the wheel in the vertical, then the centre line of the wheel in the horizontal, and finally the centre line of the wheel in the transverse (to the opposite front wheel). Having decided the angle at which the wheel

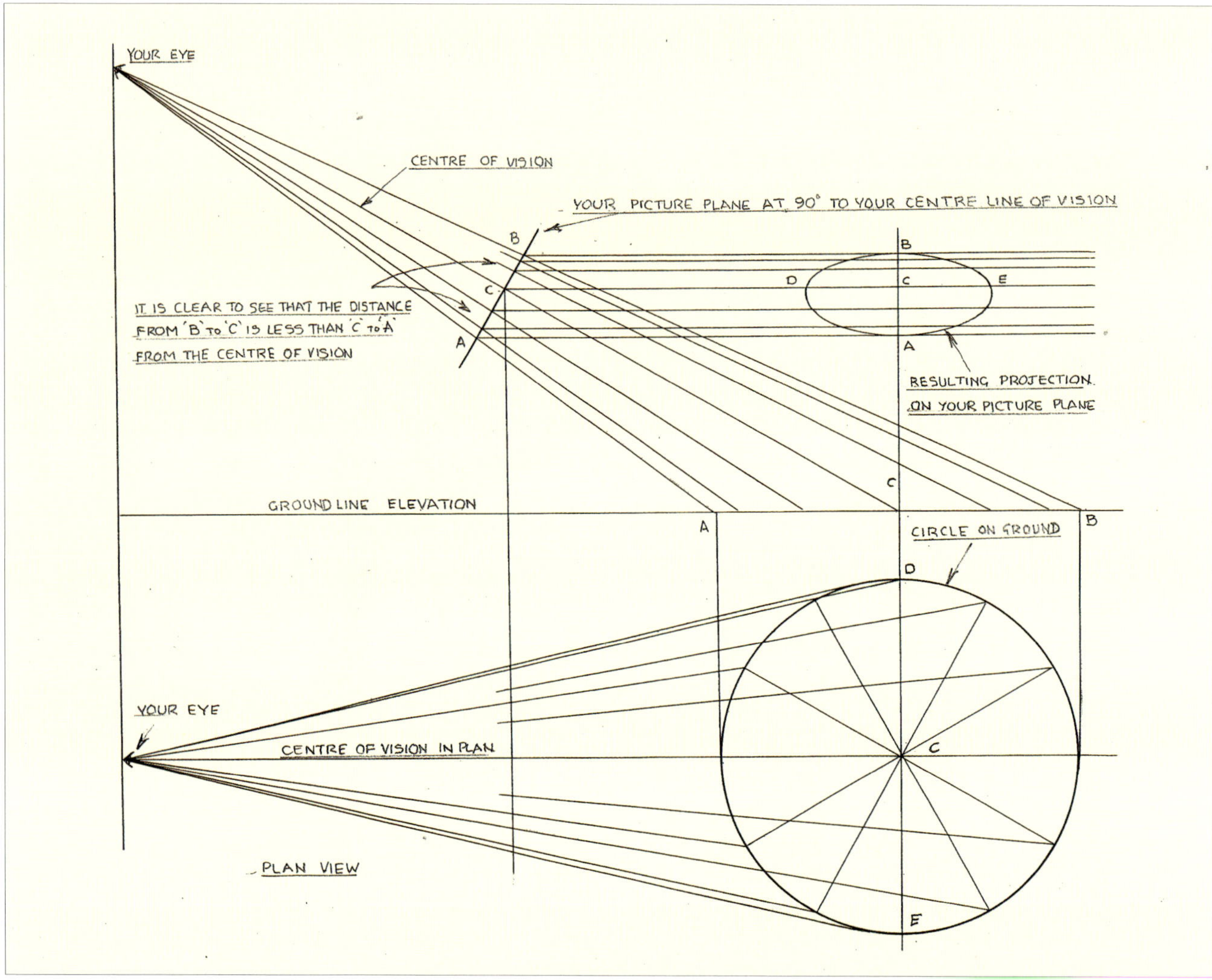

Get this perspective right and the rest is plain sailing!

will be looked at, this angle provides not a circle, but an ellipse of a certain size. It doesn't matter what size so long as it's accurate, and the only way to get that is to draw a perfect circle of roughly the size you want it to be on the finished drawing. You then divide up the wheel by drawing a series of vertical lines through the wheel, starting with the vertical line you've already drawn through the centre of the wheel. On each side of that line, you divide the wheel by drawing 6 vertical lines each side to provide 12 segments. Using the same method, now divide the wheel into 12 horizontal segments. The result is a grid over the wheel. That grid has now to be produced in perspective, with vanishing lines in the horizontal relative to the chosen viewpoint - also relative to two other points which cannot be seen, above the ground and below the ground. That in itself provides a cone of perspective. Try it for yourself by imagining being in a stand at a football match and looking down on the centre circle on the pitch. From the height at which you are looking down, that circle will appear as an ellipse. Look carefully and you will see that the side farthest from you is fatter than the side which is nearest. You will also note that the light striking the ground gives you depth. If the subject to be drawn is not within that cone of vision, which is about 60 degrees, then you are too close to the subject. In practice, that is why if you stand too close to the vehicle, you will get a distorted view of the rest of the vehicle, not to mention the wheel. The same thing happens when photographers get too close to a vehicle, you get a huge image of a wheel and a tiny scooter-like wheel in the distance at the rear of the car! The trick is to move away until the whole vehicle is in the cone of vision. And that is exactly 60 degrees. If you can arrange your space to provide that cone of vision, you will have a good perspective drawing, otherwise it will be out of perspective and to include some of the other parts of the vehicle you will have to distort them. You're then on a hiding to nothing. Get the 60 degree perspective for the wheel right, and everything else will fall into place. A final tip; check your results, and especially your ellipses, in a mirror. That's the moment of truth!'

Marooned in a high place

Getting the right perspective once landed Lawrie in deep trouble, as in up a gum tree - except it wasn't a gum tree but a whacking great crane. As he recalled it, 'I had this massive Foden lorry chassis in front of me to draw. It was so long that there was no way I'd be able to get a decent perspective unless I went up high, very high. Having explained this to the works manager, he indicated a gantry crane situated in the roof. So I climbed up to the gantry via a steel ladder set into the wall of the factory, and got into the little box on top of the crane. Perfect! Movement of the box was controlled by a man on the ground who could move me forward, backward, sideways. So I directed him to manoeuvre me to exactly the right position so I had a 60 degree cone of vision, and started sketching. There I stayed for the day, eating Eileen's sandwiches for lunch, monarch of all I surveyed. Bliss. Work was rapid, no interruptions, and I was totally absorbed in the job all afternoon. Then I noticed, almost imperceptibly, that things way down below were becoming quieter, and quieter, until there was almost absolute silence. Then the penny dropped, I was alone! Everyone had knocked off work, gone home and I was left literally high and dry, and before long, mighty worried. How to get down? To pass the time, I continued drawing, filling in little things I'd not noticed first time round. Eventually that burst of activity wore off and I became increasingly aware that I could be there all night on my own. Shouting for attention produced mocking echoes which, in time, became ever more sepulchral. Eventually, my shout was answered by a startled night watch man who had just come on duty. 'What you doin up there, then?' he shouted incredulously. When I explained and asked him to let me down, he responded 'Oh no, oh dear no, can't do that'. Thinking he was pulling my leg I asked him why not. 'Can't drive the bugger meself. Have to fetch the regular driver... if I can find him...'. Another half hour passed, by which time I was feeling decidedly hungry, cold and in need of a pee. Eventually my hero arrived, the guy who'd so cheerfully hoisted me up many hours ago. Now I could have been equally cheerful in strangling him for forgetting to take me down before going home. As an excuse for arriving home late it lacked nothing in originality '

Colourful art

The increasing use of colour in periodicals was right up Lawrie's street. He'd been doing coloured drawings for years but it was quite expensive to reproduce and was therefore used sparingly by the more penny pinching managements, which meant the majority. But with the advent of new techniques which drastically reduced the cost, and with the use of higher quality paper, the demand for coloured drawings increased dramatically. This pleased Lawrie enormously. 'If you use colour in a drawing it enables you to show different parts more clearly. For instance, you can use one colour to pick out all the transmission parts, chassis parts another colour, and so on. These really stand out, whereas in black they are virtually concealed within the chassis. Tyres can be painted in lifelike colours, with the appearance of real rubber. Max Miller, senior artist at Iliffe Press and my mentor, derived his technique from his father and this was to use black to make things stand out. Very effective it was too. Other artists used what was called Zippertone, which was ▶

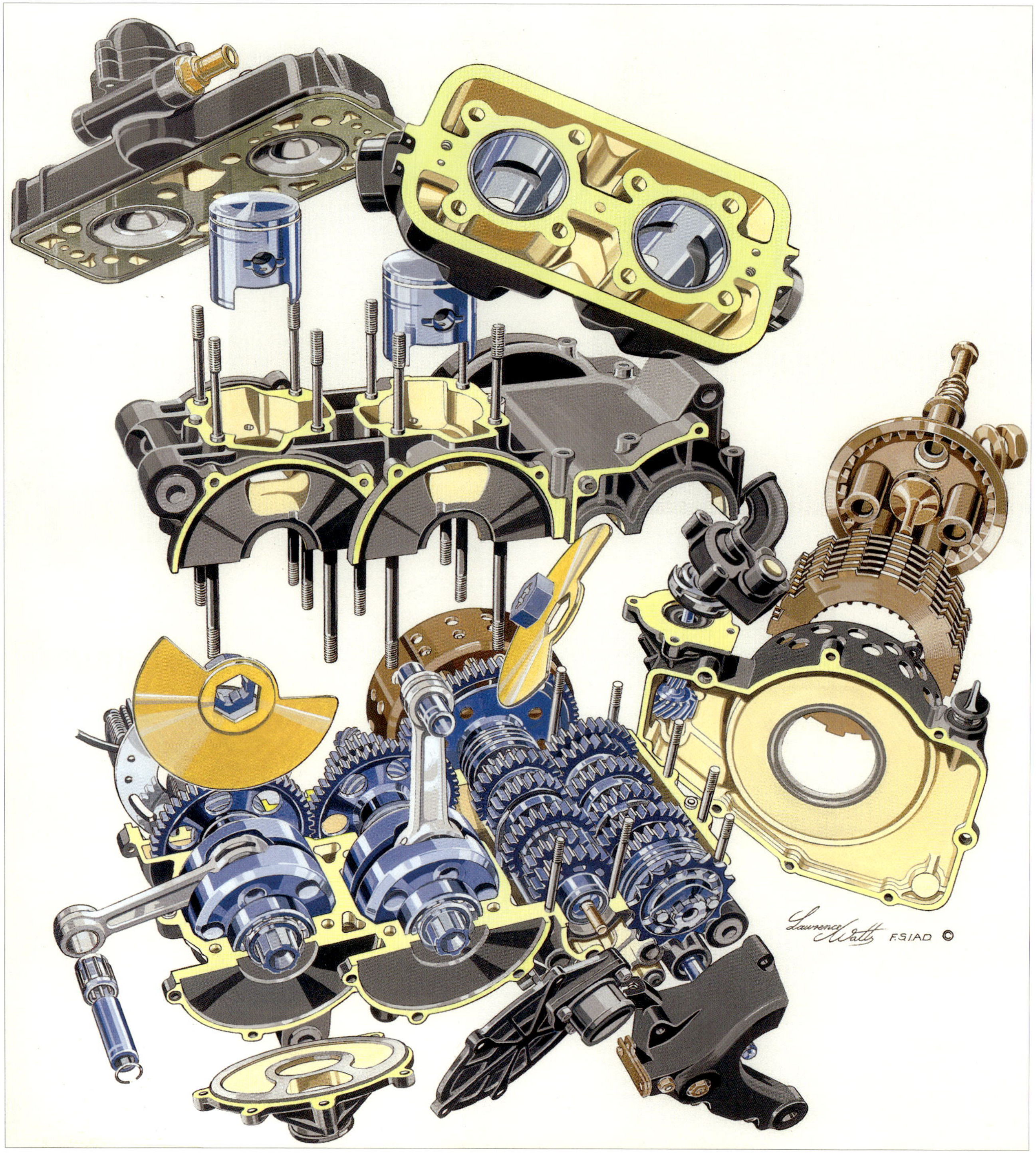

The use of colour to enhance your understanding is demonstrated in this 'exploded' view of Kork Ballington's factory KR250 Kawasaki.

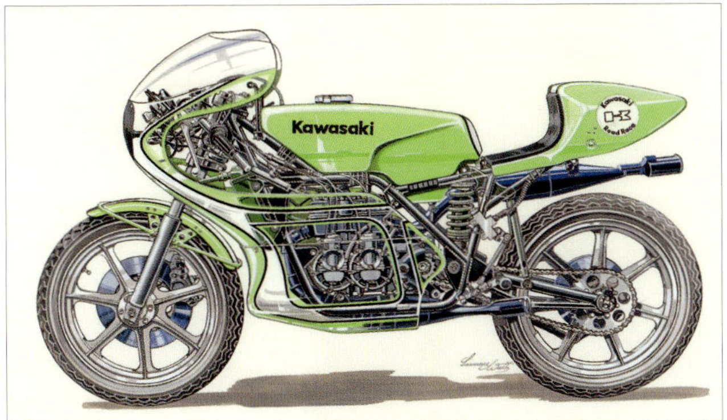

Colour is especially effective in differentiating engine and chassis components, as seen here on the Kawasaki KR250..

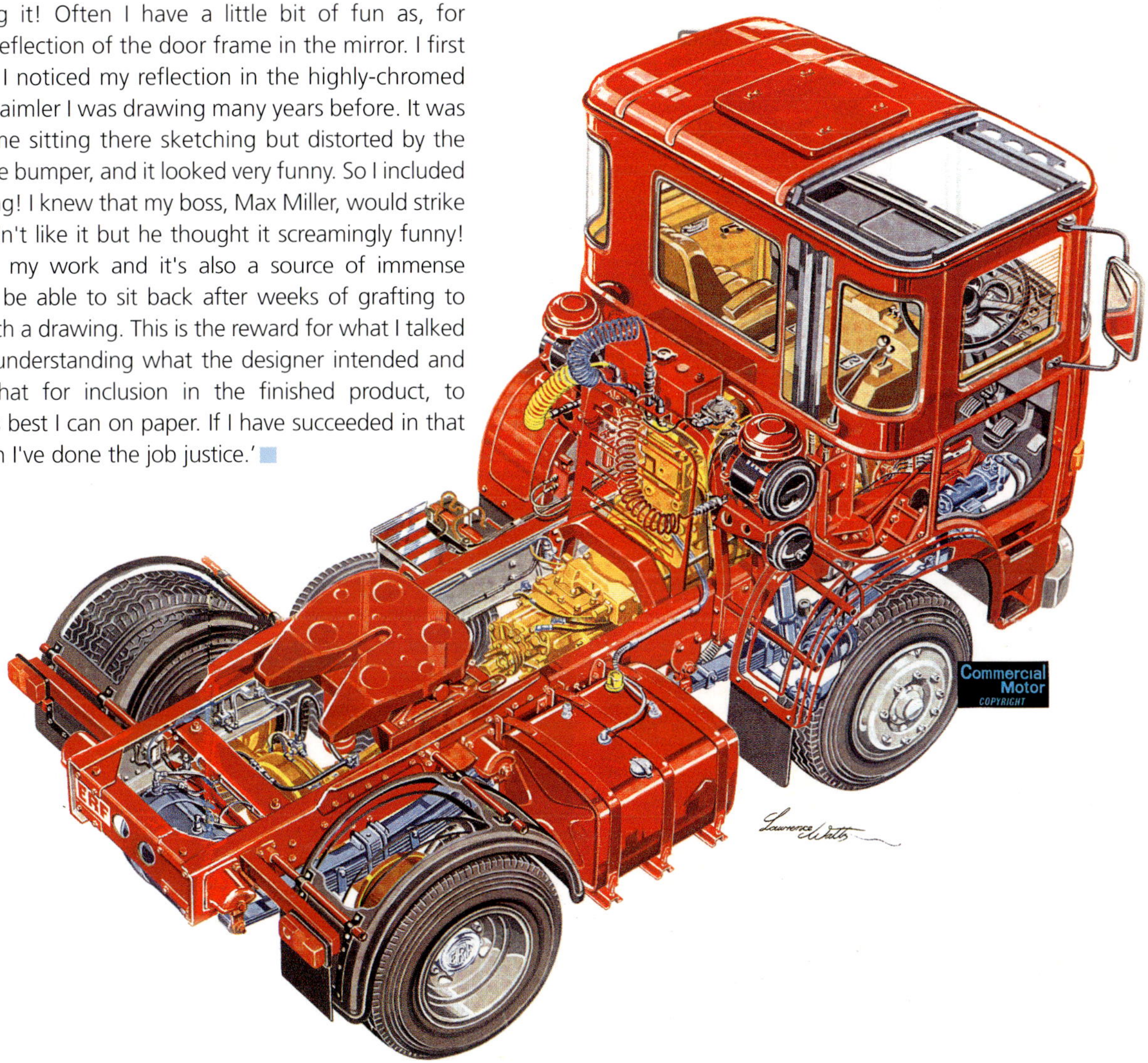

The detail work in this drawing identifies it as a Watts production; who else would show details inside the cab by means of its reflection in the driver's door mirror? Or pick out the indentations in the trailer-mounting platform in white reflections? Lawrie was inspired to do justice to the ERF by the superbly practical design.

the mechanical way of putting little dots on the drawing to create depth and perspective. Too mechanical for me, it was not art, and I preferred paint instead. So when colour printing came along, I welcomed it. The use of colour also results in a 3-D effect, lifting the overall impression of the drawing.'

Never was this more brilliantly shown than in a coloured drawing of a giant ERF tractor, which tows a huge trailer filled with our next week's groceries and behind which we join a queue of traffic for mile after mile. This pristine tractor was drawn by Lawrie in the factory with the smell of new paint still in the air. His attention to detail is stunning. On the so-called fifth wheel, which is not a wheel but a circular steel pad, tilted slightly rearwards on which the trailer is mounted, the plate has four circular hollows which presumably match with something on the trailer pad. Because the pad is all red, those hollows would not normally show up, unless you have a genius artist paint subtle white crescents on the inside rim of each hollow - where the light catches the rim to highlight the indentations. Similar treatment for the heads of nuts and bolts on the chassis and wheels, with little dabs of white highlighter showing the gloss of the new paint work before it gets a top coat of grime on the A5 or M6. The cab, seen through a cutaway in the door from an angle of 45 degrees and from a height of about 30 feet, positively bristles with details, such as the rubbers on the clutch and brake pedals, the heater panel traced in the glass of the driver's side window, the details on the instrument dials and - magnificent touch this one! - the reflection in the driver's mirror looking back on the frame of the door! Who else would go to such trouble? It must add many hours to completion of an already intricate drawing. 'The pleasure and sheer joy I get of out of painting that, it's a bonus to be paid for doing it! Often I have a little bit of fun as, for example, the reflection of the door frame in the mirror. I first did this when I noticed my reflection in the highly-chromed bumper of a Daimler I was drawing many years before. It was an image of me sitting there sketching but distorted by the curvature of the bumper, and it looked very funny. So I included it in the drawing! I knew that my boss, Max Miller, would strike it out if he didn't like it but he thought it screamingly funny! Yes, I do love my work and it's also a source of immense satisfaction to be able to sit back after weeks of grafting to completing such a drawing. This is the reward for what I talked about earlier, understanding what the designer intended and seeking out that for inclusion in the finished product, to reproduce it as best I can on paper. If I have succeeded in that challenge, then I've done the job justice.'

In 1962, Donald Campbell's 'Bluebird' was Lawrie's first dabble in colour. The subdued sepia tone matched the photo gravure printing which was then fashionable in Iliffe publications for 'special' projects, such as Campbell's land speed record car.

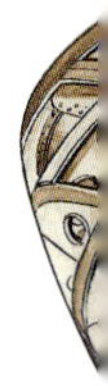

Drawn to the (steering) wheel

Unusually, an editorial artist has minimal interest in sampling the subject; it's a job to be done and when the final line has been drawn (or painted), there's a massive sense of achievement and relief. But Lawrie seems to have bonded with many of his subjects, especially racing cars (there's a surprise!) and big trucks. 'I always had the opportunity to drive these trucks and, time permitting, I always accepted the offer. I found it an enjoyable sensation driving these giants around the test track. Come to think of it, I've also driven most of the cars I've drawn. It seemed churlish not to accept such a privilege, which included the 1953 Thinwall Special, which had a Ferrari 60degree V12 engine, and its successor, the all-British Vanwall. Such opportunities, such an exciting experience! I was a racing car nut at the time and to be let loose in such famous cars, even though long since retired, was unforgettable, even though my driving was restricted to lapping the factory grounds. Luckily, it was quite a large site in Acton, West London, so I had an opportunity to change gears! I also had the opportunity of driving many of the cars tested by 'The Autocar', assisting the regular road test drivers in ferrying cars to the test tracks, sharing the driving during car v. car acceleration contests and so on. I didn't have to do that bit, it's just because I love driving and the opportunity to drive the actual car (or lorry) I'd been drawing made it all the more irresistible. Sometimes I'd get to drive a sports car before I drew it, thus giving me a driver's perspective which I usually found helpful when drawing the car.'

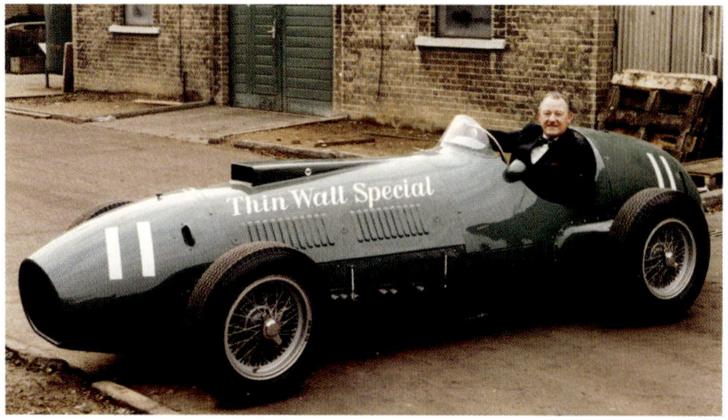

Lawrie got to drive many of the vehicles he drew. Here he samples a Ferrari-engined F1 Thinwall Special at the Acton factory. (His lap time around the factory grounds did not gain him a place in the team).

Bluebird

'Made in Coventry, right on my doorstep, Donald Campbell's land speed record car, Bluebird, was a huge task for me, then in my early days at Iliffe. Normally, my boss John Ferguson would have been allotted the assignment but, for some reason, it came to me. I tried not to let it overawe me, but the difficult part was that the car had yet to be made! Luckily, the body panels were being fabricated by the aptly named Body Panels close to me in Coventry, so visiting them as often as I liked was no problem. That's what I did each week, building up the drawing as, firstly, bulkheads were produced, then all the other bits. Luckily, I had a good working relationship with Donald Campbell and he kept me informed when something was ready for drawing. Although we got on fine on a personal level, he was inclined to be moody. One day he'd come in and wouldn't talk to you. Next day, you couldn't work because he was chatting on like a house on fire! I think that your theory about the moods may be right, that they were the result of worrying about commercial matters, the search to find sponsors, paying the never ending bills, and so on. Leo Villa, mechanic to Sir Malcolm Campbell, was also Donald's mechanic and he and I worked well together. A factor in my favour was that Leo also loved art and was very keen on what I was doing, so nothing was too much for him to facilitate my work. Also, Leo and I were working together for much of the time whereas Donald was in and out and you never quite knew when to expect him - which suited Leo and me perfectly.'

Different from virtually all of his previous drawings, Lawrie painted Bluebird in sepia, his first dabble into colour. Why such a strange colour? Well at the time, 1962, 'The Autocar' had a photogravure section (which the artists called 'photo gravy'), which itself was sepia. So Lawrie thought that he'd ▶

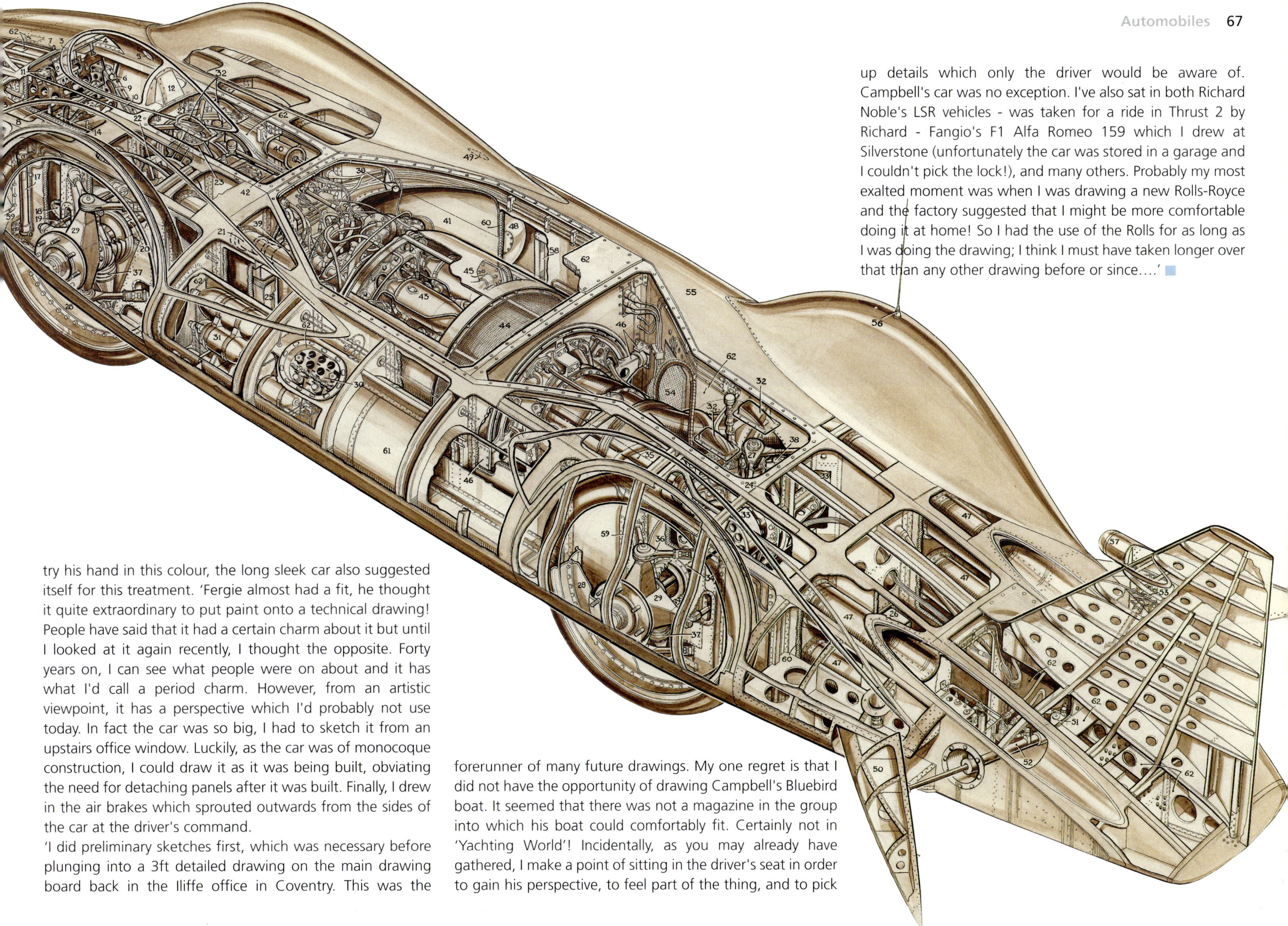

try his hand in this colour, the long sleek car also suggested itself for this treatment. 'Fergie almost had a fit, he thought it quite extraordinary to put paint onto a technical drawing! People have said that it had a certain charm about it but until I looked at it again recently, I thought the opposite. Forty years on, I can see what people were on about and it has what I'd call a period charm. However, from an artistic viewpoint, it has a perspective which I'd probably not use today. In fact the car was so big, I had to sketch it from an upstairs office window. Luckily, as the car was of monocoque construction, I could draw it as it was being built, obviating the need for detaching panels after it was built. Finally, I drew in the air brakes which sprouted outwards from the sides of the car at the driver's command.

'I did preliminary sketches first, which was necessary before plunging into a 3ft detailed drawing on the main drawing board back in the Iliffe office in Coventry. This was the forerunner of many future drawings. My one regret is that I did not have the opportunity of drawing Campbell's Bluebird boat. It seemed that there was not a magazine in the group into which his boat could comfortably fit. Certainly not in 'Yachting World'! Incidentally, as you may already have gathered, I make a point of sitting in the driver's seat in order to gain his perspective, to feel part of the thing, and to pick up details which only the driver would be aware of. Campbell's car was no exception. I've also sat in both Richard Noble's LSR vehicles - was taken for a ride in Thrust 2 by Richard - Fangio's F1 Alfa Romeo 159 which I drew at Silverstone (unfortunately the car was stored in a garage and I couldn't pick the lock!), and many others. Probably my most exalted moment was when I was drawing a new Rolls-Royce and the factory suggested that I might be more comfortable doing it at home! So I had the use of the Rolls for as long as I was doing the drawing; I think I must have taken longer over that than any other drawing before or since....'

Some Thrust!

Lawrie's motorcycling contacts resulted in one of his more famous paintings, Richard Noble's Thrust 2. As Lawrie explained, 'it happened by accident, because my mate Ken Sprayson, of Reynold Tubes, mentioned to me one day that he was building this strange vehicle in his workshop. I ought to take a look. So I did and was gobsmacked! This was something way beyond Bluebird, but the intention was the same, to gain the world land speed record for Britain. So I got on to the office in London, told them the story, they said to draw it in my spare time. When it was finished, they would see how it turned out. Additionally, Jimmy Saville had heard about the project and he had a little lad who wanted a ride in it under the banner of the television programme, 'Jim'll Fix It'. When Ken had finished the car, I went with him to Beverly airfield in Yorkshire where it was being tested and where the televised shooting was to take place. Thrust 2 was a two seater, not because Noble wanted to take a passenger with him on the record breaking runs but because he felt that a technical observer alongside him would help develop the hugely-powerful machine as it approached unknown territory around the sound barrier. When an observer was not being carried, cameras were and they provided a back up record for Richard's use later in the test facility. It was this ▶

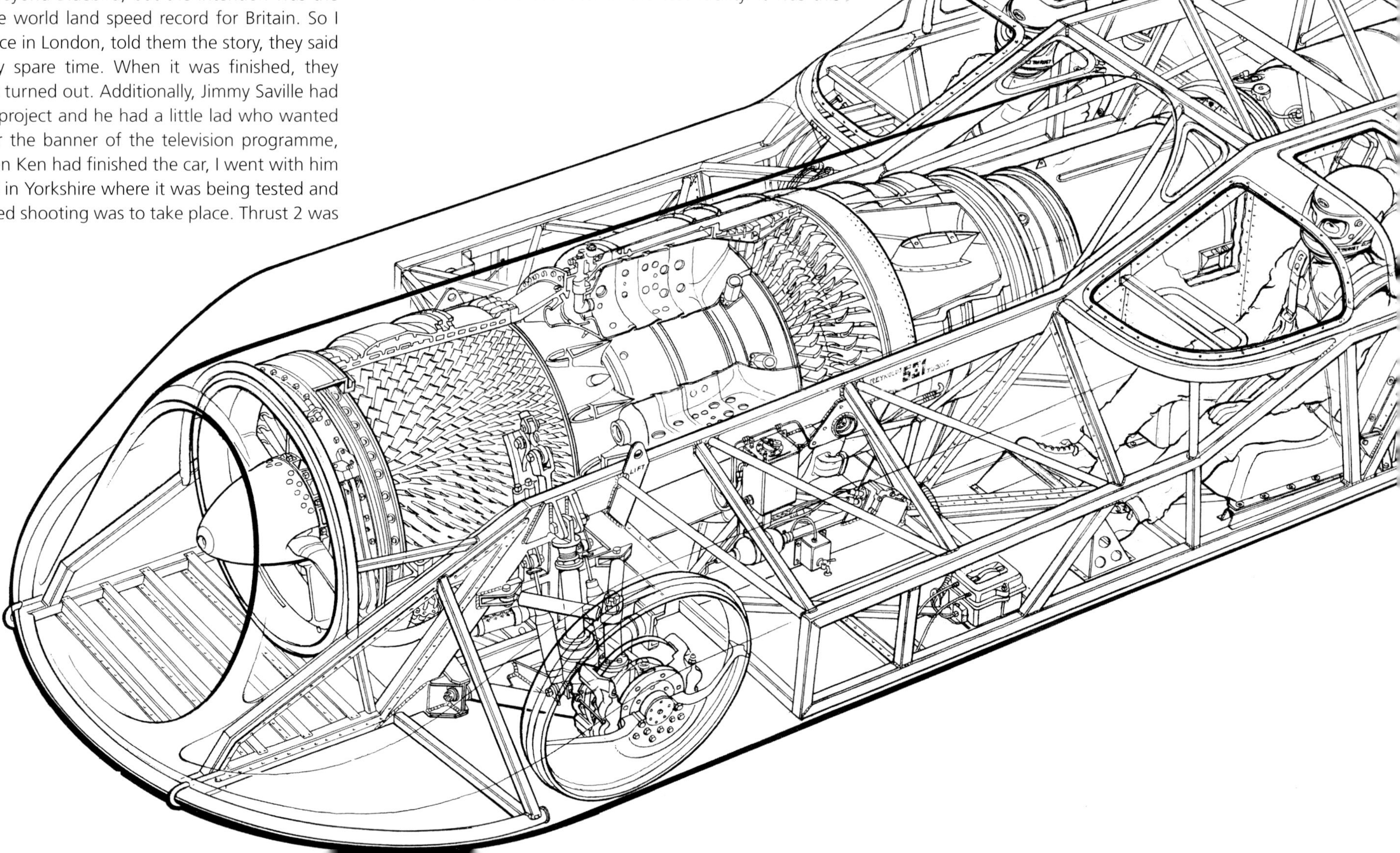

Compared with the sleek lines of 'Bluebird', Richard Noble's 'Thrust' land speed record car displays brute muscle power. The jet engine dominates the car, the driver sitting to the right of the engine for an unobstructed view of the way ahead. The second seat was installed for test engineers; on one 250mph test run, Lawrie occupied the engineer's seat.

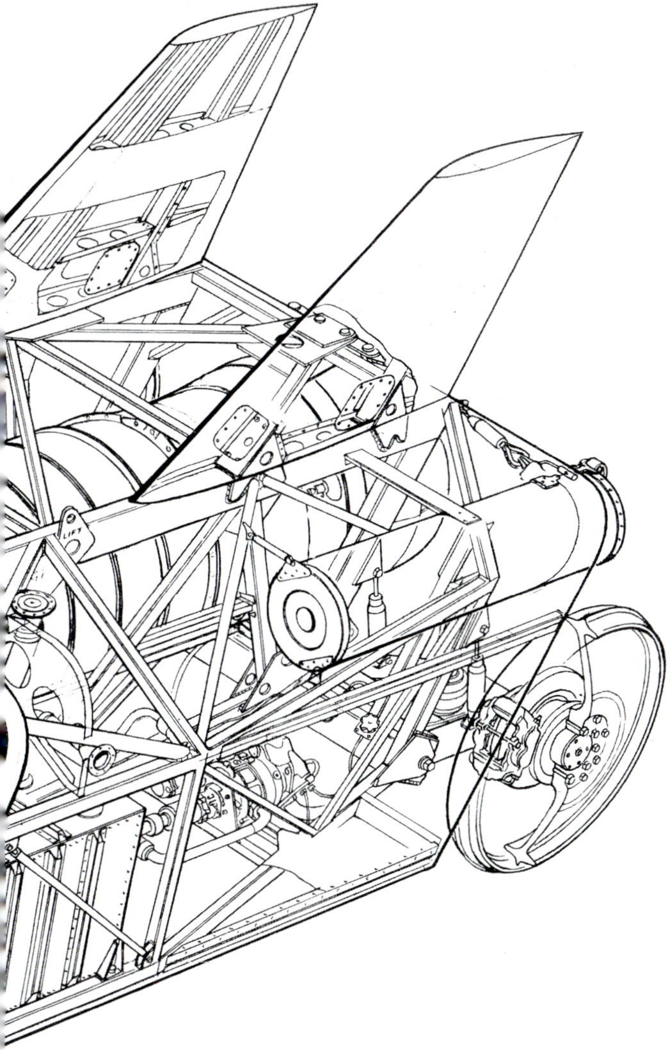

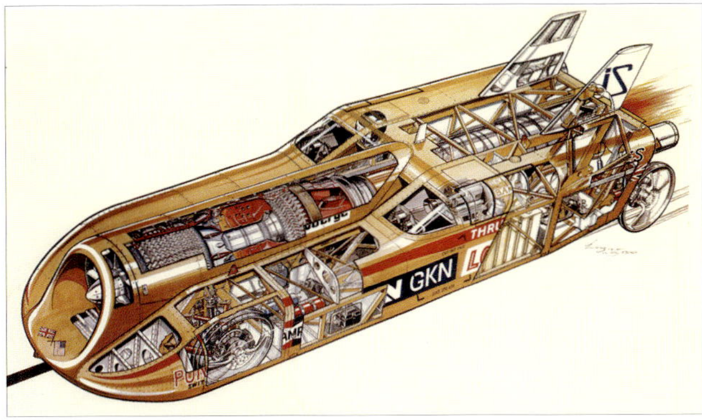

Visitors to the Coventry Motor Museum have the chance to experience over 250mph in the Thrust simulator. Both machines are on display there.

seat, positioned alongside the engine, that Saville's courageous youngster occupied during one of Richard's test runs. After the run, Richard called out to me 'How about you, Lawrie?' Like a shot I jumped in. Within a minute or so we were doing 250mph down the runway! Seems a lot, but it certainly didn't feel that fast. I'm used to 150mph in an aircraft down a runway, so this was not too different - in fact, it felt not unlike an aircraft but the biggest difference was the firm ride; false teeth wearers beware! Thrust had rubber tyres and proper suspension, but it was set up very firm and you could feel every bump. I did have a slight moment when we were nearing top speed about two thirds way down the runway. With that damned engine wittering away by my right ear 'ole, suddenly I felt warm, quite warm…'

Looking at his cutaway drawing of the engine, I noticed that he had drawn each blade inside the mighty Avon 200 Series jet engine. So I asked him to explain why he'd gone to such trouble. 'When I was talking about perspective when drawing an ellipse, I mentioned that if you don't get the segments right around the elliptical line, you won't fit all the blades in. They're not on the opposite side because of the vanishing perspective. It was quite a job, actually, because the blades have an offset to one another and as they go away from you, you have to get that offset in, otherwise they're not all going to fit into the drawing. The repetition drawing all those blades nearly drove me round the twist. I went to RAF Holton to get the engine details. Each blade is made separately then assembled by hand into a slot in the big disc of the turbine. When in position, the blade has a certain amount of movement which allows it to expand until the heat from the engine, when it fires up, causes the blade to expand securely into its housing in the disc. When all the blades have been fitted into their respective discs, that is the moment when the discs are locked up into one assembly.'

Behind the turbine blades are a series of massive gears with great big chunky teeth. Each of these teeth was meticulously drawn by Lawrie. When I remarked on this painstaking exercise, he surprised me by saying that each tooth was assembled loose. 'They have to handle some mighty loads and high temperatures. To cope with a sudden rise to high temperatures, the gears are slotted into place and actually wobble on the shaft. However, once the engine is running, rapid expansion ensures that they become a tight fit. It's important to remember the role heat plays; if everything was assembled tight, when the engine reached operating temperature it would simply fly to pieces. Incidentally, a kevlar armour-plated shield protected the driver against shrapnel in the event of a mechanical catastrophe…'

Lawrie was in at the beginning when SSC was being built by G-Force at Chichester, so he was able to build up the drawing as he had done with Bluebird, keeping pace with production. 'All the frame work, which I painted red on my drawing, was

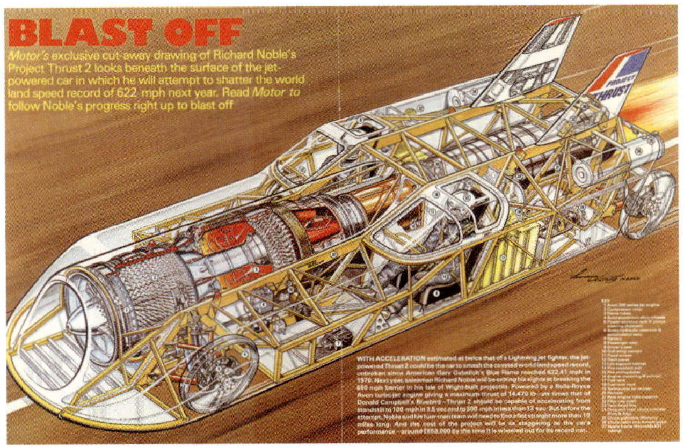

in rectangular and square section tubing. This immensely strong structure took all the load bearing. However, drawing the engines posed a problem for me as they are not mounted in parallel, nor are they parallel to the ground. In fact, they are splayed out slightly towards the rear in order to direct the thrust forwards, thereby helping to keep the vehicle in a straight line. They are also tilted up slightly at the front. So, if you drew centre lines through each engine and projected them forwards, the lines would meet some distance ahead of, and above, the vehicle. This was as a result of an aeronautical programme which called upon known data not for high-speed cars, but fighter aircraft. The reasoning was such that as they were aiming to break the sound barrier, then best aircraft practice would be involved, especially with two engines.'

Absolutely crucial was the length, and positioning, of the bodywork in relation to the tail fin assembly. Get that wrong, and the vehicle would either fly or, worse, dig in and somersault. Lawrie came up with the answers. 'There was no easy answer, they had to work by trial and error using historic Pendine Sands in South Wales, scene of many a world record attempt until it was too short, and the sand too unstable, for speeds above 200+mph. They had a model on a track at Pendine, ran it until it became stable and then scaled it up. The steering was interesting too. They used a modified BMC Mini assembly! There were long discussions about the rear end and the stabiliser, then when they tried it the suspension bottomed out! Which is why the suspension was stiffened up. Although the wheels are coupled closely together they are in fact mounted in a staggered tandem, one wheel being forward of the other. This provides wider twin tracks at the back, as shown in my depiction of the car at speed and was done to aid steering and stability.' ■

The birth of the Land Rover

Lawrie was also in on the birth of some historic cars but none more iconic than the ubiquitous Land Rover. A team of seven, led by Maurice Wilks, was charged with designing a compact four wheel drive all-terrain vehicle specifically for farming duties. The inspiration came from the sight of post war farmers buying up US army surplus jeeps for use on their fields. The Rover company could see a commercial opportunity in this and delegated the task to Wilks, who immediately formed a small design team. At the time, Lawrie was employed by Daimler but he accepted an offer to join Wilks' team, 'for the experience' he said.

He takes up the story. 'By the time the team was in place, most of the chassis had been designed so the main task for us was to make a utility body. This was indeed a change for me because I like beautiful, flowing shapes but the opposite was needed for the new Rover. So I had to adapt my concepts and think rapidly on my feet. Using straight lines instead of curves, and flat glass instead of curved, didn't appeal to me very much. Surprisingly, when I settled in I really enjoyed the job and working with such a talented team was an uplifting experience. It started with a bus journey from the old Daimler works to Rover... but everybody on the bus smoked like a chimney. I can't stand cigarette smoke, so I complained to Maurice. His response was to give me a Rover car! So those who didn't like the bus trip came with me in the company car. 'The way we operated would be impossible today, in fact it was unusual then. We split the various segments of the new vehicle between us, then we designed those specific parts for the boxy shape. Now for the unusual bit. When we had finished designing a particular item, we would go into the factory, hand fabricate and finish that part ourselves. Can you imagine that today? Of course, there was no union nonsense about who did what. The priority was to get everything made, fitted and tested. It shows how hard, and efficiently, we worked because the prototype Rover body was completed and on the chassis within ten weeks! Part of ▶

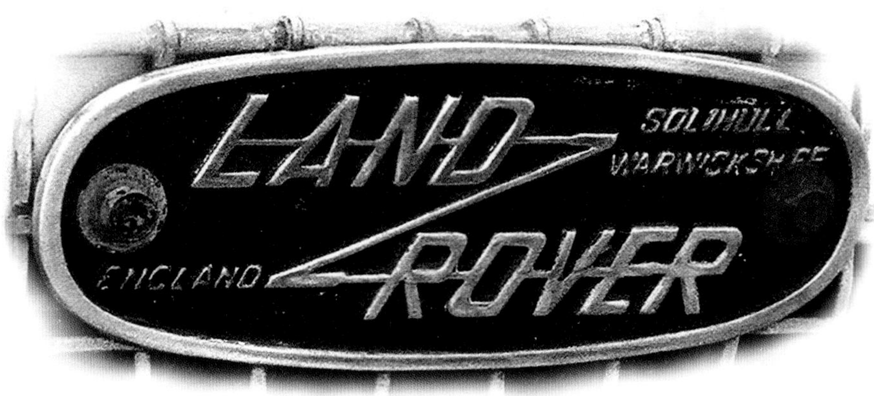

One of the original cast badges designed by Lawrie for the first Land Rover.

the reason for the swift completion was due to the fact that we were following in the tyre tracks of the original Jeep concept; we knew what that had achieved, and all we had to ensure was that our new design would be suited to civilian use. So we made room inside for four passengers facing one another on bench seats, we galvanised the entire chassis against rust, and reinforced the four corners of the aluminium body with galvanised steel for additional strength. Aluminium was used to keep weight down and also because it would not rust. In the event of damage, each panel could be unbolted and replaced quickly and easily. Another thing which expedited matters was that Rover already had a suitable four cylinder petrol engine from one of their cars which dropped straight into the compact engine bay of the new Rover. The diesel engine concept was still some time in the future.

'Then it became necessary to find a name for our new creation. Maurice had me down as an arty crafty type so he came and looked over my shoulder one day at what I was scribbling. I had just written the word 'Land' in block capitals and was thinking what to follow it with, then I scribbled 'Rover' underneath in the style I sign my signature, with my surname beneath my first name. But I didn't like it and drew a diagonal line through it, just as Maurice appeared. I apologised to him for the doodle, but to my amazement he said 'That's it, put a diagonal through it!' He then drew an ellipse around the two words, like you do when drawing attention to something in the newspaper. I spent the next day improving the drawing and then Maurice reappeared with some of his directors and took a look at what I'd done. 'We like that.' The next day, a pattern was made, followed by a casting, and that was it, which shows how quickly we worked in those days. Incidentally, it's the same Land Rover casting, which took me just a few minutes to draw, that they use today...

'Shortly afterwards, we got a Government contract to produce Land Rovers for the Ministry of Defence. They had to be in colours different to the farmers' ones; they were painted bright red so they could be visible in the fields. We did some interesting things for the Ministry. One such was a requirement for a semaphore, (a trafficator to you!) which would work without electricity. So I designed a mechanism with a large lever on the dash board which operated the foot long semaphore The lever needed to be large because the semaphore itself was so heavy. There was no bulb to illuminate the sign; the military saw no point informing the enemy which way their vehicles were about to turn!

'I was under some heavy pressure at that time, but from an outside source. I'd learnt to fly and some of the people I was flying with worked for Armstrong Siddeley Whitworth, so they wanted me to work for them. It meant me having to take an examination testing my knowledge of aircraft. This was second nature to me; I passed the exam and got a job in the design office. As I moved from Rover, I left behind the car industry and moved back into the aircraft industry.

Land Rover perspectives transferred to aircraft

'I took to Armstong Whitworth a system which I'd evolved at Land Rover. In order to make very quickly what I'd designed when working on the Land Rover, I'd also do a perspective sketch on the working drawing so that if I was not making the part then colleagues who followed me could see exactly what I intended. Rover liked the idea.The chaps in the works liked it because it saved a lot of time which in the past would have been spent trying to work out what the designer intended. It also speeded up the job. Armstrong Whitworth embraced my perspective sketch system; it was used successfully on the nose assembly of the Gloucester Meteor, which was then being built by Armstrong Whitworth. The idea proved so successful that the management formed a new department around me to produce perspective drawings for handbooks and maintenance books for the aircraft. It appeared that I was more use to them doing these drawings than designing new aeroplanes. Who was I to disagree?'

Rover and his kennel

Some years after I'd been working for Associated Ilillfe Press's office in Coventry I was asked to do a sectioned drawing of the first Range Rover. Those of my ex-colleagues who were still at Land Rover greeted me with open arms! In 1976, I again returned to Rover cars to do a sectioned drawing of the new SD1 3500 Rover for 'Motor'. What actually happened is that Rover asked me to draw the new factory in which the SD1 was being made. The commercial success of the Land Rover had given the parent company the confidence, and financial clout, to expand into a custom-made factory in which to manufacture the new SD1, a car on which Rover fortunes now rode. Recognising the significance, 'Motor' used my drawing of the factory in a double spread with the launch of the SD1, so they scored a double whammy, a new car and its factory! I put as much effort into the drawing of the factory as I did the car, with the exception that the car was sectioned and in colour and the factory was in black and white. The effort paid off because the double spread won for me Artist of the Year award for the second year in succession, the prize being a gold pen and £500, quite a lot of money in those days. When I went down to London to receive my prize, I was shocked to see my original drawing on display in Rover's showroom in Piccadilly!'

Not surprising that Lawrie's effort was a prize winner because the amount of detail in both drawings was phenomenal. By now we have become used to his style in drawing a car with its half-opened window providing a 3-D effect on the interior, sectioned panels showing the position of everything that matters. With the factory, he went overboard to show an action-man scenario, with men and machines digging holes, cranes lifting, walls being built, lorries moving materials, more than the eye can absorb on first glimpse. Sadly, I felt that the size and reproduction on the page in 'Motor' failed to do justice to the original but this did not deter the judges who were obviously wowed by the breadth and competence of the subject. Certainly, Rover were well pleased, hence their public display at HQ in London.

Lawrie's take on this was revealing. 'Looking at a selection of

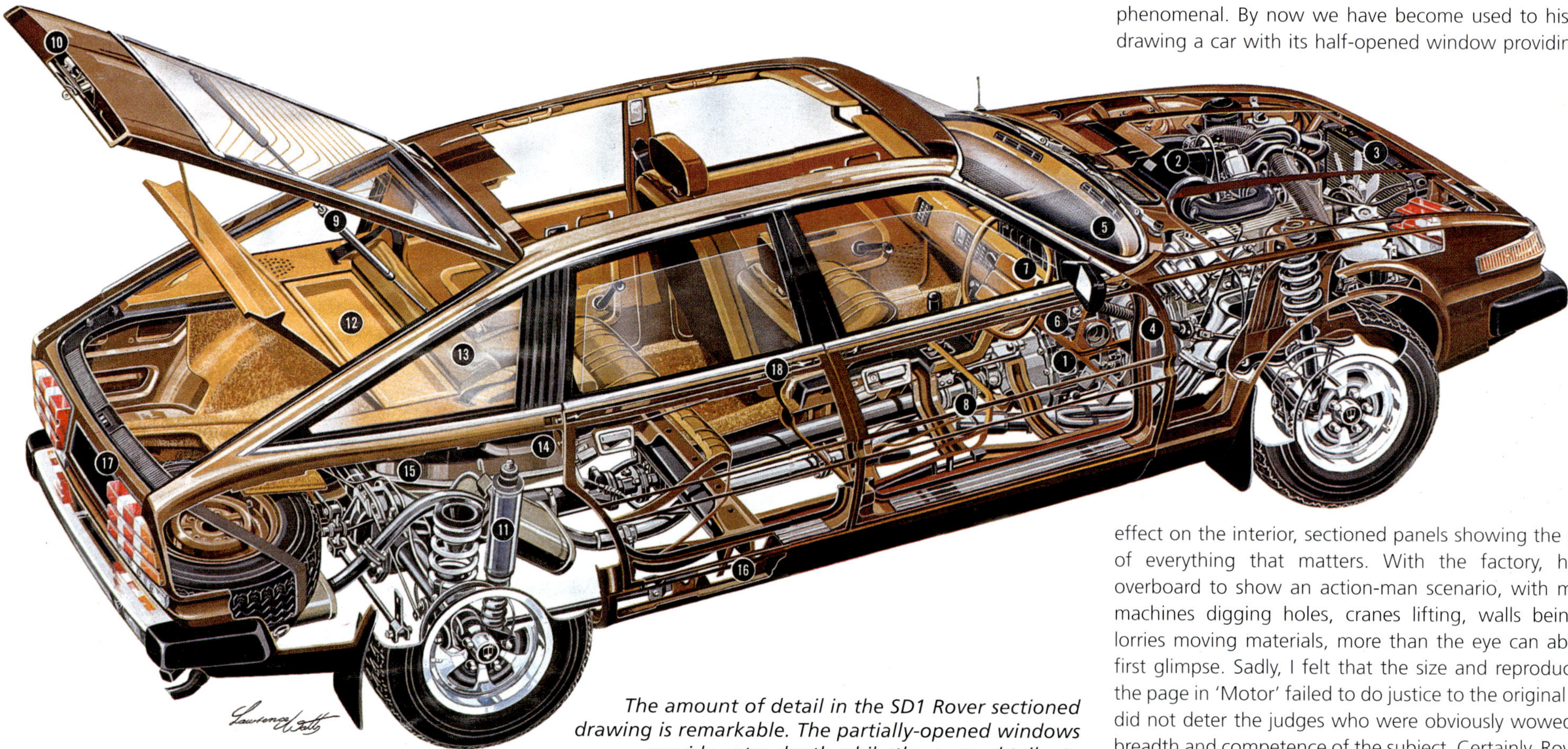

The amount of detail in the SD1 Rover sectioned drawing is remarkable. The partially-opened windows provide extra depth while the opened tailgate provides a view not normally shown. Note the heating elements in the rear window.

drawings done over a number of years, you can tell which artist did what drawing. Regardless of subject, every artist leaves his individual stamp without even knowing it! In the old days, I could tell you by the way, and the shape, who had drawn it. In my case I suppose it's not only the amount of detail I include, but other little things like reflections in mirrors, or in chrome plating, or in cutaways when you least expect them. It takes more effort, but it's worth it because you should give as much information as possible to the onlooker, and I enjoy providing something which I hope will give pleasure. That's how I really got into this drawing; the shape of the SDI appealed to me, I could see clearly what the designer was aiming for and I tried hard to convey that on paper. The judges seemed to agree with me!

'I also got to drive the car and liked it a lot, especially the model with the V8 engine, and I can see how the police came to rate it so highly. It not only had lots of torque, which made it an effortless drive, but it also had a tireless cruising speed for long journeys. And, of course, the 120+ top speed on police versions was useful for catching (or creating!) bad men. The handling was very stable and I enjoyed myself around the Warwickshire lanes. It could have been a good family car but nothing would separate me from my Volvos!' The difference that colours make in a drawing became apparent when comparing the Rover saloon and a 4x4 Range Rover, both using the same V8 engine. As would be expected of Lawrie, both vehicles were highly impressive in their stance, accuracy and attention to detail but the SDI appeared to be noticeably 'busier'. Could this be an optical illusion, I asked Lawrie? 'It's entirely due to the colours. The chocolate brown of the Land Rover did it no favours, it was too dull, but the SDI's duo-tone colour scheme cleverly accentuated the lines of the car. The same technique had been used when drawing both vehicles; firstly, pencil sketches and then, back in the office, inking in to form one big drawing.'

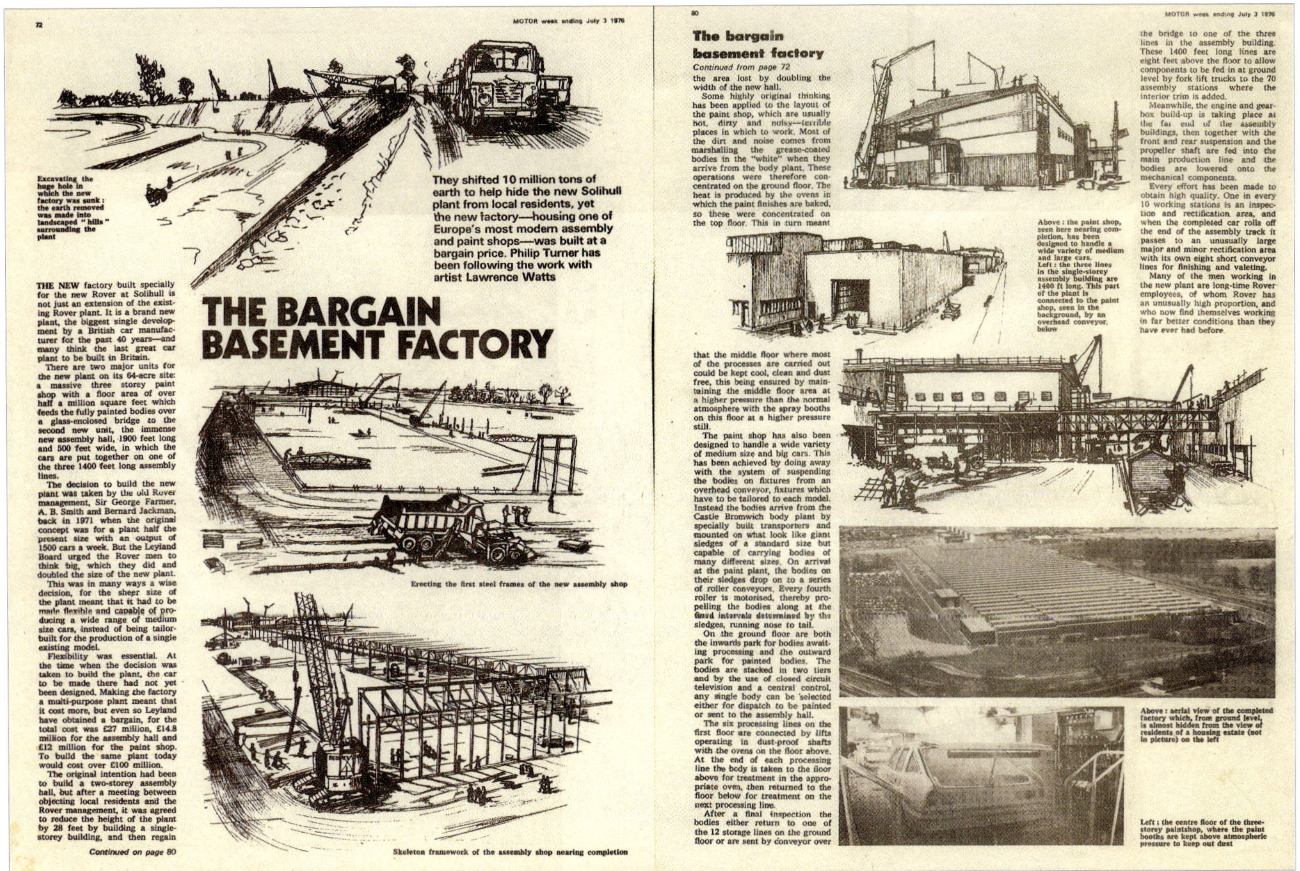

72 MOTOR week ending July 3 1976

Excavating the huge hole in which the new factory was sunk: the earth removed was made into landscaped "hills" surrounding the plant

They shifted 10 million tons of earth to help hide the new Solihull plant from local residents, yet the new factory—housing one of Europe's most modern assembly and paint shops—was built at a bargain price. Philip Turner has been following the work with artist Lawrence Watts

THE BARGAIN BASEMENT FACTORY

THE NEW factory built specially for the new Rover at Solihull is not just an extension of the existing Rover plant. It is a brand new plant, the biggest single development by a British car manufacturer for the past 40 years—and many think the last great car plant to be built in Britain.

There are two major units for the new plant on its 64-acre site: a massive three storey paint shop with a floor area of over half a million square feet which feeds the fully painted bodies over a glass-enclosed bridge to the second new unit, the immense new assembly hall, 1900 feet long and 500 feet wide, in which the cars are put together on one of the three 1400 feet long assembly lines.

The decision to build the new plant was taken by the old Rover management, Sir George Farmer, A. B. Smith and Bernard Jackman, back in 1971 when the original concept was for a plant half the present size with an output of 1500 cars a week. But the Leyland Board urged the Rover men to think big, which they did and doubled the size of the new plant.

This was in many ways a wise decision, for the sheer size of the plant meant that it had to be made flexible and capable of producing a wide range of medium size cars, instead of being tailor-built for the production of a single existing model.

Flexibility was essential. At the time when the decision was taken to build the plant, the car to be made there had not yet been designed. Making the factory a multi-purpose plant meant that it cost more, but even so Leyland have obtained a bargain, for the total cost was £27 million, £14.8 million for the assembly hall and £12 million for the paint shop. To build the same plant today would cost over £100 million.

The original intention had been to build a two-storey assembly hall, but after a meeting between objecting local residents and the Rover management, it was agreed to reduce the height of the plant by 28 feet by building a single-storey building, and then regain

Continued on page 80

Erecting the first steel frames of the new assembly shop

Skeleton framework of the assembly shop nearing completion

80 MOTOR week ending July 3 1976

The bargain basement factory

Continued from page 72

the area lost by doubling the width of the new hall.

Some highly original thinking has been applied to the layout of the paint shop, which are usually hot, dirty and noisy—terrible places in which to work. Most of the dirt and noise comes from marshalling the grease-coated bodies in the "white" when they arrive from the body plant. These operations were therefore concentrated on the ground floor. The heat is produced by the ovens in which the paint finishes are baked, so these were concentrated on the top floor. This in turn meant that the middle floor where most of the processes are carried out could be kept cool, clean and dust free, this being ensured by maintaining the middle floor area at a higher pressure than the normal atmosphere with the spray booths on this floor at a higher pressure still.

The paint shop has also been designed to handle a wide variety of medium size and big cars. This has been achieved by doing away with the system of suspending the bodies on fixtures from an overhead conveyor, fixtures which have to be tailored to each model. Instead the bodies arrive from the Castle Bromwich body plant by specially built transporters and mounted on what look like giant sledges of a standard size but capable of carrying bodies of many different sizes. On arrival at the paint plant, the bodies on their sledges drop on to a series of roller conveyors. Every fourth roller is motorised, thereby propelling the bodies along at the fixed intervals determined by the sledges, running nose to tail.

On the ground floor are both the inwards park for bodies awaiting processing and the outward park for painted bodies. The bodies are stacked in two tiers and by the use of closed circuit television and a central control, any single body can be selected either for dispatch to be painted or sent to the assembly hall.

The six processing lines on the first floor are connected by lifts operating in dust-proof shafts with the ovens on the floor above. At the end of each processing line the body is taken to the floor above for treatment in the appropriate oven, then returned to the floor below for treatment on the next processing line.

After a final inspection the bodies either return to one of the 12 storage lines on the ground floor or are sent by conveyor over the bridge to one of the three lines in the assembly building. These 1400 feet long lines are eight feet above the floor to allow components to be fed in at ground level by fork lift trucks to the 70 assembly stations where the interior trim is added.

Meanwhile, the engine and gearbox build-up is taking place at the far end of the assembly buildings, then together with the front and rear suspension and the propeller shaft are fed into the main production line and the bodies are lowered onto the mechanical components.

Every effort has been made to obtain high quality. One in every 10 working stations is an inspection and rectification area, and when the completed car rolls off the end of the assembly track it passes to an unusually large major and minor rectification area with its own eight short conveyor lines for finishing and valeting.

Many of the men working in the new plant are long-time Rover employees, of whom Rover has an unusually high proportion, and who now find themselves working in far better conditions than they have ever had before.

Above: the paint shop, seen here nearing completion, has been designed to handle a wide variety of medium and large cars.
Left: the three lines in the single-storey assembly building are 1400 ft long. This part of the plant is connected to the paint shop, seen in the background, by an overhead conveyor, below

Above: aerial view of the completed factory which, from ground level, is almost hidden from the view of residents of a housing estate (not in picture) on the left

Left: the centre floor of the three-storey paintshop, where the paint booths are kept above atmospheric pressure to keep out dust

The new factory in which the SD1 was to be made was also drawn by Lawrie. The same attention to detail is evident throughout.

The lost Gypsy

'What I call the lost Gypsy was a utility vehicle designed by Austin to challenge the Land Rover in 1958. 'Power Farming' invited me to Austin to do a cutaway drawing for the magazine. At the time, I was working mainly for 'Autocar' and due to the lapse of time, I suppose, my employer forgot that I was one of the design team on the original Land Rover 10 years before. So there I was, in the opposition's camp, to draw Austin's answer to the Land Rover! Of course, my employer was now Iliffe Press so there was no conflict of interest. Anyway, nobody at Austin was any the wiser and I got stuck into the job of drawing the Gypsy.' Had Austin merely copied the Land Rover, hoping to refine that successful design, was the obvious question. But Lawrie refuted this. 'The chassis was certainly all Austin, with some ingenious little additions, but the square-ish body was distinctly similar in looks and functionality to the Land Rover. One of the Gypsy's engineering differences concerned the suspension and steering - the steering rack comes through the front suspension arms, while the rear is also different from the Land Rover. Originally, the Land Rover was designed to replace the ex-WW2 American Jeeps which farmers had snapped up and worn out. Austin had presumably the same utility market in mind with the added bonus of potential army and overseas contracts in mind. Whatever, they never sold in the same quantities as the Land Rover, either at home or overseas, and eventually the Gypsy (or Champ in military guise) faded from the scene.'

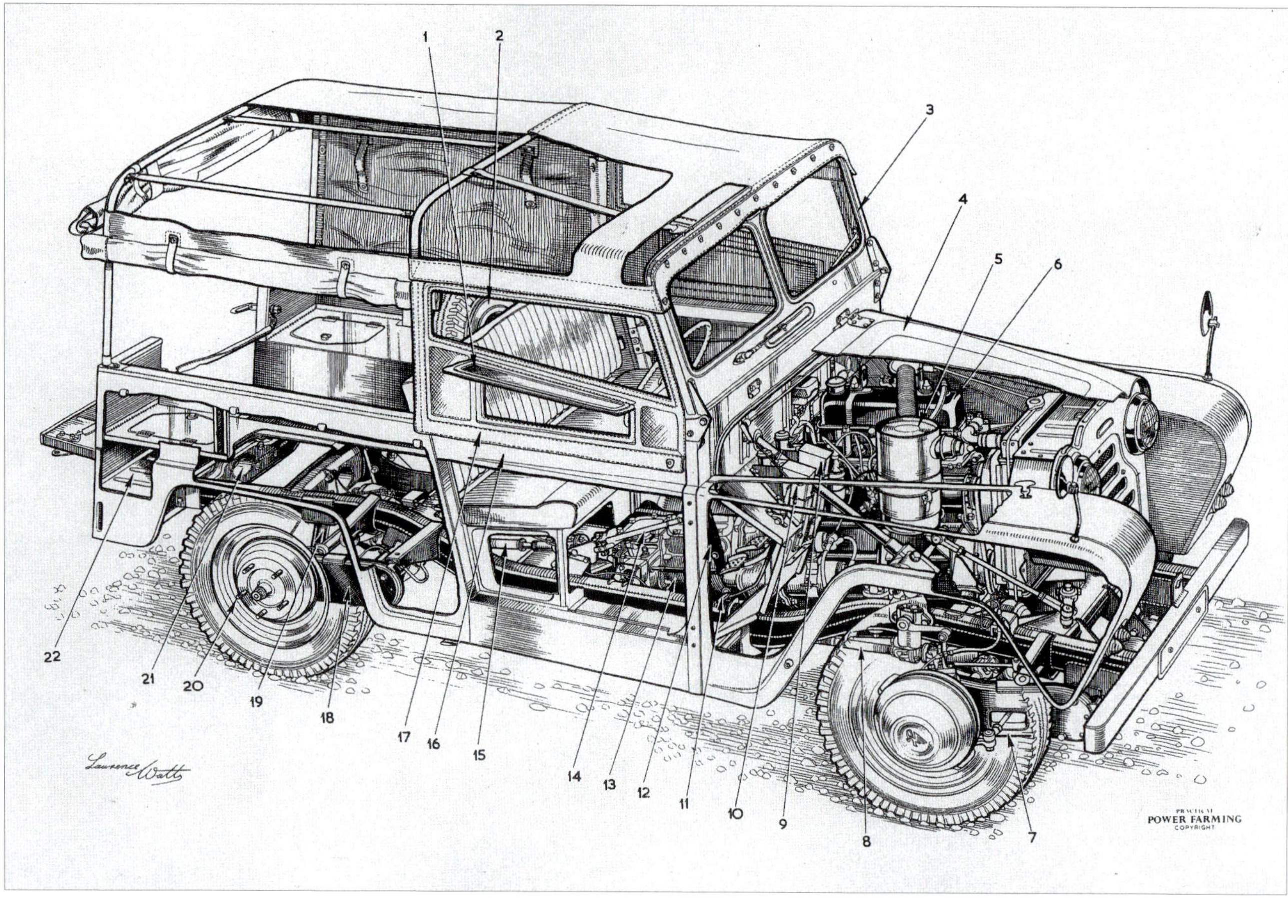

Though visually similar to Rover's Land Rover, Austin's Gypsy differed in mechanical design.

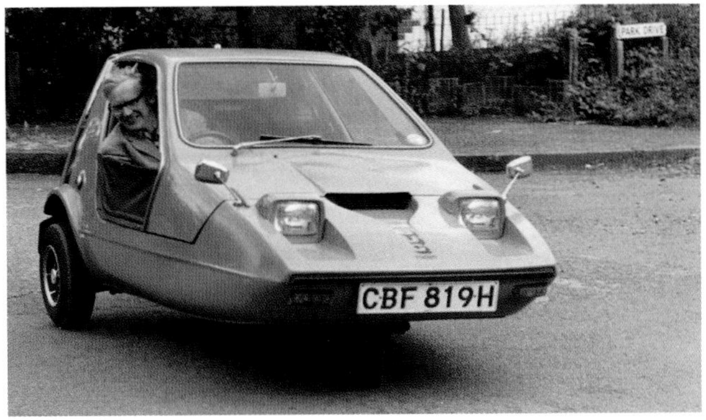

Flying Bug; the author leaves Mick Woollett's home in the Bond Bug which they shared for a trip to the Dutch motor cycle GP at Assen. The car was comfortable, speedy, great fun to drive and attracted attention wherever it went - especially on two wheels in the hands of some of the GP riders at Assen...

Orange Bug

None of Lawrie's colour drawings can compare with his superb sectioned view of the striking Bond Bug, an oft-derided three wheeler of the early seventies. Its wedge shape, covering the single front wheel, high squared-off back, and bright orange finish made the Bug one of the most recognisable shapes on the road. My colleague Mick Woollett and I borrowed an early production one in order to cover the Dutch motorcycle grand prix at Assen. We loved it! Not to put too fine a point on it, Mick knows his cars and can be quite critical yet he too became enthusiastic about the Bug. Aided by an excellent power-to-weight ratio resulting from the use of a glass fibre body, the all-alloy ohv engine gave some hint of its racer origin in the pre-war Brooklands Austin 750 single seaters as it produced surprisingly rapid acceleration and a comfortable 70+mph cruising speed. I recall 80mph on occasions. It was also very economical, averaging in the region of 60mpg, in spite of being thrashed for much of the trip. The 'flying wedge' aerodynamics were such that the little three wheeler was rock steady at speed. Comfortable, too, with reclining foam-rubber moulded one-piece seats. Generously-sectioned tyres on black and polished cast-alloy wheels ensured that we had plenty of grip on corners and it was only when we got to Assen and started giving demo rides to the racers that the outer edges of the performance envelope were explored by such as Peter Williams. Displaying his trademark 'butter wouldn't melt in my mouth' innocence, he tried to see how much of the circuit he could travel on two wheels, not only on the circuit, but also around the local traffic which had by now reclaimed their roads for normal usage...An impressive demonstration of the Bug's ease of control and manoeuvrability (do not repeat this experiment at home). The sole grouse of both Mick and myself concerned the lack of luggage space, restricting each of us to a small holdall (and no duty free).

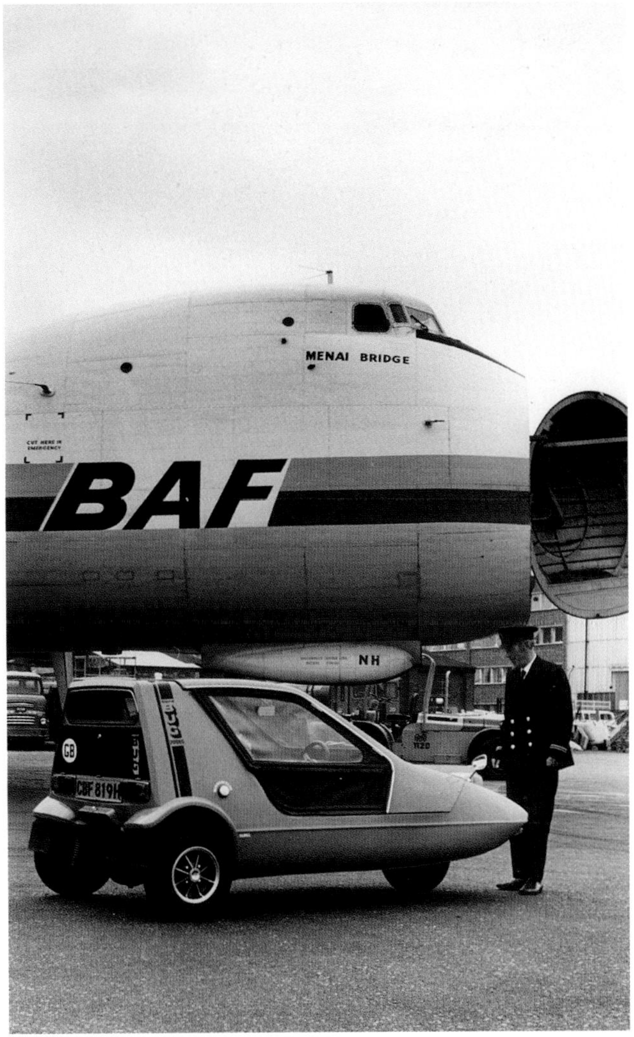

Dwarfed by a Bristol Freighter, the Bug awaits loading for flying across the Channel. The aircrew were smitten by the little vehicle... as the occupants were by the sheer size of the Freighter.

'I liked the Bug too, it was a fun car, fun to drive and I was amazed that it did not catch on,' chipped in Lawrie. 'Once Bob Currie and I were out in it and on a fast left hander, I somehow managed to lift the inside rear wheel. I had to restrain Bob from climbing out of the non-existent window as he was impersonating his racing sidecar passenger trick, hanging over the side to restore balance! After swapping drives, we both agreed that the Bug was great fun to drive and as such, must have a bright future. How wrong can you be?'

Another factor which convinced Lawrie, and Bob, that the Bug would appeal to mechanically-minded folk such as motorcyclists, was the manner in which the rear of the car was designed. 'That rear axle, from the original Austin racing car, was exposed to view, for instance. I felt that's the type of thing which would appeal to motorcyclists with their liking for working parts and those who wanted something different ▶

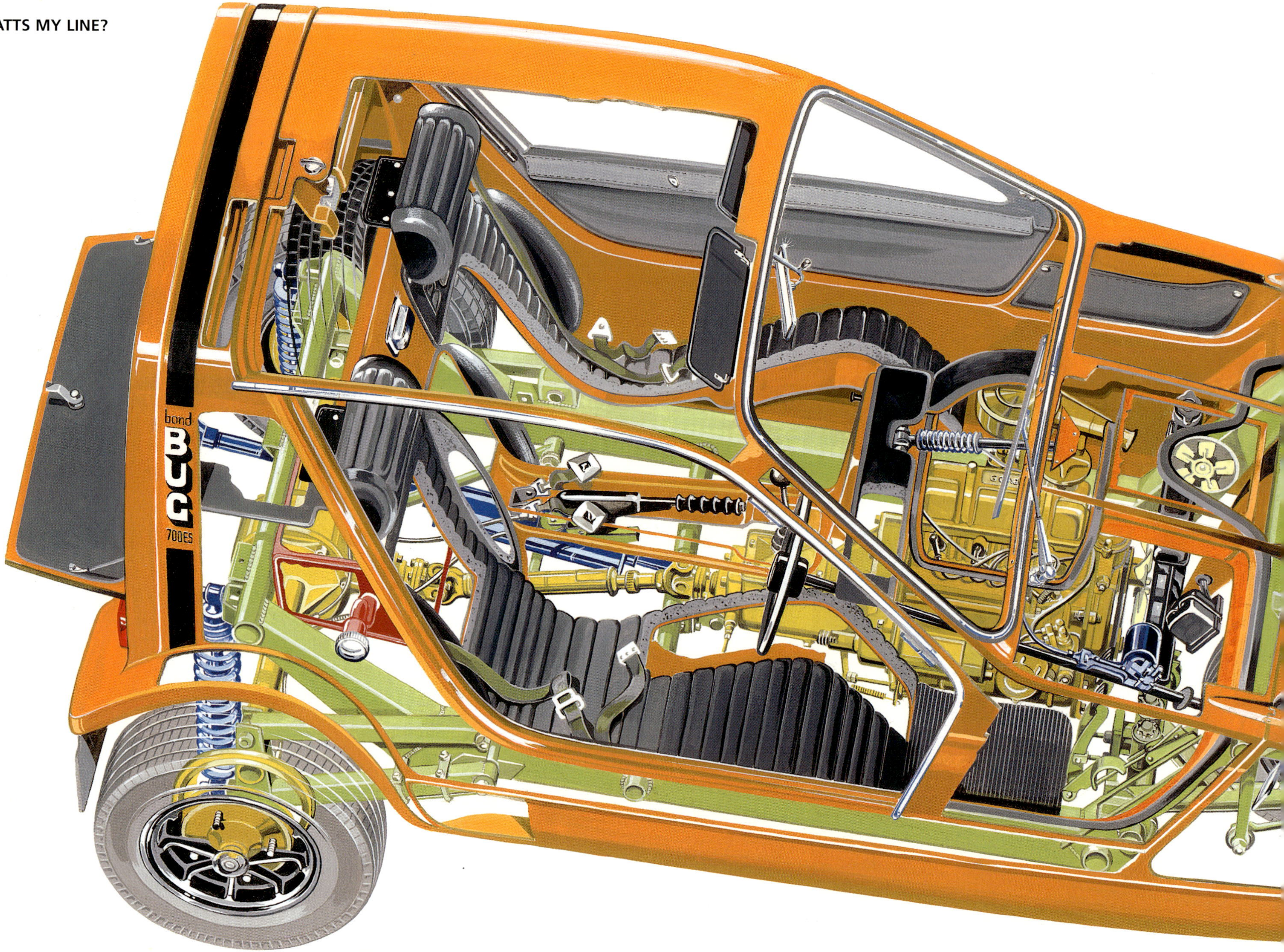
bond
BUG
700ES

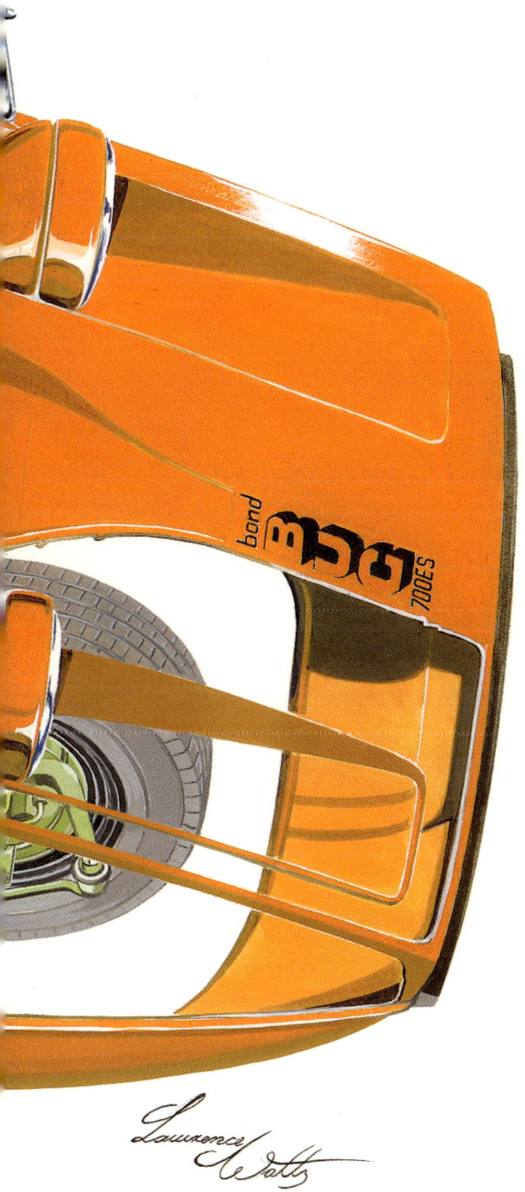

to drive without the hassle of gaining a car licence. We all agreed that the Bug had potential. Apart from minimal luggage space, the only drawback is that those comfy reclining seats were not adjustable fore and aft. For me that's a problem 'cos my shorty legs needed to get nearer the pedals. Only by stuffing cushions behind my back could I move forward.'

Other than the fact that the Bug was a three wheeler, with a distinctively-styled glass fibre body on an orthodox rectangular-section chassis, there was little of ground shaking technical interest in the design. But Lawrie has displayed yet again in this drawing of a modest vehicle his artistic skills and enthusiasm by including the most extraordinary detail and workmanship which one would expect in a prestigious vehicle such as a Rolls-Royce or Aston Martin. All the Bug's chassis is in green, the transmission in brown, and the steering gear in black all of which 'lifts' the drawing, bestowing an almost three-dimensional effect. Lawrie had first executed a black and white drawing, but then he decided to do the Bug justice in a colour painting, using brushes only; no pencils or pens. The result shows clearly the benefit of a colour medium because the freedom it allows enables even the smallest details to be identified in colour. In black and white they would have been lost. For example, by using a very fine brush Lawrie has been able to paint such tiny details as the welding of the various brackets on the chassis, or even the welding of the chassis itself and his cutaway of the moulded seats shows clearly that the medium is foam rubber, with hundreds of tiny dots identifying the rubber foam. It doesn't stop there either because he replicated exactly the padded non-slip seat covering!

Greater contrast could not be imagined between the lowly Bug and a Formula 1 World Championship-winning Tipo 159 Alfa Romeo. For a racing and sports car series in the mid 60s Lawrie was commissioned to draw Juan Manuel Fangio's 159 Alfa. The car was stored in an anonymous Silverstone garage and to draw it that's where Lawrie took up residence for ten days. The same attention to detail displayed in the Bug painting was shown in the sectioned drawing of the Alfa but with the accent on power and speed, the focal point on the Alfa was the magnificent straight eight cylinder engine packed under that long tapering bonnet. So much was there that Lawrie had a field day with his passion for detail. Italians like their metal and a casting is not just a casting, it must look like a work of art, with sculpted ribbing around the exterior, or strengthening gussets cast instead of fabricated, that sort of thing. All grist to the pen of Master Craftsman Watts. 'If the Italians saw the need to make it, ▶

then I'll draw it. No problem.' And there it is, exquisitely reproduced in its original colour.

The Alfa chassis seemed to be comprised more of lightening holes than metal, all of which Lawrie faithfully reproduced. Even the welding on the gusset plates on the chassis is shown, as is the alloy welding on the fuel tank. There's an indent in the tank, needed for clearance of the rear suspension; faithfully shown is the typical neat alloy welding around the indent! Wheels, the bane of many an artist but not this one, are shown complete with their magnificent finned brake drums, with cutaways to show the internals. Needless to say, each spoke is individually drawn, polished, and laced into the polished alloy rims at the correct angle too. What an immense amount of work! For an encore he's not only drawn a line of cooling louvres in the raised engine bonnet but has also painted little highlighters on each louvre to show that it's raised to catch the airflow. Using a magnifying glass will reveal even more details…

For me the final flourish is the manner in which he's painted the Alfa Romeo insignia in red and green onto each knock-on hub nut. Showing off? 'Yes! I so much enjoyed doing this drawing, of such a beautiful and historic car, that I got carried away - to the point that if there had been a loose nut in sight, I'd have drawn that as well!'

Even today, he wonders how he managed to do it. 'If I were asked today to do this drawing I'd have to think twice,' he admits. 'But, yes, of course I'd do it, it's the sort of challenge I like. But back in 1964, such challenges were helping me to establish myself, and this one in particular helped me up the ladder because it won for me the Editorial Award that year, one of five in succession I have won. But the biggest challenge was to do the best job I could. If it then won an award that was a bonus, especially if it was awarded by my peers.'

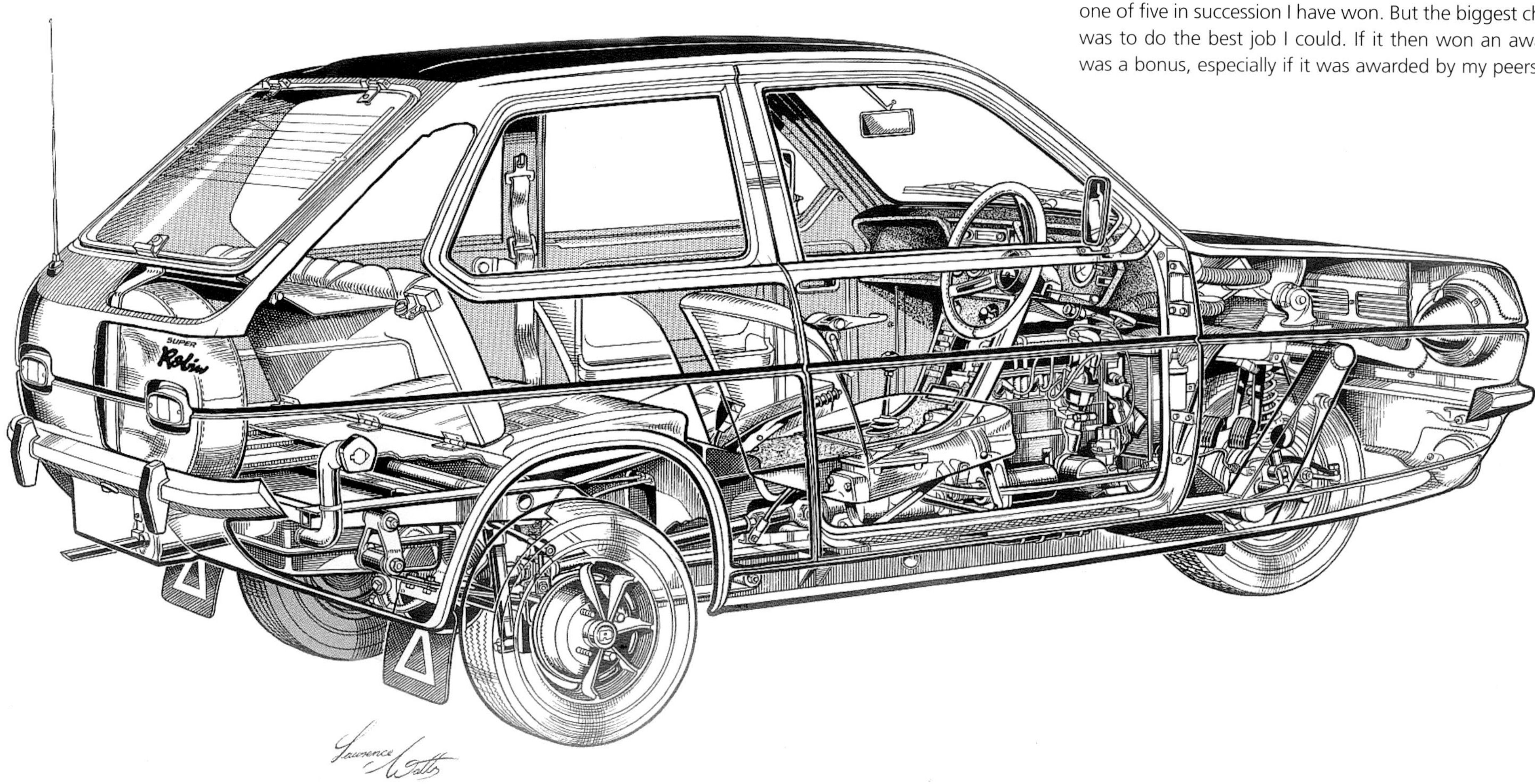

Saloon version of the Reliant Robin carried four persons in relative comfort. Robins enjoyed popularity because they could be driven on a motorcycle licence.

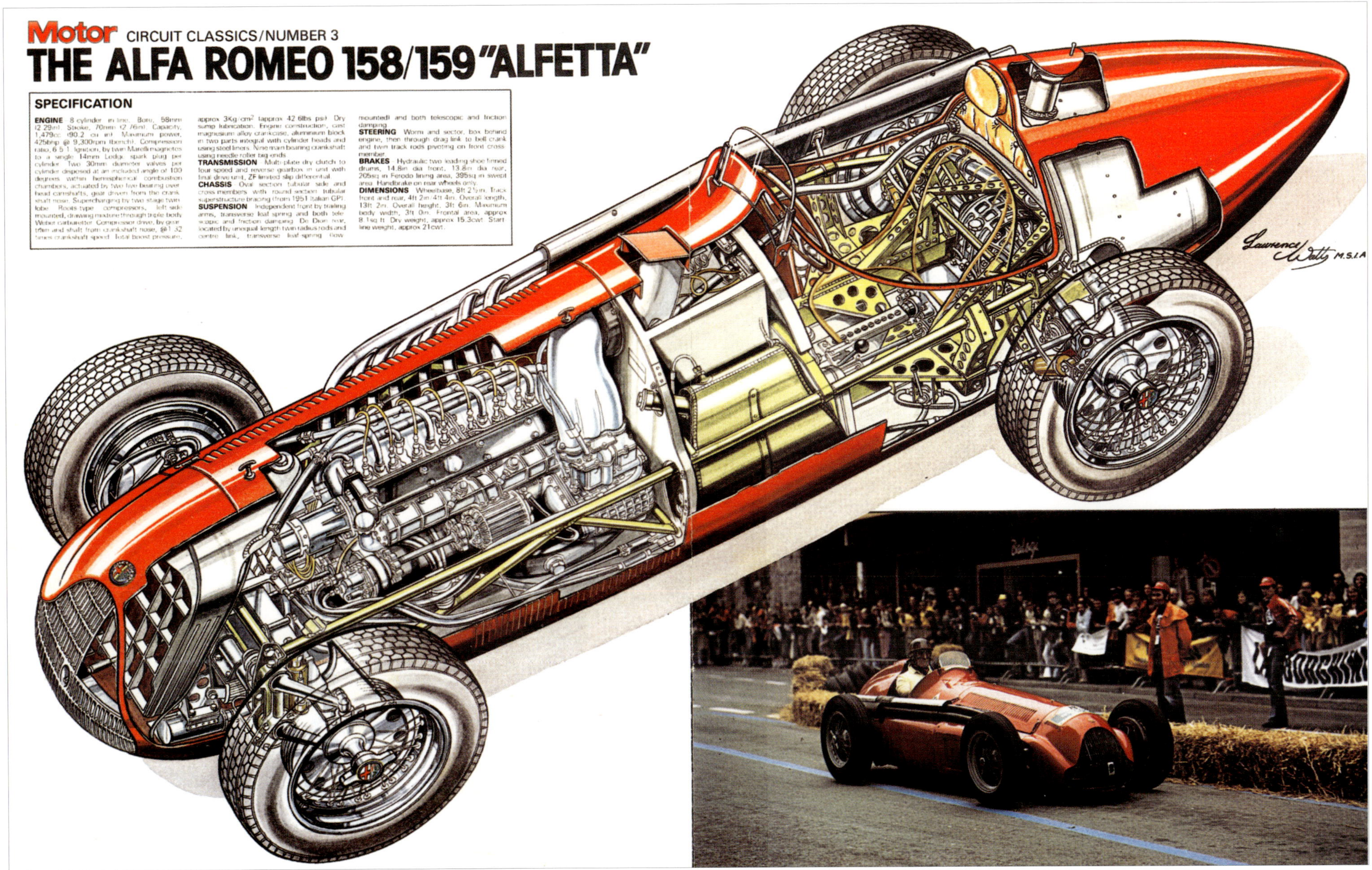

Motor CIRCUIT CLASSICS/NUMBER 3

THE ALFA ROMEO 158/159 "ALFETTA"

SPECIFICATION

ENGINE 8 cylinder in line. Bore, 58mm (2.29in). Stroke, 70mm (2.76in). Capacity, 1,479cc (90.2 cu in). Maximum power, 425bhp @ 9,300rpm (bench). Compression ratio, 6.5:1. Ignition, by twin Marelli magnetos to a single 14mm Lodge spark plug per cylinder. Two 30mm diameter valves per cylinder disposed at an included angle of 100 degrees within hemispherical combustion chambers, actuated by two five bearing overhead camshafts, gear driven from the crankshaft nose. Supercharging by two stage twin lobe Roots-type compressors, left side mounted, drawing mixture through triple body Weber carburettor. Compressor drive, by gear train and shaft from crankshaft nose, @1.32 times crankshaft speed. Total boost pressure, approx 3Kg/cm² (approx 42.6lbs psi). Dry sump lubrication. Engine construction, cast magnesium alloy crankcase, aluminium block in two parts integral with cylinder heads and using steel liners. Nine main bearing crankshaft using needle roller big-ends.

TRANSMISSION Multi-plate dry clutch to four speed and reverse gearbox in unit with final drive unit, ZF limited slip differential.

CHASSIS Oval section tubular side and cross members with round section tubular superstructure bracing (from 1951 Italian GP).

SUSPENSION Independent front by trailing arms, transverse leaf spring and both telescopic and friction damping. De Dion rear, located by unequal length twin radius rods and centre link, transverse leaf spring (low mounted) and both telescopic and friction damping.

STEERING Worm and sector, box behind engine, then through drag link to bell crank and twin track rods pivoting on front cross-member.

BRAKES Hydraulic two leading shoe finned drums, 14.8in dia front, 13.8in dia rear, 205sq in Ferodo lining area, 395sq in swept area. Handbrake on rear wheels only.

DIMENSIONS Wheelbase, 8ft 2½in. Track front and rear, 4ft 2in/4ft 4in. Overall length, 13ft 2in. Overall height, 3ft 6in. Maximum body width, 3ft 0in. Frontal area, approx 8.1sq ft. Dry weight, approx 15.3cwt. Start line weight, approx 21cwt.

Judges in the annual 'Artist of the Year' award were influenced by the incredible amount of detail in this drawing of Juan Manuel Fangio's F1 Alfa - another award for Lawrie!

Tom Walkinshaw's Le Mans effort

A man of strong views and few words, Tom Walkinshaw lived up to the popular image of the dour determined Scot that he was. A clever and innovative mechanical engineer who was also a successful racing driver in his own right, his race track successes benefited his Jaguar car dealership in Coventry and his engineering business at Kidlington, near Oxford. Lawrie knew him only by reputation, but he admired him enormously for his single-mindedness and courage in putting his money where his mouth was in the highly specialised world of sports car racing. 'I was a bit apprehensive when sent to meet him,' admitted Lawrie, 'because of his reputation, but it soon became apparent that I'd nothing to fear. Contrary to everything I'd been told, he was chatting away amiably within a few minutes and was anxious to assist me in any way possible. I was there to draw his Le Mans winning V12 sportscar and he provided all the background information that I needed to get on and draw the car. It seemed that Jaguar did not want to know him until after he'd won Le Mans, with a young Martin Brundle as one of his drivers, but then Jaguar was all over Tom like wet paint. He had proven that taking a stock 6 litre V12 engine, developing it to produce 630-bhp at 7,000 rpm, and installing it in an aerodynamically proven package, you could break the monopoly that specialist firms such as Porsche had enjoyed for many years. That suited Jaguar, so they mounted a successful return to Le Mans with their own team of cars, but it was Walkinshaw who deserved the credit for lighting the touch paper. And that's all the inspiration I needed to make what I feel is one of my better drawings.'

Walkinshaw offered Lawrie a drive in the car at an airfield near the TWR base at Kidlington, Oxfordshire. Years later the memory still produces a happy grin from Lawrie. 'Quite an experience! Within a short time, and helped by the advice of an engineer along for the ride - and to keep an eye on me! - I'd become familiar with the car. It was user-friendly, not surprising when you think how long the drivers had to spend controlling it at speeds of up to 200mph in the mixed weather traditional to Le Mans, and through the long hours of darkness. Having put the engineer at ease, I think, it was time to give the car some welly and I was amazed how effortlessly it rocketed up to, er, high speeds, then lost speed almost as rapidly when the powerful brakes came into use. I could have played with it all day but using up someone else's tyres and time tempered my enthusiasm and I regretfully handed it back. What a car! And what a bonus for only doing my job.'

Again, Lawrie's beady eye for detail is evident throughout, even down to the dzus quick-release fasteners used to retain the engine cover. 'I was impressed by the practical details,' recalled Lawrie, 'so I tried to do them justice by including everything that the designer put in. For instance the whole back end was quickly detachable. The engine was hung from the main bulkhead, facilitating a rapid change, if need be. As might be imagined, fabricating the exhaust system for 12 cylinders took some doing, but it was all beautifully tucked away.' Responsibility for the aerodynamics was farmed out to a specialist, the Imperial College London's wind tunnel facility. March Engineering supplied the five-speed racing gear box which drove through a Borg and Beck clutch.

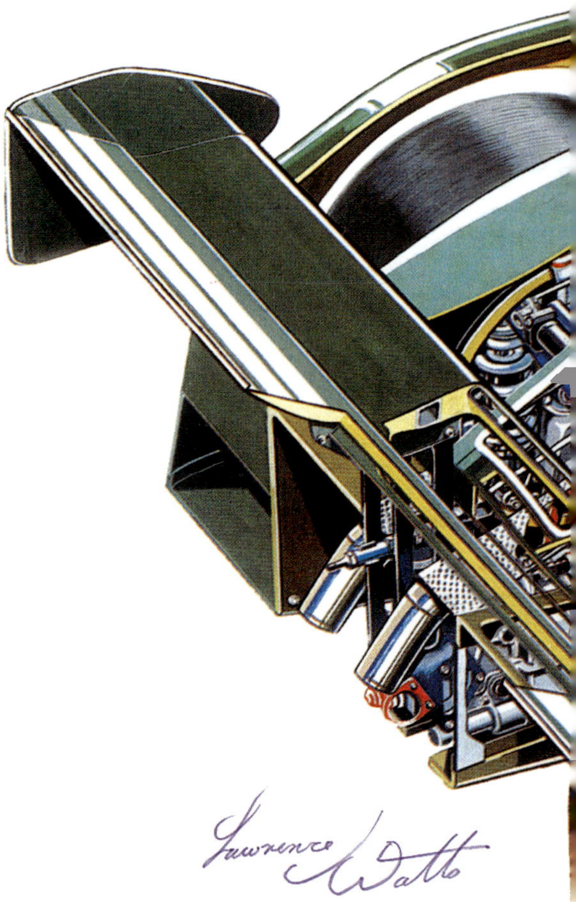

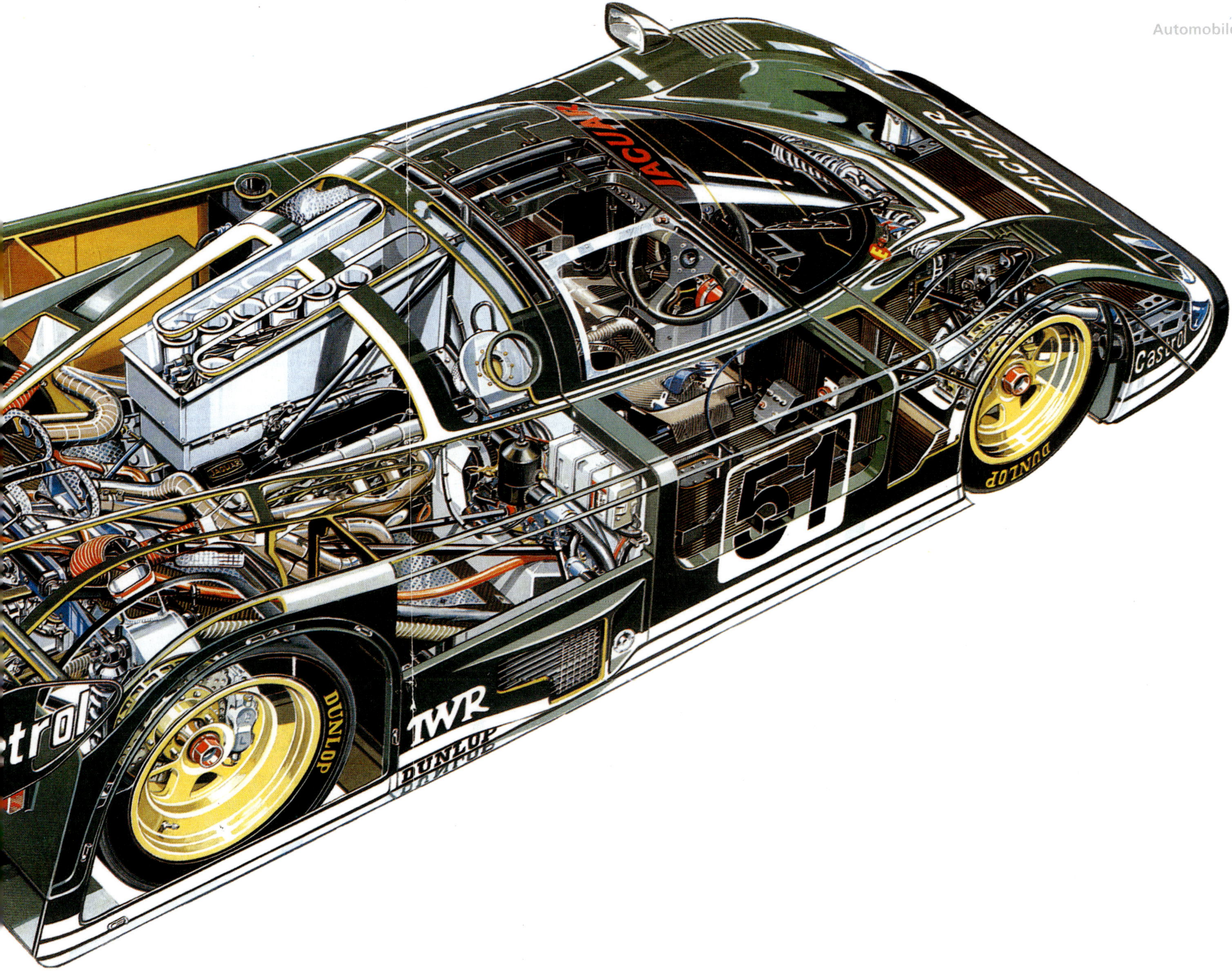
JAGUAR
Castrol
51
DUNLOP
TWR
DUNLOP
DUNLOP
trol

Rover-BRM Le Mans turbine engine

Over 40 years ago, Rover entered a gas turbine-powered car in the 1965 Le Mans 24 Hours race. Yes, a gas turbine car from Rover! Hardly seems credible but the Coventry car maker at that time was at the cutting edge of technology, as NSU had been with its rotary Wankel petrol engine. It is believed that the Wankel inspired Rover to seek an answer in their turbine. Together with BRM, at that time a leading Formula 1 racing car manufacturer, Rover built a Le Mans challenger to prove the reliability of their new gas turbine engine, having already introduced an experimental car two years earlier. This time it was to be for real, a fully fledged race car with streamlined body, lights, all the gubbins necessary for maintaining 100mph+ average speeds for 24 hours at Le Mans.

Lawrie was asked to draw the 1965 engine, which was about the same size as an equivalent car unit and produced 145bhp. As he explains, 'Rover's intention was not so much to go car racing but fuel and reliability testing. Gas turbines were normally used for aircraft where they excelled at ▶

Reprinted from **Autocar** *16 July 1965*

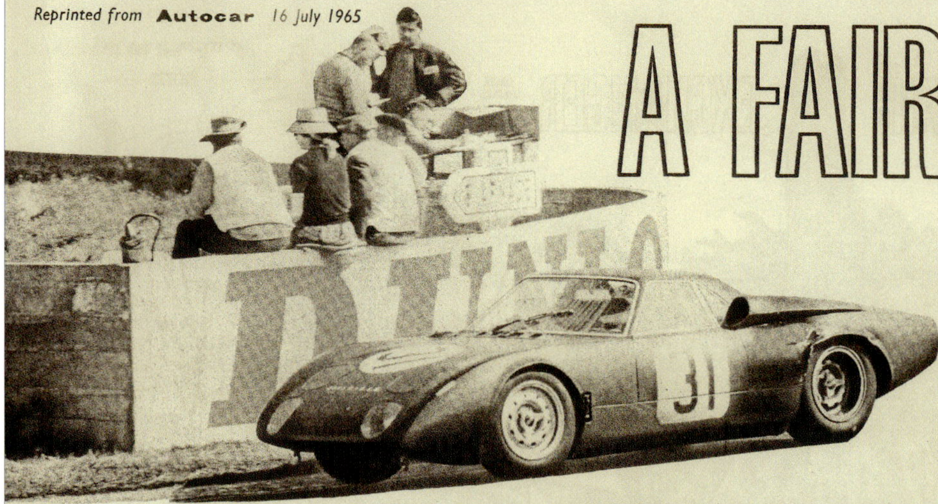

A FAIR EXCHANGER

by EDWARD EVES

ANALYSIS OF THE LE MANS ROVER-B.R.M. TURBINE ENGINE

IT was historically important that the first gas-turbine car to go motor racing in earnest—its appearance at Le Mans in 1963 was experimental—should finish its first race. That the particular race should be of 24 hours duration was a measure of the confidence that the Rover Company and the Owen Organization had in the Rover B.R.M., and the result was proof that the propulsive unit was not as delicate a flower as some would have us believe.

In the present state of the turbine art, however, when aircraft engines are achieving overhaul lives of thousands of hours, there *should* be nothing remarkable in a gas turbine engine running satisfactorily for 24 hours. However, besides operating reliably, jet aircraft have to earn their keep by giving fuel economy. Despite poor specific fuel consumption (the amount of fuel consumed for each horsepower produced), turbine engines can do this because they work well at high altitude where the air is cold and its resistance is low, so that aircraft to which they are fitted can fly very fast. This way they achieve good m.p.g. figures in spite of heavy fuel consumption on a time basis.

A road vehicle cannot fly high, and has to operate at the same speeds as other traffic most of the time, so the turbine engine fitted to a car has to stand on its own feet as an economical power producer in competition with the well-developed piston engine, if it is to be a practical proposition. The only way to achieve this is to make it more efficient by converting a bigger proportion of the "goodness," the calorific content, of the fuel into energy. Since most of the waste calories come out of the exhaust in the form of heat, the most obvious way to effect a saving is to return this waste heat to the work cycle. While this sounds simple enough, it is difficult in practice, and the main re- →

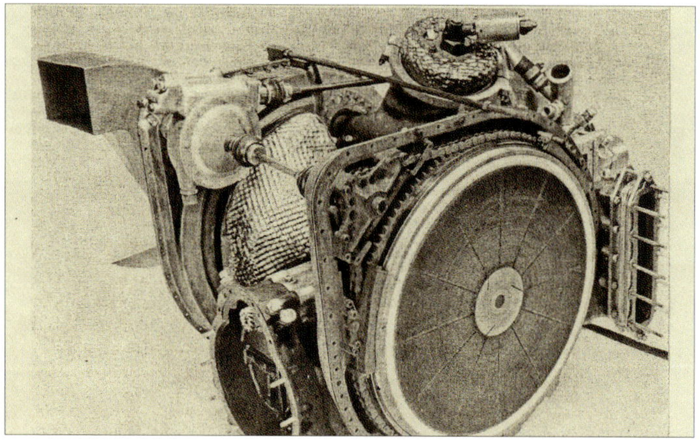

Complete engine with the heat exchanger covers removed to show the chain drive. Although the heat exchanger matrix is wound continuously in manufacture, it is made up in segments afterwards. Total heat absorbing area is 1,300 sq. ft.

2 AUTOCAR, 16 July 1965

A FAIR EXCHANGER

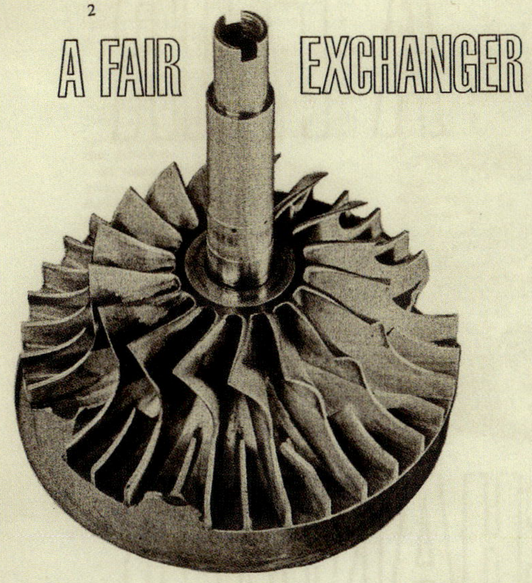

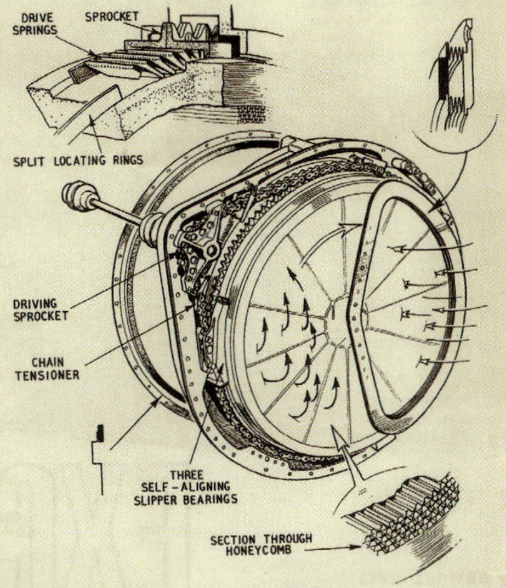

Left: Damaged compressor turbine of the Le Mans Rover-B.R.M. Even with this amount of damage, the car was able to average almost 100 m.p.h. Right: Details of the heat exchanger element and its mounting

YEAR	1963	1965
Designation	2S/150	2S/150/R
Type	2-shaft non-recuperative	2-shaft regenerative
Power Output	150 b.h.p. at 15°C	145 b.h.p. at 15°C
Pressure Ratio	3·92 : 1 at 65,000 r.p.m.	4 : 1 at 65,000 r.p.m.
Fuel Consumption (race)	0·82 lb/b.h.p. hr 6·97 m.p.g.	0·57 lb/b.h.p. hr 13·52 m.p.g.
Air Mass Flow	2·1 lb/sec.	2 lb/sec.

search effort in automobile gas turbine development over the past few years has been to develop satisfactory low-cost regenerative heat exchangers.

So for turbine engineers the main interest in the Rover-B.R.M. run at Le Mans was in the fuel consumption figure. The final placing of the car in the Index of Thermal Efficiency, which put it on level terms with piston engines, was proof that, so far as fuel economy is concerned, the gas turbine for road use had come of age. If damage to the compressor turbine of the Le Mans car had not reduced engine efficiency, it is highly likely that the car could have finished high in the overall classification, as well as in the Thermal Index.

The credit for this success goes fairly and squarely to the Corning glass-ceramic rotary heat exchangers used, which were the outcome of nine months intense co-operative effort between Rover and the American company. Incidentally, Corning have been quite astounded by the degree of technical frankness which the Rover Company brought to the project, and in return they say quite happily that this, and the urgency of the project, telescoped development of the Cercor heat exchanger by at least a year.

To see just how a heat exchanger benefited its fuel consumption, one must first understand the workings of the 2S/150 Rover axial gas-turbine. As with all gas turbines, the basic principle is to draw in cold air, energize it by heating it, and pass the resulting hot gas-air mixture through a turbine wheel. Expansion of the gas inside the enclosed engine casing increases its velocity to such an extent that turbine wheel speeds in the region of 40,000 r.p.m. can be achieved. The efficiency of the engine can be measured as the temperature rise between the inlet and the turbine.

The twin-spool Rover unit comprises essentially two sections. First in order of function is the gas-producing unit, basically a simple jet engine consisting of an air compressor, combustion chamber and turbine. This section delivers gas through a sheet-metal casing to an axial-flow power turbine which drives the rear wheels of the car through reduction gearing. There is no mechanical connection between the two sections, so the power turbine with its associated static guide vanes is, in effect, a gas torque convertor. For this reason no clutch is needed between the output shaft and the wheels, giving a very simple transmission system.

The complete engine and heat exchanger unit is as bulky as a piston engine of equivalent horsepower

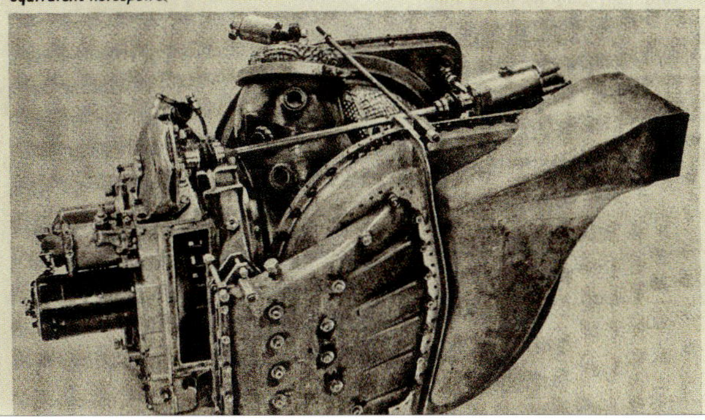

producing power reliably over very long periods of time and distance, while using comparatively little fuel in ideal conditions at high altitudes and low resistance due to the thin air up there. On the ground, those conditions no longer applied and some drastic re-thinking was required to provide an engine at least as economical as a petrol-powered vehicle if the gas turbine was to become a practical proposition. Incidentally, they probably had the Wankel in mind as it was notably thirsty, though a high performer in all other respects (as it was 40 years on....)'. What Lawrie was asked to draw was the culmination of an intensive development programme which involved the enthusiastic co-operation of a major American industrial giant, the Corning Corporation, which designed and produced glass ceramic rotary heat exchangers. This proved the key to managing the high temperature gases produced by a small jet engine which fed an axial-flow power turbine which drove the rear wheels through reduction gearing.

'It was quite different to anything else I'd drawn,' remarked Lawrie, 'so I had to digest carefully everything which I was told just so I could understand how it worked. One of the beauties of the system is that there were fewer working parts than in a four stroke petrol engine, everything seemed to go in rotation rather than up and down. As it looked a likely candidate for the engine of the future, one had the feeling of being involved in an exciting new project, and I'm all for that!' ▶

JTOCAR, 16 July 1965

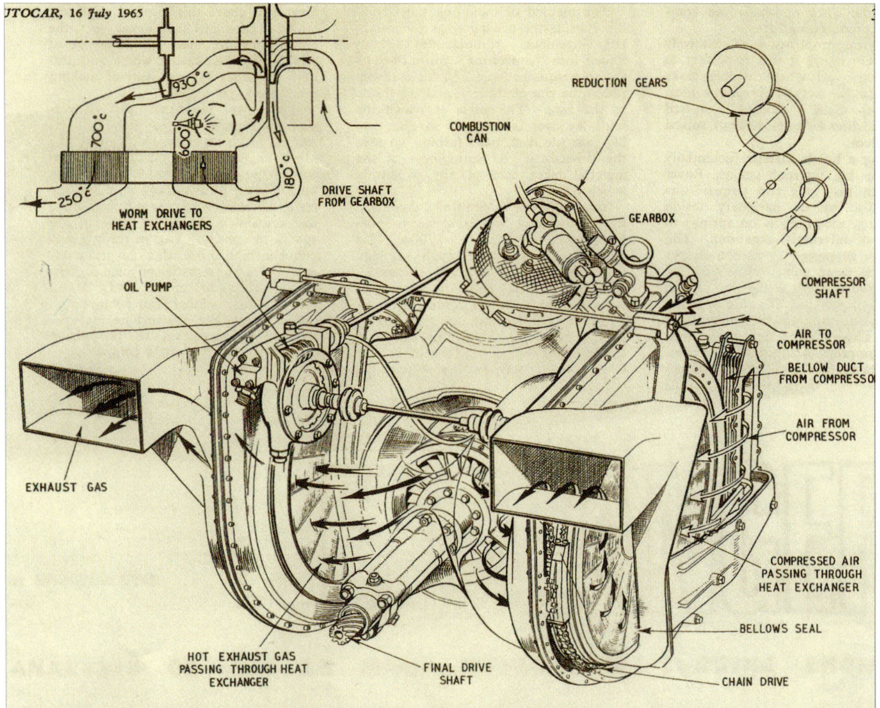

This drawing shows the location of the Cercor heat exchanger elements paddle-wheel fashion alongside the turbine unit. The insert depicts the path of the air and hot gases through the engine and disc

In spite of a damaged compressor turbine, the car lived up to Rover's expectations by finishing Le Mans at an average speed of almost 100mph at a fuel consumption of nearly 14mpg. This was a marked improvement on the 6.97mpg which the car achieved two years earlier on its debut. However, it was one thing to achieve these speeds coupled with this economy under the controlled conditions of a race circuit, but altogether harder challenges faced the myriad problems of controlling the unit in everyday traffic, the chief of which was lag in throttle response during acceleration and deceleration. Associated problems of heat dispersal also provided major headaches and, as development funds mounted, even Rover finally had to admit that in its controlled environment, the gas turbine car had a huge potential, the high street was not to be its scene, even in the Swinging Sixties.

Taking the gas-producing section in more detail, air enters the centrifugal compressor through guide vanes and is fed outwards through a sheet metal annular chamber to the combustion chamber, which is an open-ended can fabricated in Nimonic 90 alloy to withstand the intense heat. Only part of the air from the compressor is used for combustion; the remainder is by-passed round the can and mixed with the hot gas afterwards to cool it to a temperature of 930 deg.C.—the most the turbine blades can withstand. After passing through the power turbine the gases exhaust to atmosphere.

It follows that if some of the heat from this exhaust gas could be fed back into the gas-producing section, between the compressor and the combustion chamber, less fuel need be burned to achieve the required gas temperature at the power turbine.

Rover started off by using a static type of exchanger, rather like a big radiator heated by the hot gases and through which the compressor output passed. This type is heavy and expensive. Most experimenters have now turned to the rotary type of heat exchanger, consisting of a disc or drum of heat-retaining material turning slowly in the hot exhaust gas and conveying heat to the gas-producing section mechanically by inserting the hot disc in the air flow. This pattern has more to offer because it is compact and, being continually cooled by the compressor air, can be made from less expensive materials. Until recently these rotary exchangers have had stainless steel matrices, like that used by Chrysler.

However, in Corning Cercor, which is a sintered ceramic bound together with glass, the Rover Company found an ideal material. On the face of it, glass or ceramic seem unsuitable, because both are insulators. But this property is in fact a virtue, making it possible to build up a steep temperature gradient through a moderate thickness of the material and so save bulk and weight.

The Cercor exchangers are like a honeycomb in section, and are built up by winding a plain strip and a corrugated strip of paste-like unfired ceramic round a core, rather like a roll of corrugated cardboard, until the necessary diameter is reached. The resulting disc is then fired in a furnace, like pottery. After cooling, the faces are ground flat and the disc is ready for use. The materials are cheap, so is the manufacturing process, although a lot of know-how was acquired during development. I was told that in production the discs need cost no more than £10 each, which compares more than favourably with a stainless steel static type with £80 worth of raw materials alone.

In the Rover-B.R.M. two of these Cercor discs are used, mounted alongside the engine like paddle wheels. Sheet metal ducts pass the hot exhaust gases through the ceramic matrix and looped ducts feed compressor air through the discs and into the compressor turbine burner casing.

Damage to the heat exchanger matrix of the Le Mans car, caused by a foreign body, was only slight, and the element is quite serviceable for future use. Quite possibly the heat exchanger saved the engine from destruction

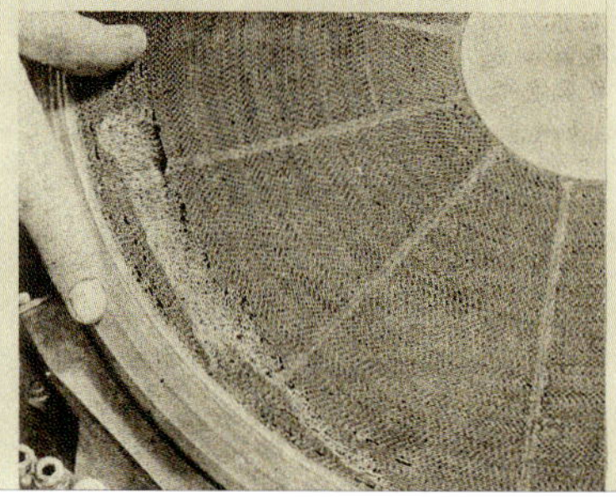

Mechanical problems were relatively simple. Obviously it was important to make effective seals where the ducts make contact with the faces of the Cercor discs. Flat sheet metal "mouths," pressed against the discs by metal bellows, solved this problem.

In using a brittle material potentially subject to big internal strains, Rover decided to play safe and suspend the Cercor disc by its periphery inside a steel ring, mounting it on springs to take care of differential expansion. The rings were supported by carbon slipper bearings at three points and driven by furnace chains, teeth being cut on the periphery of the ring for this purpose. The heat-exchanger drive is taken from the usual accessories gearbox at the front of the compressor turbine, by a compound spur gear train. The speed reduction is 3,000 to 1 to give a disc rotation speed of 15 r.p.m.

The method of mounting the Cercor discs inside the driving rings is particularly ingenious. Nimonic 75 sheet is folded into "concertina" springs which are then squashed flat and inserted between the rim of the disc and the inside of the ring. The angle at which the folds lay over is arranged so that they dig into the disc, pawl fashion, to take the drive, and the springiness of the material takes care of any expansion which occurs.

Simple figures show just how efficient the glass-ceramic heat exchanger is. Turbine outlet temperature is 700deg. C.; after the gases pass through the 3in. thick disc this is reduced to 250deg. C. Thus the disc has absorbed 450deg. C. Air temperature at the compressor outlet is 180deg. C. and it comes out of the other side of the Cercor matrix at 600deg. C., a temperature rise of 420deg. C. It will be seen that only 30deg. C. of temperature have been lost, indicating a heat transfer efficiency of 93 per cent. Over and above this level of efficiency, which is far higher than metal types, the ceramic matrixes have the advantage of staying flat when heated, which considerably simplifies the problem of making satisfactory gas seals.

While we feel that this Rover-Corning exercise has put the small gas turbine on to an equal footing with the piston engine in terms of fuel consumption at the expense of bulk and weight, but, without in any way sacrificing the main advantages of the turbine, there still remains a lot to be done in the sphere of control. Lag in throttle response in both the acceleration and over-run phases was a matter of some concern to the drivers of the car at Le Mans. If a low-cost solution can be found to this problem, the gas turbine may become real opposition to the piston engine, and more than a competitor to the Wankel. ■

Aerodynamics on Formula 1 cars

A lecture on wind tunnel design on Formula 1 cars at the Royal Aeronautical Society provoked some pungent comments by Lawrie on the type of aerodynamics on F1 cars a few years ago. 'What I've seen on Formula 1 cars looks like schoolboy aerodynamics. You'd never get away with it in aviation! I went to the lecture to learn about the problems of the enormous area occupied by the cross section of F1 tyres. This area occupies well over half the frontal area of the car! When you consider that the top of the wheel is travelling faster than the bottom (one is the rolling radius, the other is the radius attained by the fling of the material), somewhere there are molecules catching themselves up as they rotate around the circumference of the tyre. What's happening is that the air the vehicle is travelling into at around 200mph is being crushed between the ground and the rotating tyre. Meanwhile, at the top of the wheel, the air is being flung forward to meet the oncoming air and is turning into massive turbulence about 18 inches above the top of the tyre. Additionally, the wheel's rotation at very high speeds creates its own turbulence. This pushes air ahead of it, building a wall of compression ahead of the car which affects it before the air even reaches the car! This turbulence is a massive problem, an aerodynamic nightmare, which was my reason for raising it with the lecturer. Unfortunately, he skated around the subject because, as I understood it, they had not gone into this area of aerodynamics at any depth. What they had achieved was based largely on results of wind tunnel testing and a rolling road but, sadly, none of this involved investigating the air turbulence created by wheels rotating at high speeds. So my question remains unanswered at the time of writing, and Formula 1 cars may end up as an aerodynamic compromise because an aerodynamacist has been greatly limited by the regulations imposed upon him, hence the barge plates and other bits tacked on to try and make the air go where the aerodynamicist wants it to - Suck it and See Science! These add-ons create drag, which absorbs power, so the end result is an unsatisfactory compromise. To illustrate how much power is being absorbed to push the mass of a Forumla 1 car through the air, there are now production motorcycles which are as quick in a straight line, but which use a fifth of the power to push a drastically slimmer and lighter profile through the air.' One thing which did astonish Lawrie, and other pilots at that lecture, was the rule which forbade Formula 1 drivers to adjust their front or rear wings while the car was in motion. This could only be done while the car was stationary in the pits. In comparison, pilots would be unable to fly without moveable ailerons!

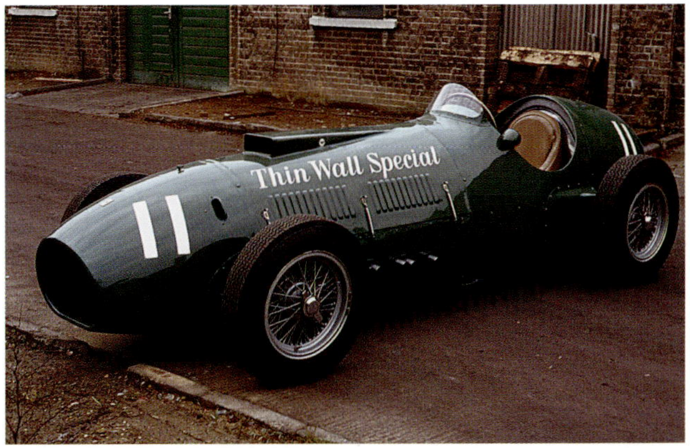

Aerodynamics had yet to come into the equation when this F1 Thinwall Special was created in the 1950s.

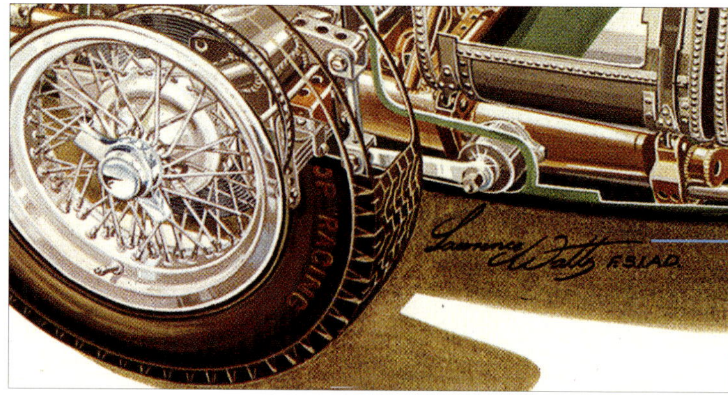

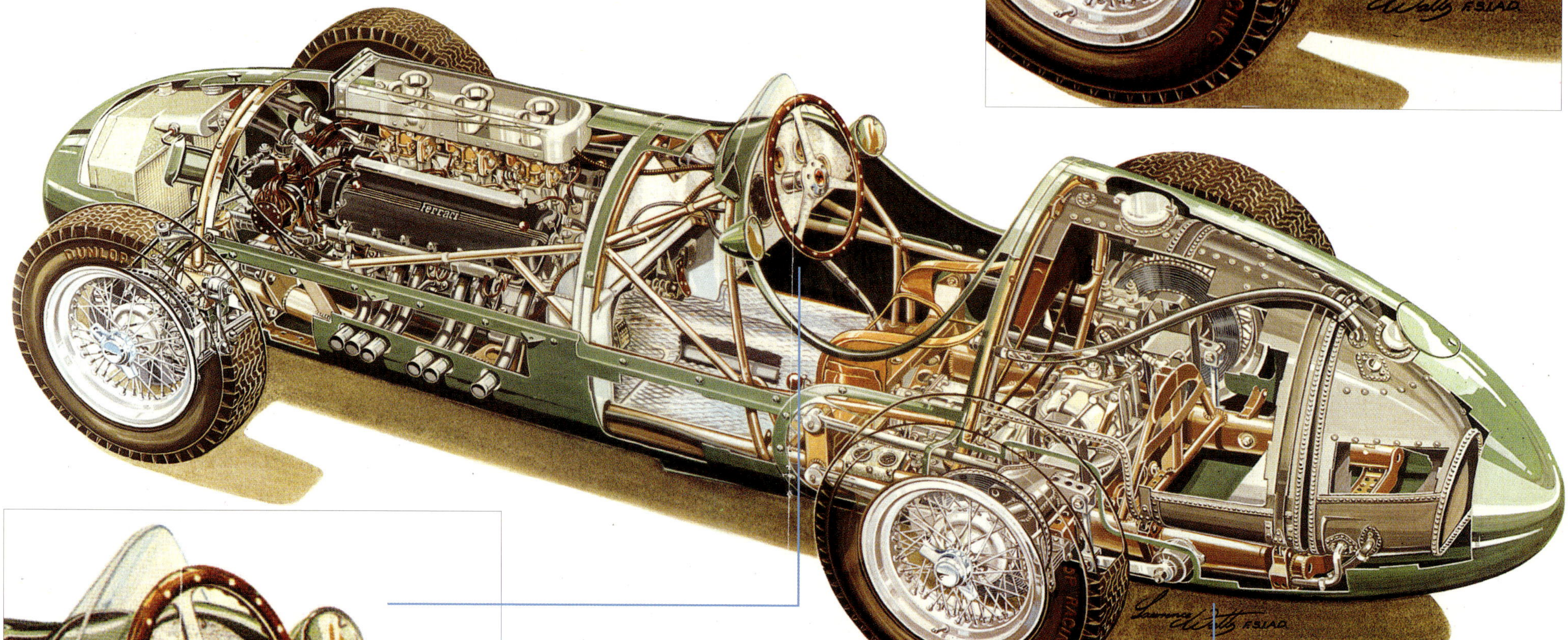

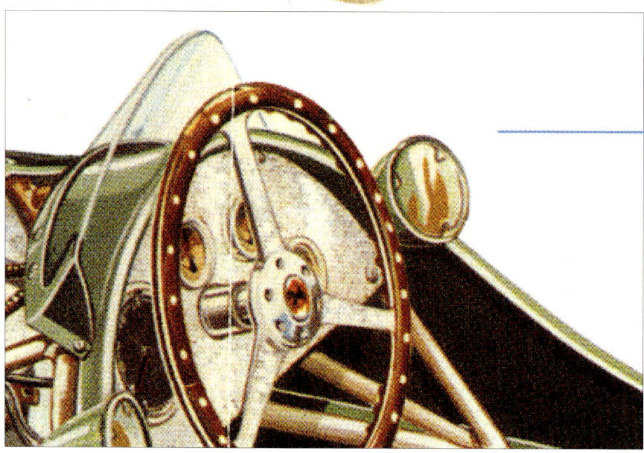

The Thinwall Special with the lid off. Power unit is a V12 Ferrari which Lawrie has not only drawn but dutifully named on the cambox. One of his most exacting tasks, but ultimately rewarding, was drawing each spoke at the right angle in each wheel and in the correct perspective. Note the welding on the aluminium-alloy petrol tank.

Vanwall and Thinwall

In the 1950s, there were two Formula One Vanwalls; Ferrari-powered, the first was called the Thinwall Special and was built by industrialist Tony Vandervell one of whose businesses was making Thin Wall plain bearings, of which Ferrari was a customer. So they lent him a V12 cylinder engine with which to contest the new Formula 1. Encouraged by his initial successes with mainly British drivers, Vandervell developed his own engine using the highly-successful double overhead camshaft 499cc Norton racing engine as a slave unit. From this was evolved his liquid-cooled two and a half litre car which, with drivers Stirling Moss and Tony Brooks, won the first Manufacturers' World Championship in 1958. This was the car which Lawrie had been asked to draw for a 'Motor' series on sports and racing cars in the late 50s at Vandervell's factory in Acton, west London.

'No problem gaining access to the car, it was in the entrance hall to the factory,' recalled Lawrie. 'Everyone connected to the car was most helpful, removing panels so I could draw the engine and transmission, anything which needed drawing. Nothing was too much trouble for them. And no problem sitting there in the hall of this nice big airy factory, no more uncomfortable than sitting in a museum! One thing led to another and they said, why not draw the Thinwall? That's the car which started it all and it's stored here. So the 'Motor' agreed and I returned to draw the Thinwall. Here's the drawing to prove it.'

When I mentioned to Lawrie about the Norton inspiration for the Vanwall's engine, it reminded him of another motor cycle engine which inspired a car, but one which proved an abject failure. He explains, 'I seem to remember that Edward Turner used three Triumph 650cc engines, minus gear boxes, as the basis for the prototype V8 engine in his Daimler SP250, known as the Dart until forced to drop the name because it had first been registered to Chrysler. I drove the Dart while assisting the 'Autocar' test team and it was a waste of time! It was a real tit-shaker; the whole scuttle vibrated badly at certain speeds and it didn't handle at all well. Frankly, it was never any good and it was one of the few cars I've been asked to draw but declined. As I've said before, I need to be interested in what I'm drawing and I couldn't lift a pencil to even start on the Dart.'

Harry and the Arabs

Invited back to Jaguar by Autocar's one-time Technical Editor and Jaguar designer, Harry Munday, to see the new XJS model, Lawrie went out to lunch with the directors. Chatting over coffee, Harry asked Lawrie what he thought of the new model. He replied, 'Put tracks on it and you could sell it to the Arabs!' There followed a deathly silence, eventually broken by someone talking about the weather.... 'It was not one of my greatest moments,' admitted Lawrie. 'Before opening mouth, I should have engaged brain and realised firstly, that the Arab/Israeli war was at its height, secondly some of the distinguished company around the dinner table were Jewish, though I'm not sure if Harry was. Anyway, he didn't speak to me for some years afterwards. Then, out of the blue, he telephoned me and invited me around for a drink at his home in Kenilworth, only a few miles from Coventry. Over drinks he told me that he'd regretted freezing me out, he realised that there was no malice in what I said and the whole silly business had been a storm in a tea cup. Thereafter, we used to meet for a sherry every Wednesday until he died. He was a good bloke, great company, we had some marvellous discussions over the years.'

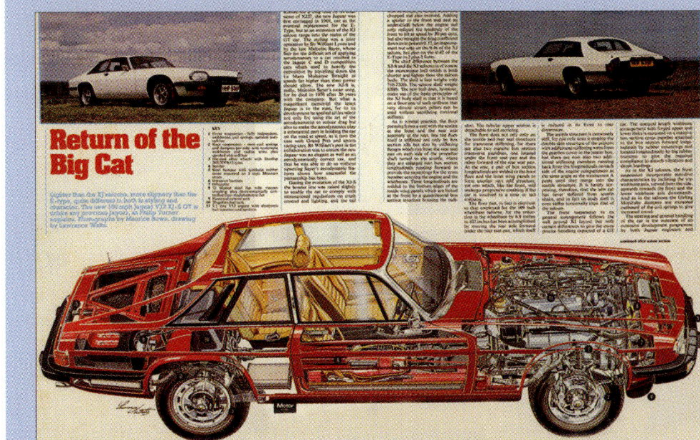
Return of the Big Cat

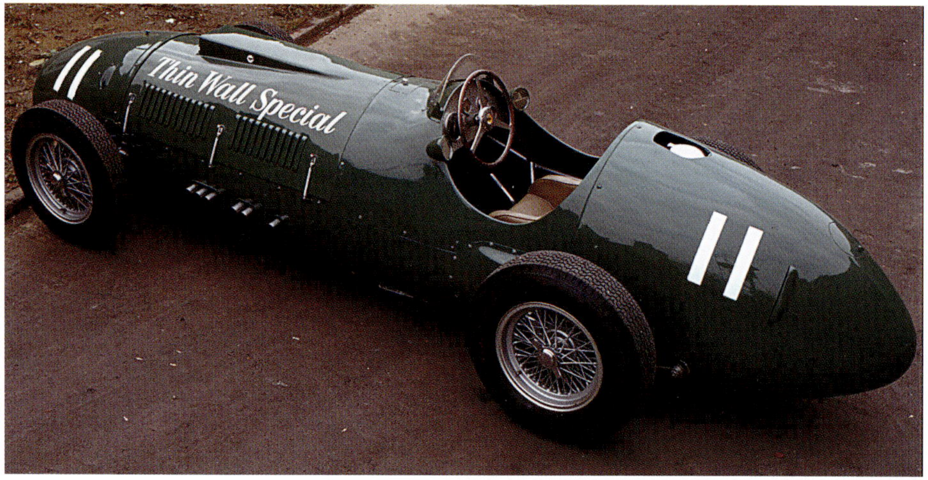

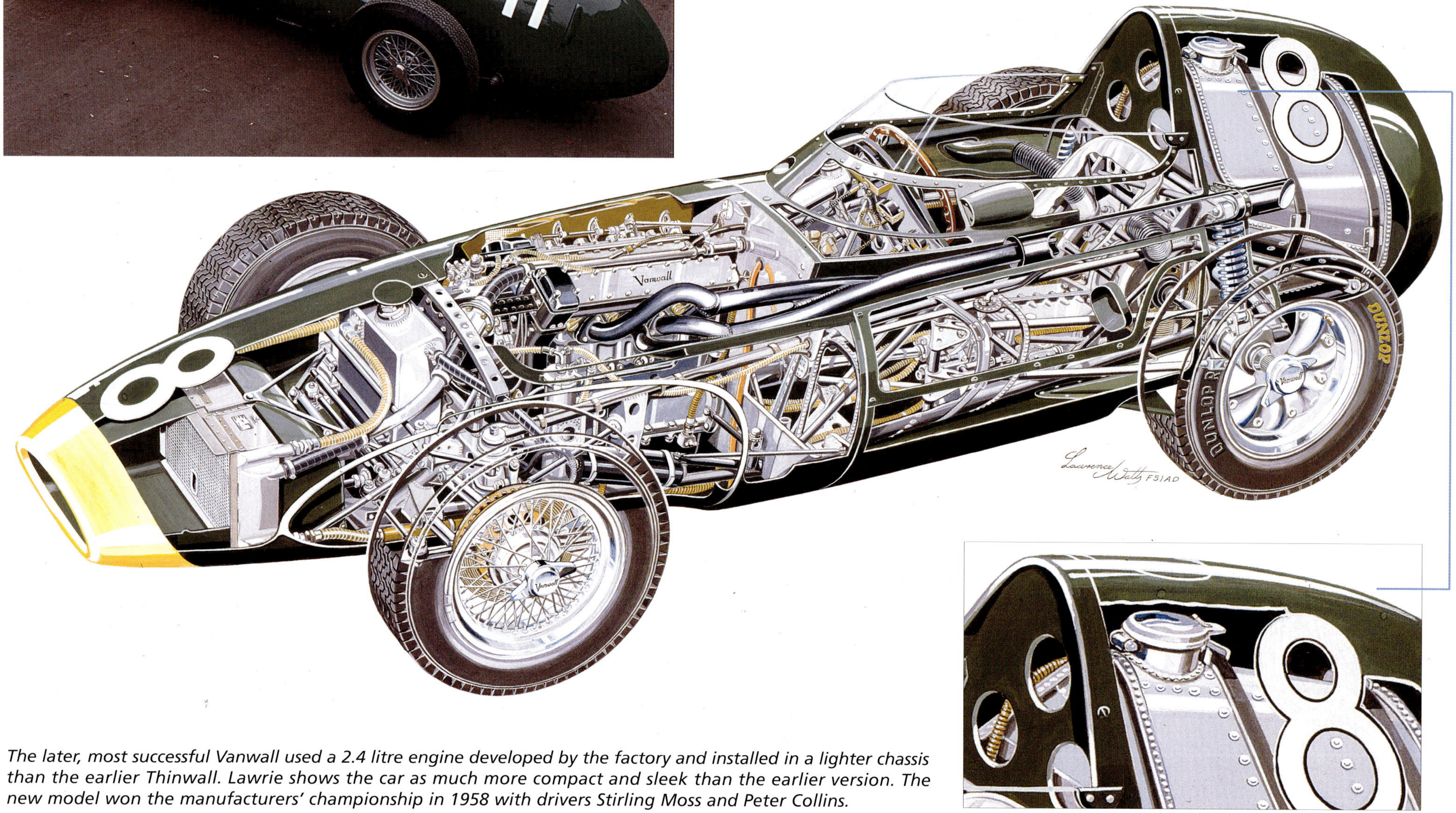

The later, most successful Vanwall used a 2.4 litre engine developed by the factory and installed in a lighter chassis than the earlier Thinwall. Lawrie shows the car as much more compact and sleek than the earlier version. The new model won the manufacturers' championship in 1958 with drivers Stirling Moss and Peter Collins.

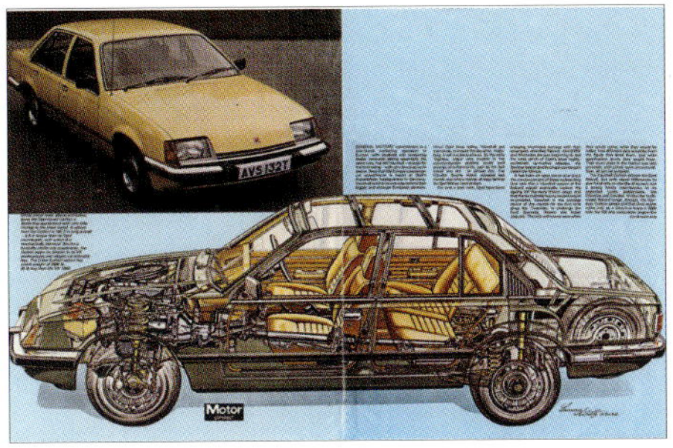

Music while they worked

'One of the happiest places I've ever worked in a production factory was at Luton, yes! at Vauxhall Cars. I shouldn't have been there, this was within the orbit of London office, but for some reason in 1978 they sent me to Luton to draw the new Carlton. For me, a good move, because the atmosphere was just right. It was one of the cleanest, precise, most organised factories I'd ever been in. Employees worked in teams, like at Volvo, and they adopted the Volvo way where teams of men built a car from beginning to end, their enthusiasm expressed in constructing the car. Another interesting thing, which affected me deeply, was that one of the team would start singing and then everyone else joined in, just like a choir. All down the assembly track, it was the same, teams singing their hearts out. An unbelievable atmosphere! I think that this happiness reflected itself in the product; the Carlton, which turned out to be an excellent car, was taking Vauxhall into the big league in hot pursuit of Ford. Vauxhall were keen to rid themselves of the image problem which they had had for years, and having a happy labour force - and you had to be happy to sing like that! - meant that Vauxhall got a better product. This, in my book, is an important factor in the success of the Carlton. Alas, it would not last forever, because in the 80s, strikes crept back in and it made me very sad to see how a once-happy workforce should now be on picket duty outside the very gates through they marched singing just a few years before...

'The Carlton was a nice car, very quiet, beautifully trimmed, comfortable, and it drove very well. I liked it a lot. What did it owe to its German Opel cousins? Well I'd say a fair bit, mainly because they used the same General Motors European design centre and commonality of parts was beginning to play an important part in new car designs, a practice now universally adopted throughout the industry, thanks to GM.

'Shortly afterwards I was off to Frankfurt to draw a new small Opel, called the Corsa. I stayed overnight at the hotel, spent only a day in the works to get all the information I could on the new Corsa and then back to the UK and finish the drawing in the office.'

Time constraints meant that the drawing was done in black and white. By this time Lawrie had evolved a back-up system, using a camera to photograph things which he wanted to draw. He uses his camera as a 'picture notebook'. Using these photographs back at base enabled him to complete the Corsa drawing within a week. The detail is quite astounding. The driver's seat, on the left hand side (remember this is an Opel), has a cutaway which shows not only the interior springs in the seat but also the foam backing which is identified by seemingly millions of little pin pricks! The covers on the nicely-shaped Recaro-type seats have even been drawn in their original patterned material. A cutaway in the rear panel shows the rear suspension, the single separate spring and a hydraulic unit. On the driver's seat, Lawrie has even shown the catch which releases the seat back, allowing entry or exit to a rear passenger. The boot lid is open to show the heater element in the rear window, the hydraulic strut which supports and also controls movement of the boot lid - not to mention the cord which supports the rear parcel shelf. And on the shelf is a little speaker for the radio! As Lawrie pointed out, 'This is where the camera was useful because I might not have spotted this at the time I did the original sketches. So it's a very important departure. Because I'm telling a story to customers, those looking at the finished drawing want to see the full picture. It's my job to show it to them so that's why I include every little detail. There's a spin off using a camera. Although I'm focussed on a certain item, there will be other parts in the background, perhaps on the assembly track, and this record can assist me. So it's quite useful to have a camera, even though the factory personnel have always been very helpful, providing whatever I wanted and with the freedom to go anywhere. In this case, the Corsa was not yet on the market, in fact it was being set up for production in Spain. Any information I gleaned with the camera was of considerable help.'

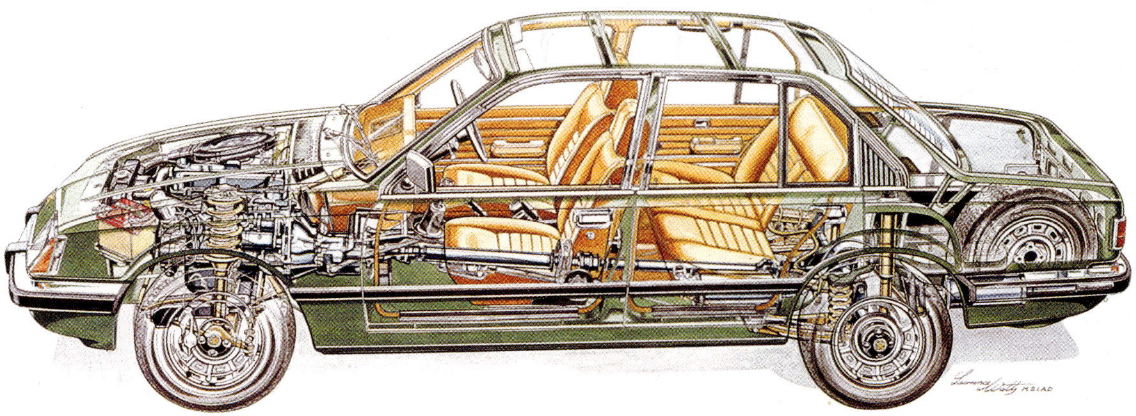

Vauxhall Corsa

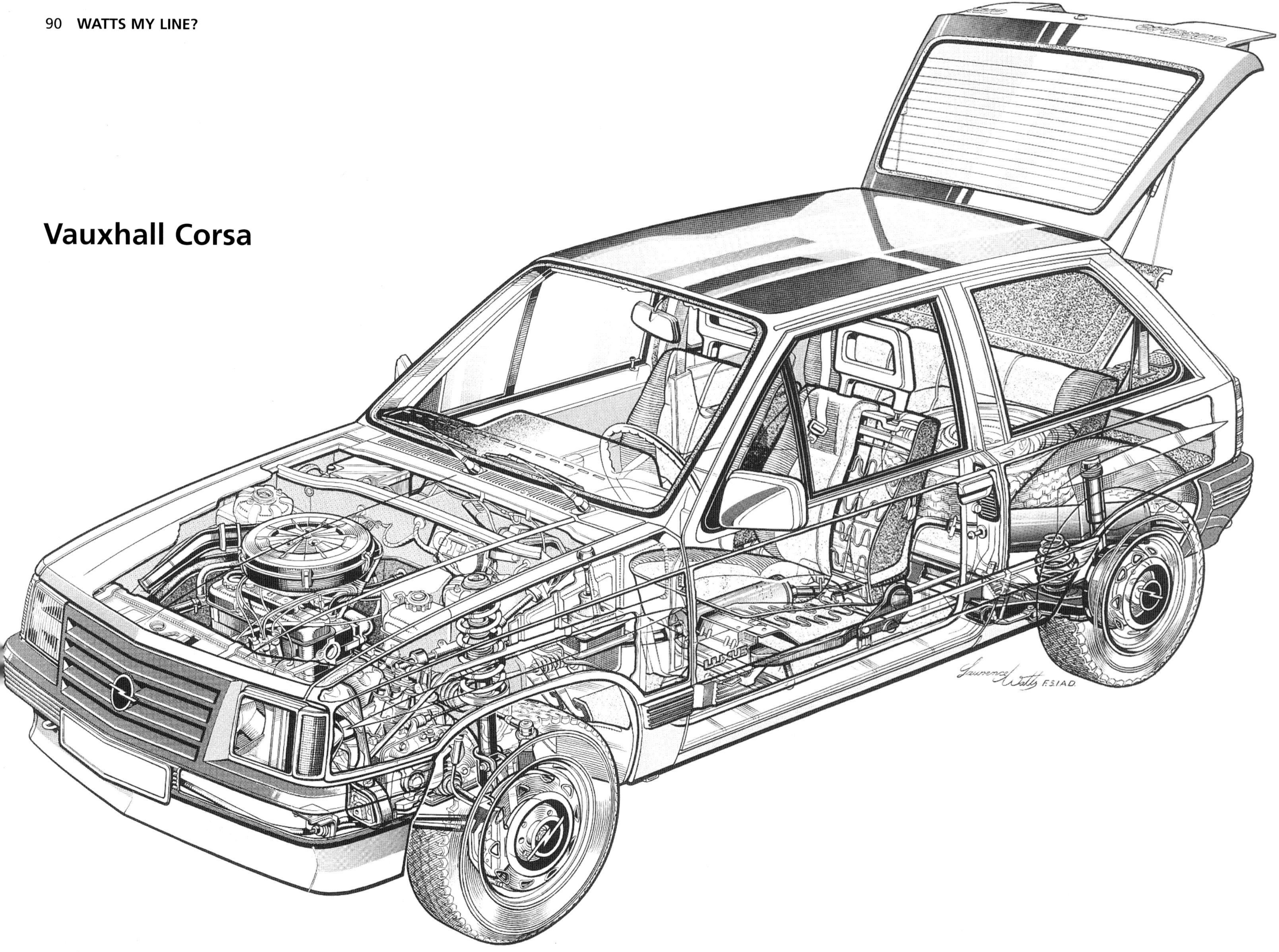

Peugeot 205

A direct competitor to the Corsa, which Lawrie drew a couple of years later, was Peugeot's little 205. Comparison of the drawings shows clearly the respective designers' approach to the same problem; how best to accommodate a driver and three passengers, and their luggage, in roughly identical-sized cars. Lawrie's cutaway of the rear suspension on both models shows that Opel gained valuable luggage space by mounting their rear hydraulic units horizontally and the springs separately, whereas Peugeot retained their spring and damper units mounted vertically in the rear wheel arches, thus sacrificing considerable luggage space. 'These two are contemporaries,' said Lawrie, 'and it was noticeable to me when drawing them how much more fussy one is than the other.' I remarked on how much more enthusiastic Lawrie had been in his rendition of the Corsa than the Peugeot, as exemplified in his stripping out the seat to show how it's made, and in the rear end detail. In comparison, he's not shown the rear parcel shelf on the Peugeot, he's not lifted the rear seat squab to show the fuel tank beneath, leaving the impression that he was not as involved with the French car as he was with the German. That I found odd, as the Peugeot was built down the road from Lawrie at Ryton. 'An interesting comment,' he replied. 'If you look at the Peugeot you'll see that there's a dark black shadow under the car, but there's none under the Corsa, and I'm at a loss to know why. It also looks as if I've not signed the drawing but look closer and you'll see that I have all but hidden the signature in the black shadow, almost as if I didn't want to claim ownership! It's very strange, because I've not noticed this before - thank you for asking! - but I think we've stumbled onto a psychological conundrum. Perhaps I didn't want to draw that car! It was a job which, now that you've brought it to my notice, I'd done without as much enthusiasm as I usually did. One of life's little mysteries!'

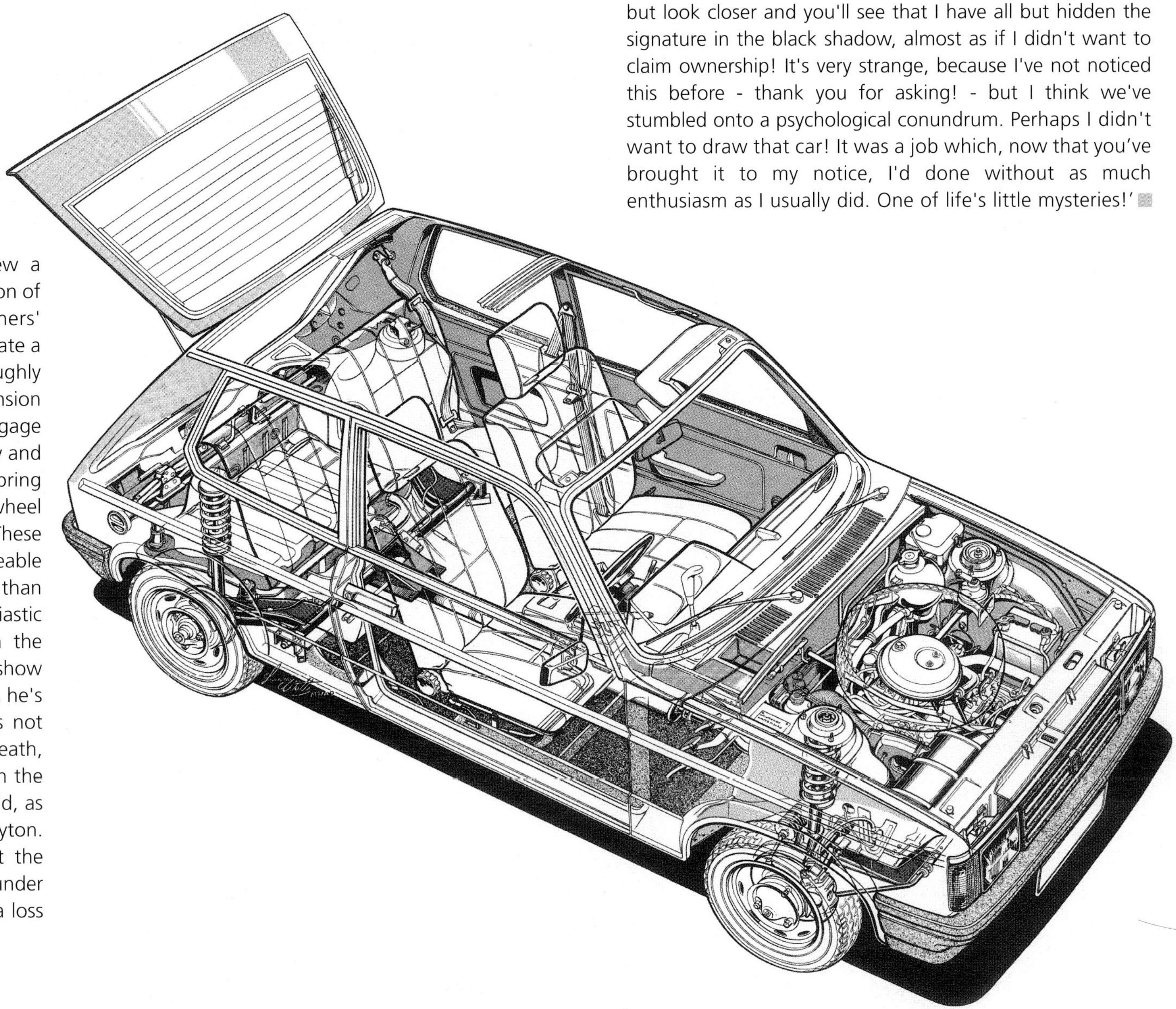

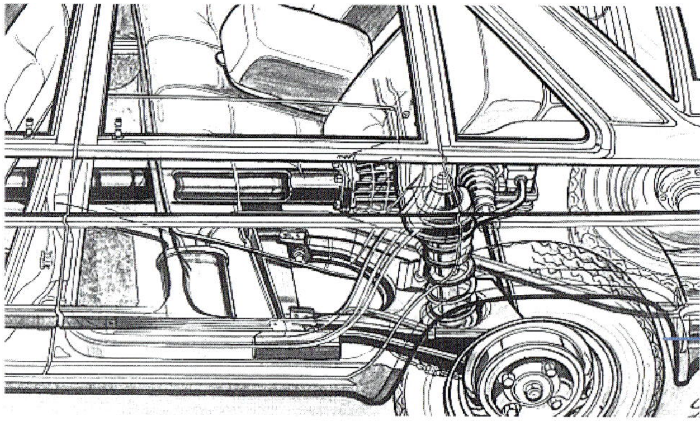

Talbot Tagora

Not a landmark name, or model, in the British industry yet in the early eighties this old-established Coventry manufacturer Talbot produced a 2.3 litre four cylinder model, the Tagora (which also had a 2.6 litre V6 version), which impressed Lawrie with a unique technical feature. The longer he spent drawing the car, the bigger its impact on him. 'I thought that it was a solidly built car, roomy, practical, providing an ideal means of family or business transport. It was smooth, too, no noticeable vibration from the engine and the transmission was very quiet too. I drove the Tagora after I'd completed the drawing and it lived up to my initial impressions, but it failed to light the fires in any great number of buyers. Perhaps the styling was too bland for them, but I thought that it looked quite smart.' The technical innovation which so intrigued him was that the front engine and the drive shaft was one solid piece. The drive shaft was enclosed within a large tube, so that the geometry of the chassis was determined by the fact that the rear axle was joined onto the gear box. The idea behind this highly unusual layout was that the whole mechanical assembly could be put together on the production track and the body could be dropped on from above. 'It was a cracking good design., yet it was not a commercial success. I was disappointed, and I know the manufacturer was.'

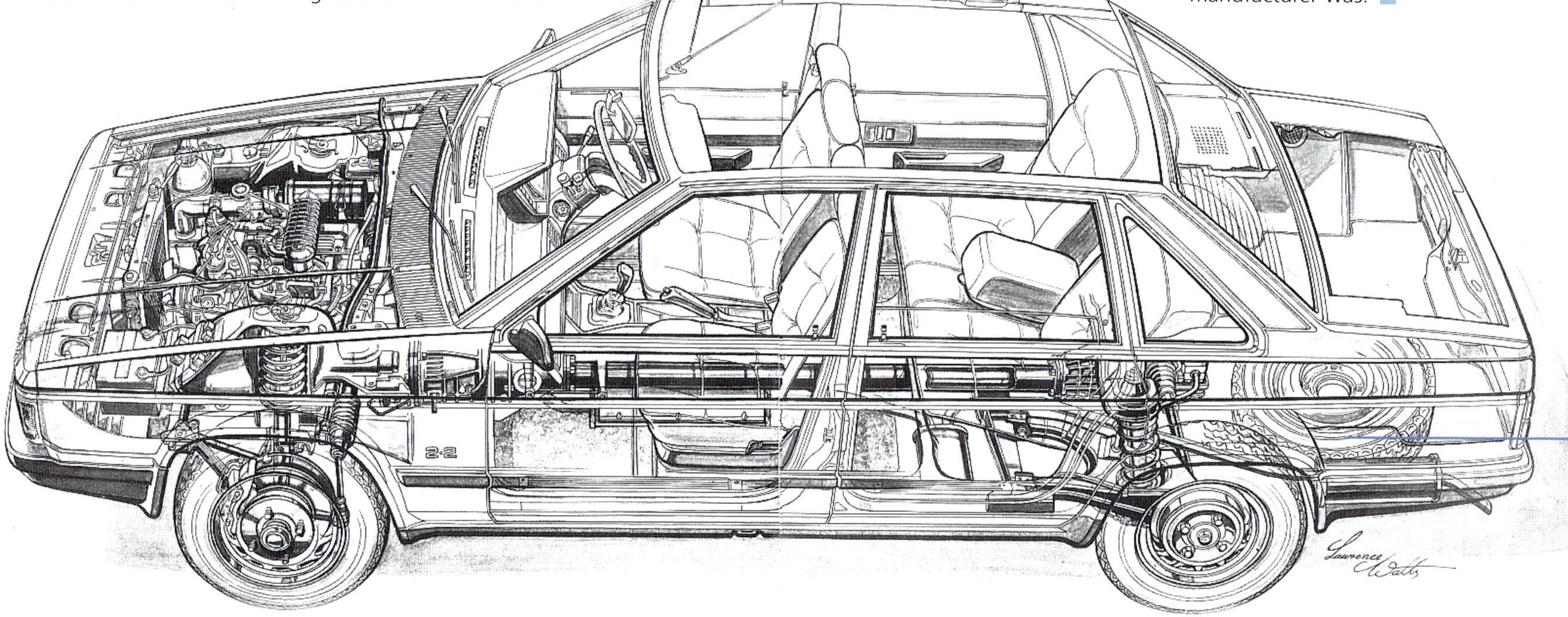

Top: Unusual feature of the Tagora was the propshaft housing which was bolted to the engine at the forward end and to the gearbox and axle at the rear. The whole mechanical assembly was put together on the production track and the body dropped on from above.

MGB GT V8

'This was fun! As you see, there's a lot of detail. It was drawn down at Abingdon in a nice friendly atmosphere. A great car, this one was pre-production and hand built. I had it for a couple of days beforehand, so was familiar with it and identified lots of details before I'd started drawing. This probably made for a better drawing, certainly I liked the finished product. And the car was a dream to drive, with an engine more powerful than you could fully use, thanks to the lightweight three litre V8. To put more weight over the rear wheels, I put a large paving slab in the boot - crude but effective! The car was built quite swiftly at Abingdon, MG was able to get it into production fairly quickly and inexpensively, and it sold well at home and abroad.'

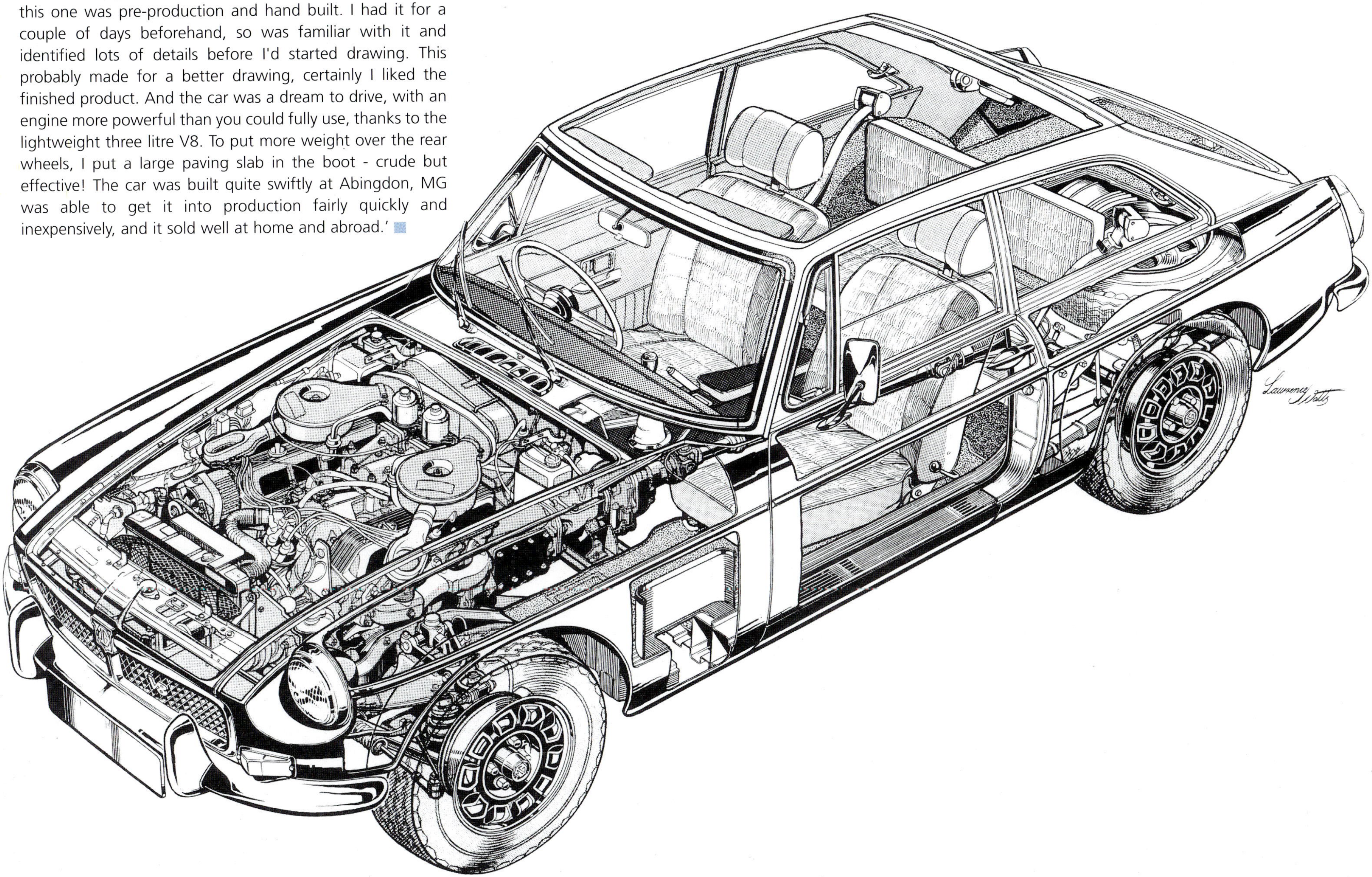

Audi 4WD

One of Lawrie's most outstanding drawings, which won him Artist of the Year in 1985, was his coloured drawing of Audi's sensational 1981 five cylinder 2.2 litre turbocharged four-wheel drive World Championship-winning Quattro. Submitted by his employer, Associated Iliffe Press, the drawing encapsulated everything which gained him the acclaim of his peers. 'It was done as part of a series for '4WD Magazine' and appeared in their December 1984 issue', recalled Lawrie.

The 30 degree three-quarter frontal viewing angle has the front wheels set on left lock in order to show the brake and ▶

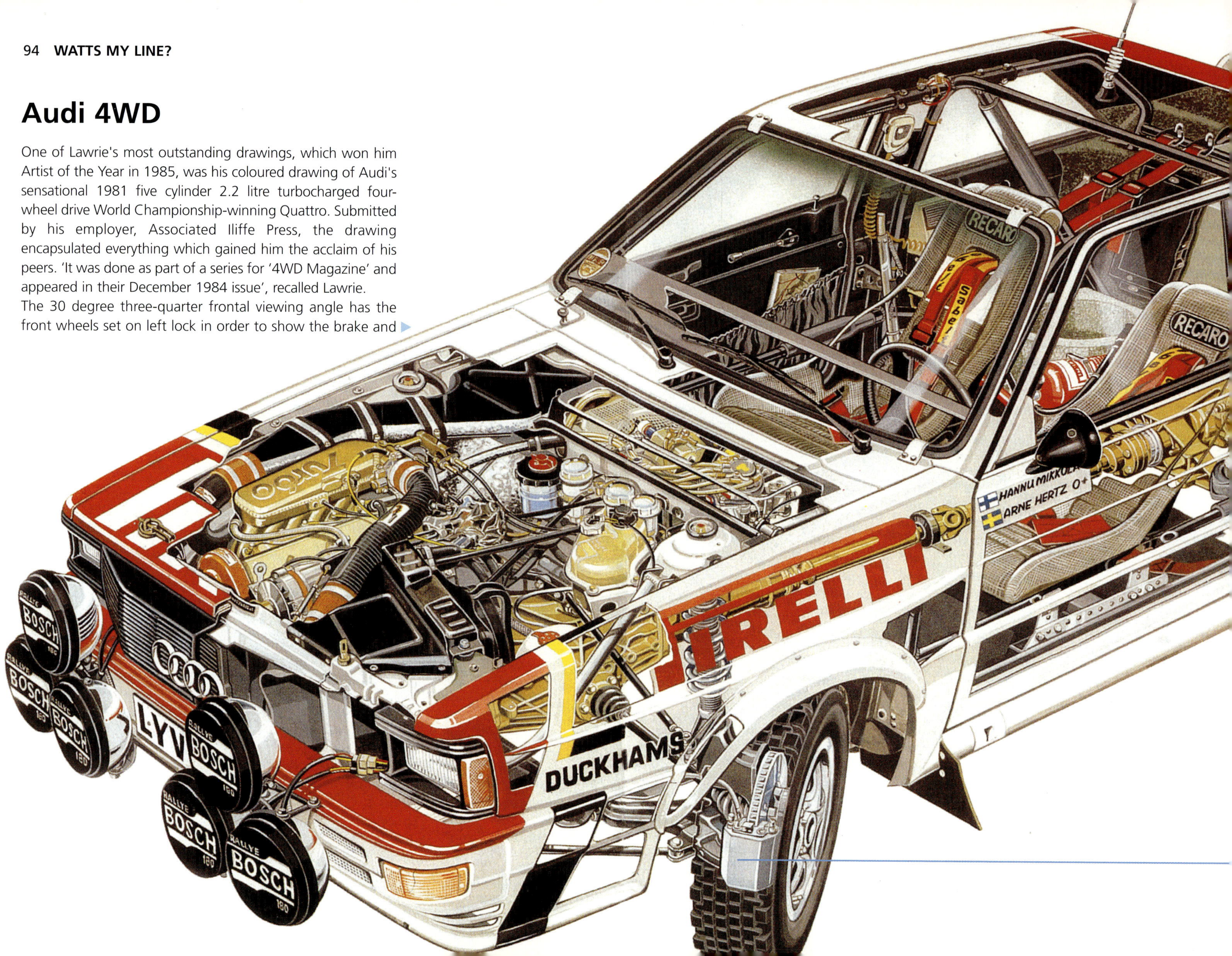

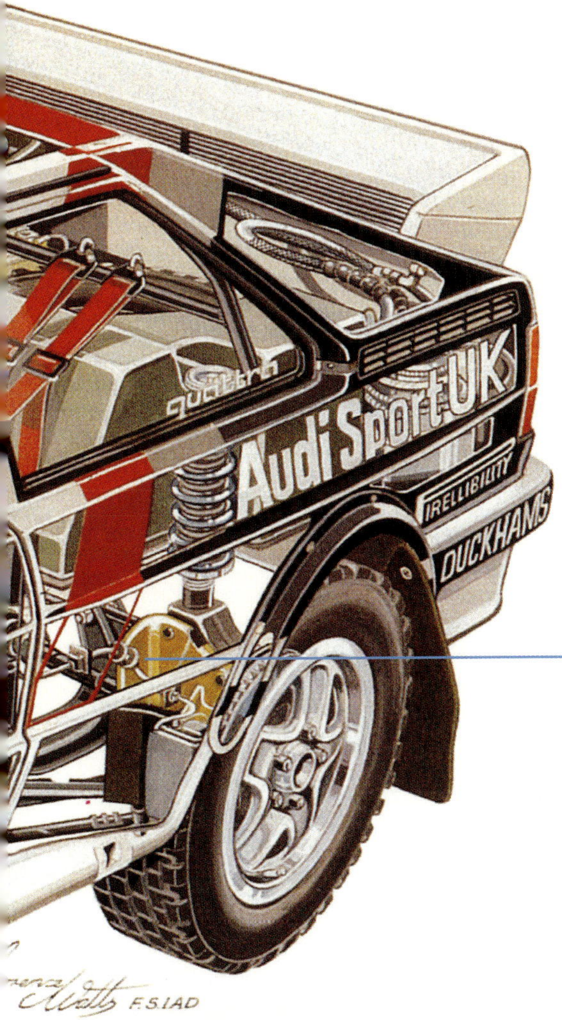

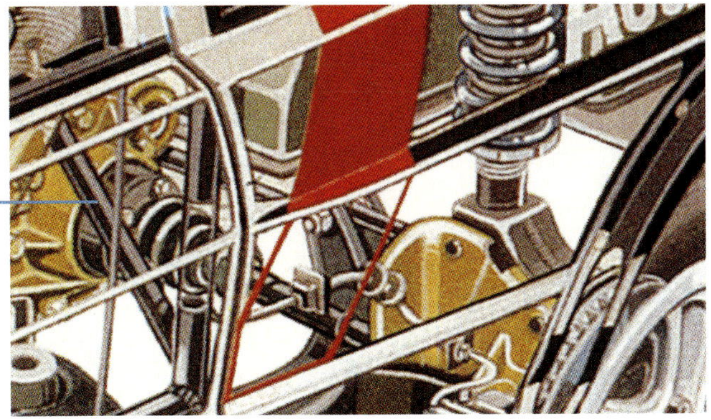

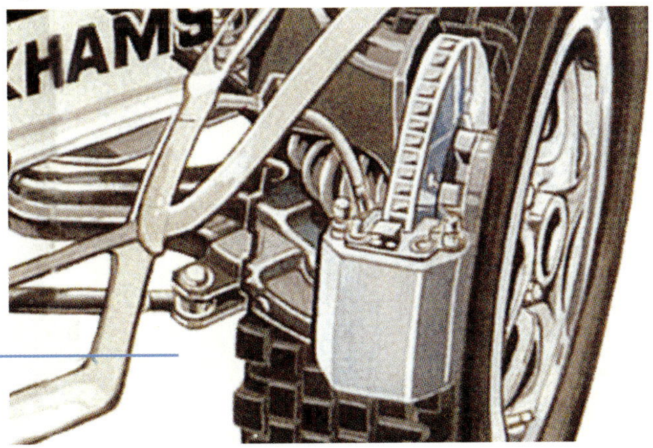

one of the drive shafts, and has cutaways in the bonnet and sides to enable the special rally equipment in the driver's cabin to be shown to best advantage. 'It was such a beautifully functional car, with ground-breaking design which was successful from it's inception, that it caught my imagination. Because of this I did go to town and made sure that I included as much as possible to instruct the viewer', admitted Lawrie. 'Everything about it, down to the smallest detail, was impressive. Which is why I was honour bound to do the car and its designers justice. This included not only the radical items, such as the four wheel drive and the obligatory roll cage inside the car, but also the tiny details which contributed in their own way to the outstanding success of the car.' Tiny details? Behind the front wheel he shows the rivets securing the special cutaway mud flaps! Also the rivets holding on the 'eyebrow' extensions over the wheel arches, necessary to allow clearance for the chunky rally wheels and knobbly tyres. Not satisfied with this he meticulously drew every knob on the rally tyres before he grew tired, saved himself effort, and sectioned the tyre and wheel. Why? He wanted to illustrate the vented brake disc and show one of the front drive shafts entering the left-hand hub. Below this is the massive lower fabricated wishbone and radius arm, viewed through 'windows' which he drew in the front mudguard and bodywork. By skilfully making the bodywork appear translucent, yet retaining its shape, he showed the coil spring and hydraulic suspension strut, all massively strong to withstand the terrible hammering they were to receive in international rallying in the hands of soon to become legendary, the flying Finn, Hannu Mikkola, and Norwegian co-driver, Arne Hertz. Use a magnifying glass and you can see the details better, but the 'out takes' alongside the main drawing show what care and precision Lawrie took with detail work. Where another artist would have been content merely to show the front drive shaft disappearing towards the gear box, Lawrie used precisely the right shade of grey on the drive shaft gaiter so that you know it's rubber. And you can also count the number of bolts securing the end plate to the gear box! Not only that, the cast magnesium end cover of the box is shown with its strengthening ribs. How do you know the cover is magnesium? Only the best for Audi! So why is it coloured gold? The actual colour on the box could have been gold but one of the reasons for Lawrie's innovative introduction of coloured drawings was to allow him colour-code related items. So here we can identify the transmission and engine components because they are coloured gold. Note the raised TURBO on the air intake cover? Lawrie reproduced the letters exactly as they were stamped into the cover, with delicate touches of white replicating light catching the raised lettering. He even showed fluid in the windscreen washer reservoir, with red cap, in front of the bulkhead. Through the cutaway P of Pirelli, you can glimpse the gold coloured prop shaft, with the universal joint about to pass between the driver and passenger seats. Most of the door has been cut away but for one small, but highly significant, panel. That's the one which shows the driver and co-driver's names, each nationality indicated by minute flags, followed by their blood groups. (Rallying can be dangerous). ▶

The obligatory fire extinguisher is located between the two seats, by the elbows of driver and co-driver, ready for use. Mounted on robust light-alloy frames, whose members are bolted together for ease of replacement in the event of damage, are two moulded Recaro race seats. Notice the difference between them? Mikkola liked a reclining position for a straight-arm relationship with the steering wheel, but Hertz needed to be more upright to enable him read the route notes and maps, seen here dropped on Mikkola's seat ready for the off. But how about this for detail? In the pocket of the navigator's door is a map book which, if you look closely, has an illustrated page of the route! The book is retained in place by an elasticated plastic pocket which you can deduce by the bulge which Lawrie has drawn - feel that rucked shiny plastic! Clipped to the door pillar beside the navigator's head is a radio headpiece, with the red extendable cable clearly shown. Those long adjustable straps from the tubular cross member above the fuel tank pass through slots in the seat backrests over the occupants' shoulders and are anchored beneath the front of the seats. Rally cars and punishment are synonymous, so safety harnesses are essential in securing driver and navigator, especially when flying over jumps ('yumping') or, if the flying becomes unintentionally inverted... in which case the occupants are left hanging not by their bootlaces (as in the old days), but by their safety harnesses.

Even though the rear axle and suspension is normally out of sight, Lawrie ensures that you, the viewer will see whatever is there through the magic of his pen and paint brush. Not content with showing the massively strong castings for the gear box, differential, and suspension, but he's also included the hydraulic pipes for the brakes. A massive aluminium fuel tank is fabricated to fit directly over the rear axle and into the boot space.

Mounted outside the boot is a large aerodynamic wing. Except it's not. The top is formed as a wing, to provide down thrust and therefore grip at speed, but it doubles as a box section to house a massive full-width radiator for cooling engine oil. The clue to this is inside the car where Lawrie has drawn in the bottom right corner of the rear window, a steel braided high pressure hose pipe which carries oil to the radiator and cool oil back to the engine at the front. Why not have a front radiator, as in a normal car? Firstly, a radiator of that size was needed to cool the oil of a highly tuned, hard working turbo-charged engine (a synthetic oil had to be developed for the ultra-high temperatures of the turbocharged engine). Secondly, even if there was room for it at the front a radiator of this area would have been vulnerable to flying stones and even minor shunts. Besides which, the hot air would have exited over the engine. All of these problems were solved by mounting the cooler in the optimum position at the rear where it collected cold air flowing over the roof of the Audi and down the rear window. Hot air from the cooler wafted into the wake of the car.

Lawrie's treatment of the glasswork is breathtaking. He's left the windows in place but, in showing the interior, he's drawn everything exactly in position. By using a misty white spray vertically, he's suggested a reflection of light in a panel on the clear glass of the driver's side window. But, and this is the really sneaky bit, the 'reflection' panel follows the sloping line of the door as it leans in to join the roof. His treatment of the windscreen shows that it's tinted because everything viewed through the screen is a subtle shade darker. For example, compare the colour of the navigator's seat with the driver's and the deeper red of the safety straps. But where the top of the navigator's Recaro seat projects above the top of the windscreen, the colour matches the driver's seat! As with the side window, Lawrie uses light reflections on the glass to indicate that the windscreen is still in place. And he's also used a graduated tint to subvert glare at the top of the screen. That Audi were proud of their Quattro is further indicated in the bottom corner of the rear window where they have etched onto the glass the word Quattro. So Lawrie dutifully reproduced it!

At the front of the car you will, of course, have noted the six chromium-plated Bosch spotlights. If you look closely you'll see that they are all slightly different. That's the result of Lawrie focussing them! The lower left front is focussed on the left hand side of the road, the one beside is focussed on the road surface 50 metres or so in front and the one above is set for distant vision to enable the driver find his braking and turn points. Same sequence for the right hand set of lamps. Instead of showing the lamp leads disappearing into the grille, Lawrie has drawn the quick-release fasteners, complete with their colour coding. No detail spared! Not surprising that the judges made him Artist of the Year. He bequeathed us this masterpiece which, years after it was drawn, still provides so much pleasure just to pore over it with a magnifying glass, seeking some new details to marvel over. All executed by a craftsman at the peak of his powers. ■

Snakes alive! Testing was often full of surprises - in this case a black adder found in the boot of the Cambridge.

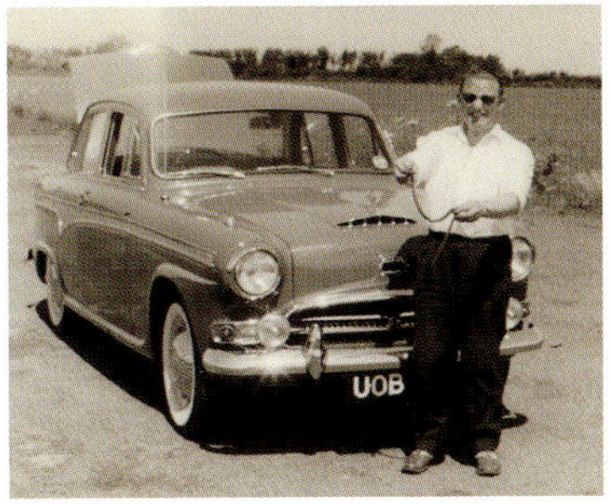

Road tester Loll

When time allowed, Lawrie assisted the road testers on 'Autocar' and later, 'The Motor'. 'Reliable drivers were needed because group testing of cars was now becoming the norm with cars usually outnumbering available drivers. So I was welcomed aboard, especially as I was known to the guys and we all shared a common employer, so there were no insurance difficulties. I could also provide an informed impression of any car I drove, so that was a bonus. Truth is, I just loved a blast in an interesting car! That's how I found myself heading to Warwick one day to collect an Austin Healey 100 from Donald Healey himself at the factory. Soon after I set off for the testing site at Brize Norton in Oxfordshire, I realised that the car was a bit lively at the back end. Incidentally, Brize Norton was probably the main US Air Force base in this country and yet they closed the airfield for our performance testing! They detailed a few uniformed guys to look after us and they were absolutely delightful. A sergeant asked how much one of these cars cost. Do you know what he drawled? 'Well then, I could work for three months and then I could buy one.' I'd never before encountered that manner of thinking. If I work, then I can afford it.... nothing about borrowing, just work, earn, buy. Illuminating!

'The road test team comprised chaps from London office, and myself with usually six different cars. We'd take it in turns to drive all the cars, later giving the editor a written report to help him collate a consensus of opinions on each. Anyway, after lunch we did our acceleration tests, car v car. I lined up my Austin-Healey level with my oppo in a Sunbeam Alpine. On the signal to go, I floored the accelerator, the Healey wheels spun uselessly and the car pirouetted 360 degrees amid a cloud of burnt rubber! Then I sussed it, straightened the car and hurtled down the road after the fast -disappearing Alpine. When we got to Brize Norton, my oppo said 'That was very impressive, Lawrie!' Little did he know that I had had no option! That's when we realised that, in spite of each of us having a passenger, whose task was logging figures off the performance equipment, that we would have a problem getting grip from standstill with these high-powered sports cars. The watching Americans saw the problem, found some concrete slabs and placed them in the boot of each car. The extra weight did the trick. On one occasion we raced all six cars abreast down the runway, just to see how they would compare, and to put on a show for our friendly hosts. Meantime, all my fellow artists were hard at it down in London with their pencils....'

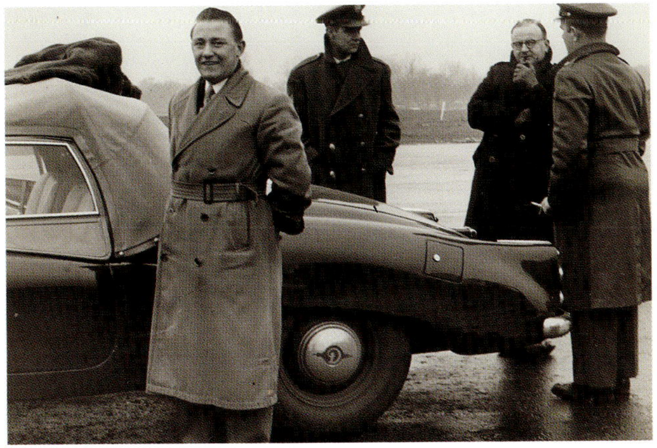

A keen driver, Lawrie filled in his spare time helping the road test teams at 'The Autocar' and later, 'The Motor'. He is checking in an early Daimler SP250 at Brize Norton airfield in Oxfordshire, then (as now) occupied by the US Air Force.

Double booking

Wherever people work there will, inevitably, be problems of a personal nature, a clash of personalities perhaps, professional jealousies, or just plain dislikes. Artists, perhaps they have more in common than others, don't seem to suffer over much from human frailties. But one guy did have it in for Lawrie from very early on, even before he met him, and this continued for many years; not usually a problem with Lawrie in Coventry, the other guy 120 miles away in London. Until the day of the double booking... Lawrie had instructions to draw something, found out by chance that this other guy was also booked, but anxious not to rattle the cage, Lawrie made Peter Willis, his boss, aware of the double booking. 'No problem, Lawrie, you draw it. The other guy's doing a computer drawing.' To which Lawrie replied that while the other guy was setting up the computer co-ordinates he, Lawrie, would have the damn thing finished. When the drawing was finished, painted and delivered to the office, the computerised drawing had reached the basic drawing stage! So the firm ended up with two drawings, the score standing at Artist Drawing 1, Computerised Drawing 0.

Six cylinder Triumph

One of the last engines designed by the Triumph group in 1977 was in fact intended for the new Rover models, superceding the Rover 2000. Built in Triumph's Canley factory, the new six cylinder overhead camshaft unit, in 2.3 and 2.6 litre forms, was not a warmed over old unit but a completely new design to replace Triumph's 2000 and 2.5, and also Rover's 2.2 unit. The new engine sizes were differentiated by the 84mm stroke of the 2.6 engine, against 76 mm for the smaller unit. An interesting departure was the use of German Mahle heavy-duty strutted pistons, with solid skirts, and the combustion chamber formed partly in the piston crown, features clearly shown in the drawing. In contrast, the 2.3 engine used English Hepworth and Grandage W-slot pistons. Both piston types had two compression and one scraper rings. On the 2.6 litre engine, the die-cast aluminium-alloy cross-flow head contained the remainder of the squish area complementing that already formed in the piston crowns.

Another interesting feature was the use of valve gear originally designed for the Triumph Dolomite. In this, the same lobe of the overhead camshaft operates both valves; on the inlet side it operates directly on an inverted bucket above the valve, whereas on the exhaust side it opens and closes the valve by means of cast iron rockers.

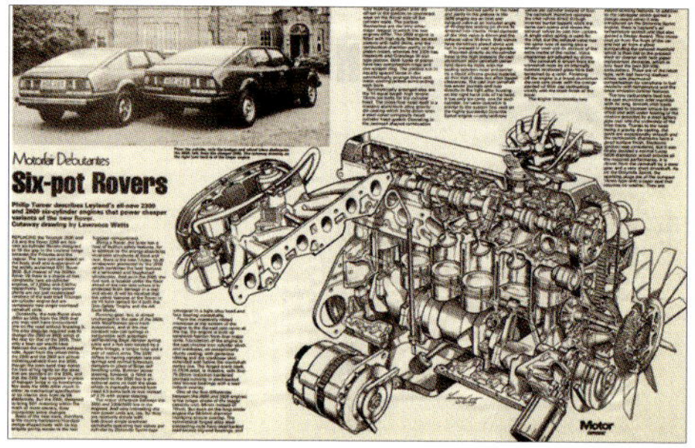

In both forms, the new engine filled quite a few existing models in the Triumph and Rover ranges, with no external clues as to which engine was fitted, other than twin tail pipes through which came a distinctive snarl. This drawing won Lawrie Artist of the Year award in April, 1978.

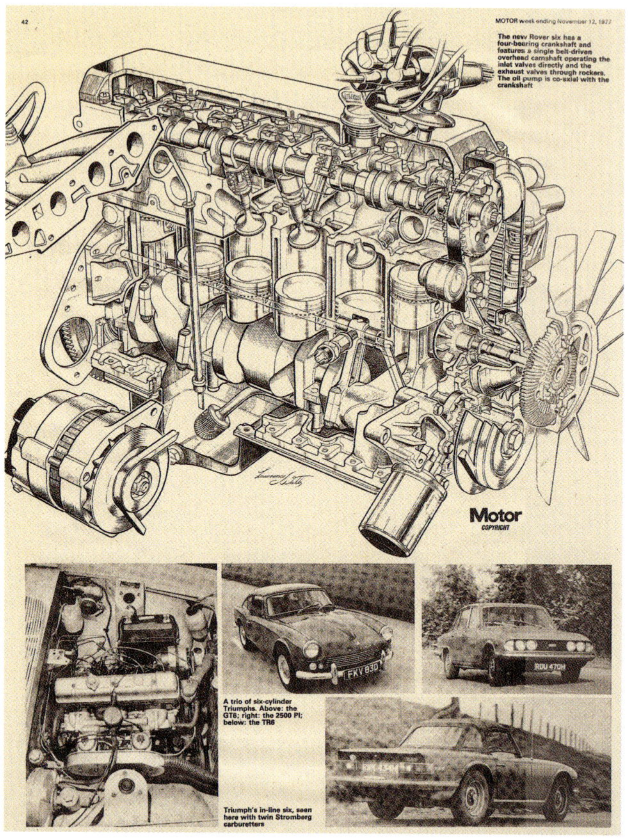
MOTOR week ending November 12, 1977

The new Rover six has a four-bearing crankshaft and features a single belt-driven overhead camshaft operating the inlet valves directly and the exhaust valves through rockers. The oil pump is co-axial with the crankshaft

Motor

A trio of six-cylinder Triumphs. Above: the GT6; right: the 2500 PI; below: the TR6

Triumph's in-line six, seen here with twin Stromberg carburetters

The Triumph six-cylinder engine was fitted in all of the cars illustrated in the magazine article. Most models were successful and popular in motor sport events, an answer to Rover's sporting saloons.

Foden

There was not a lot wrong with the mechanicals of British trucks but driver comfort had not been high on the list of priorities, manufacturers tending towards utility and running costs. Drivers were considered well looked after if wind and rain were kept out of the cab, heaters were for cissies, overnight stops were in inconveniently located doss houses. The advent of the Common Market changed all that. Continental long haul trucks appeared on UK roads, their equipment up front put many of our lorries to shame, and continental drivers could cover greater mileages in comfort before kipping down overnight in specially-designed living quarters behind the cab without wasting time locating accommodation. All of this helped cut operating costs, the sacred cow of the transport industry. Inevitably, British hauliers demanded better designed long haul tractors and lorries forcing UK manufacturers to respond or lose business. Hence the arrival of air-conditioning and driver-friendly cabs to rival the best of the continental products, along with labour saving features such as tilt cabs and hydraulically-operated gadgets to improve efficiency and cut operating costs, all of which kept Master Artist Watts busy drawing.

In one drawing of a new Foden Lawrie illustrated an innovative feature, the tilt cab. The entire cab hinged forward allowing easy access, not only to the engine and mechanicals beneath the cab, but also to the back of the driving instrument panel - an area normally inaccessible. At the same time, Foden introduced a new air circulation system the direction of which Lawrie shows by directional arrows to the windscreen and side windows, with cooling air being ducted through channels in the roof. 'Cabs of big tractors get hot so it's vital to keep them cool and this system achieves that by expelling hot air through the roof,' explained Lawrie.

New air circulation system on Foden's tilt-cab shows how the air is circulated over and around driver. Also facilitates access to the engine for repairs or maintenance.

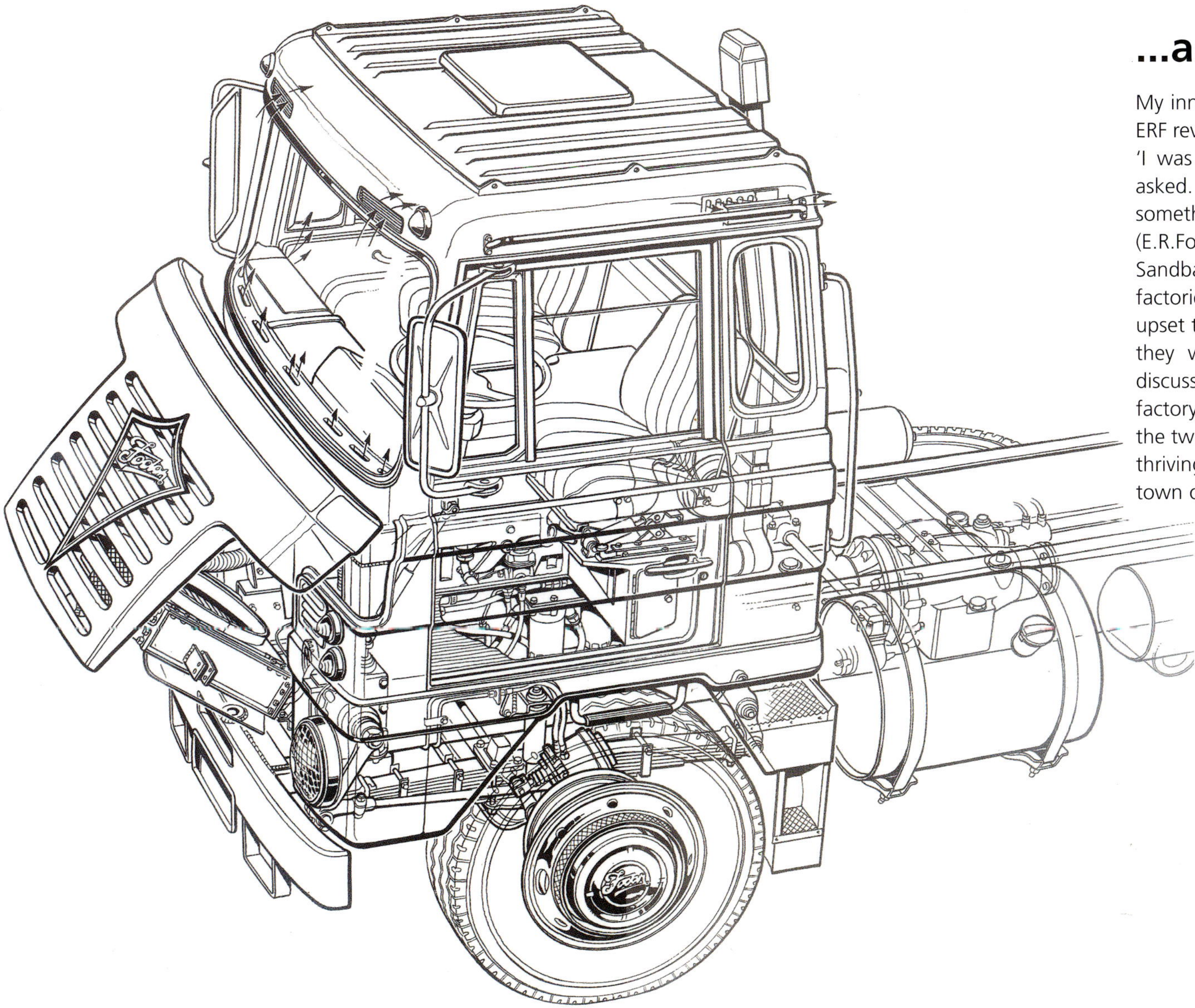

...and ERF

My innocent query as to the difference between Foden and ERF revealed a little bit of transport history. Lawrie explained, 'I was curious myself about the difference, so one day I asked. It appears that they were the same family but something happened and they split up, one to form ERF (E.R.Foden) and the other remained in the original factory at Sandbach. This was embarrassing as I had to work in both factories at different times. So I had to be very careful not to upset the equilibrium, and to appear as if I didn't know that they were in opposition. Most importantly, I could not discuss with either what I'd recently drawn at 'the other' factory. In fact, there was little significant difference between the two products. Yet, at that time, both firms seemed to be thriving, both operating out of the same small Cheshire town of Sandbach!'

Bedford five tonner

Bedford offered a range of chassis upon which you could fit whatever body you wanted.

Drawn for the trade magazine, 'Bus and Coach', the idea was to show this five-ton chassis from an unusual side angle so that potential customers would see details of the load platform. 'This would help them choose the most suitable body to fit on that platform,' explained Lawrie of the unusual perspective, rather than his normal higher position and sharper angle. 'In addition, the drawing also shows clear access to the cab, access through the floor to the engine, and a flat platform for whatever body was chosen - nothing protruded above that top chassis line. It was the responsibility of the body fitters to drill the massive channel members which form the chassis, and to fix mounting brackets for attaching the body to the chassis. Complete versatility was the vehicle's forte, you could fit almost anything on it, within reason. Everything was made in Bedford's factory, even the massive channel-sections which formed the chassis backbone.'

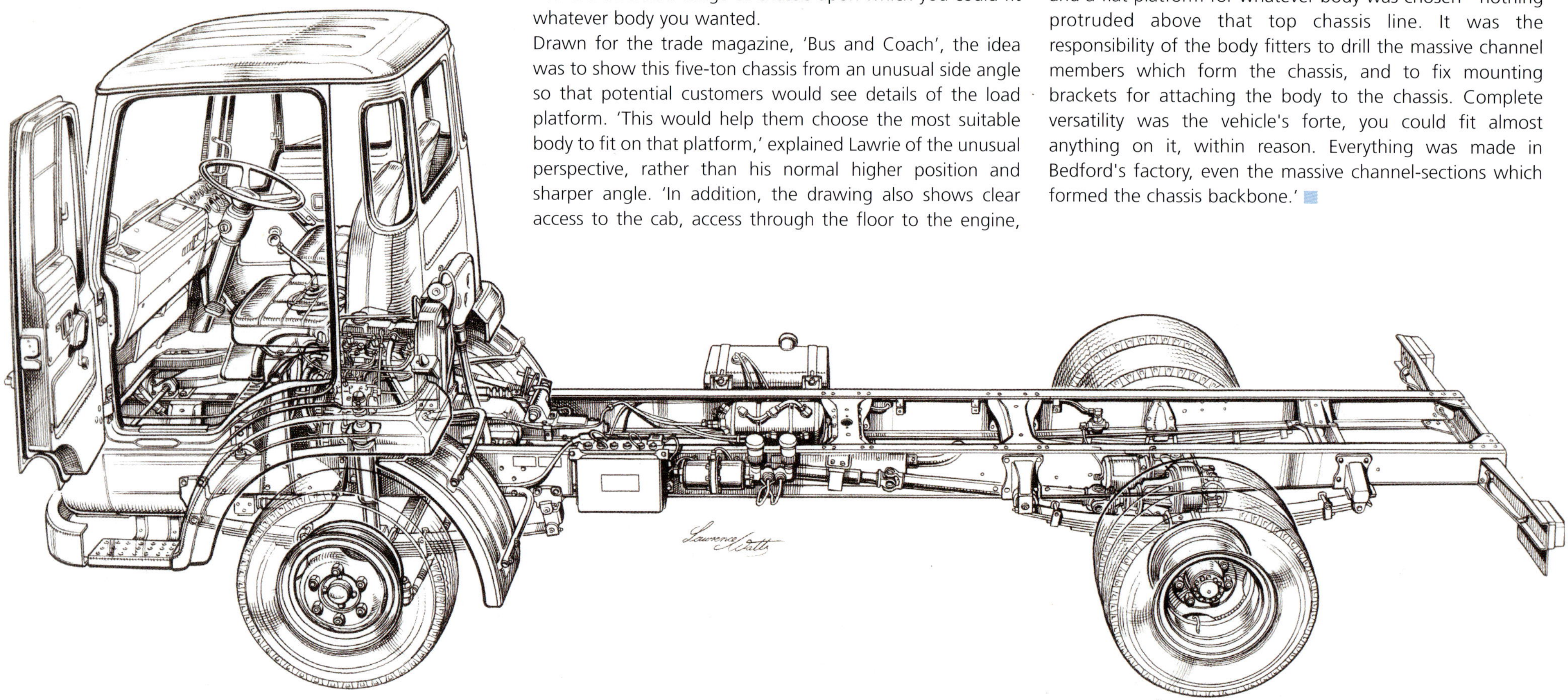

Lawrie enjoyed drawing trucks, their simplicity of purpose and ruggedness appealed to him, as exemplified in this Bedford design which was supplied ready for body fitting.

Wilson self-changing gearbox

One of the most difficult automotive challenges Lawrie faced was the Wilson automatic gear box, not as automatic as we interpret it today, because with the Wilson you pre-selected a gear and the box automatically changed it. Lawrie was called to draw the Wilson box when it was fitted to the Armstrong-Siddeley Sapphire car. Executed in 1964, the black and white drawing is remarkable because it shows a seldom-seen piece of machinery which was popularised on British luxury cars of that era. Its origins go back even earlier. I remember when I was a sprog sitting on my father's knee as I steered his Talbot up the driveway of our home in Ireland in the late 1930s. That had a pre-selector box.

The modern hydraulic-electrical system came about because in pre-motorway days in the late 1950s and early 1960s, British cars were earning a reputation overseas for being under geared. To counter this, an overdrive was designed and offered as an optional extra on upmarket models, especially those exported overseas. Even sports cars, such as the Austin-Healey 3000, benefited from having an overdrive fitted. It made for an effortless drive on the embryo UK motorways and the European and American motorway networks where, at the time, speed limits did not exist. As fully automatic drives developed and became standard equipment on many upmarket models, so demand for the pre-selector box receded.

But 40 years ago it was quite different, as Lawrie explained, talking me through the drawing, as he remembered it. 'The rectangular box on the bottom right of the drawing is the shuttle box, which contains a labyrinth of tubes with shuttles which travel back and forth according to the oil pressure on which side of the shuttle it is. That electronically engages the spring-loaded clutches. There's a clutch ring, rather like a brake band, around each set of sun and planet gears. Whichever one is locked on, it manifests itself through the planets - so called because they rotate around the sun wheel - into the sun wheel which is attached to the shaft. This gives you the drive shaft speed. That is then transmitted to the torque converter, or fluid flywheel, which is what it used to be called, and which does the driving. The gears are controlled by hydraulics.' From casting marks, which Lawrie faithfully reproduced, it appears that at least some of the transmission was made by AP, Automotive Products, one of the biggest, if not the biggest, in the automatic gear box business today.

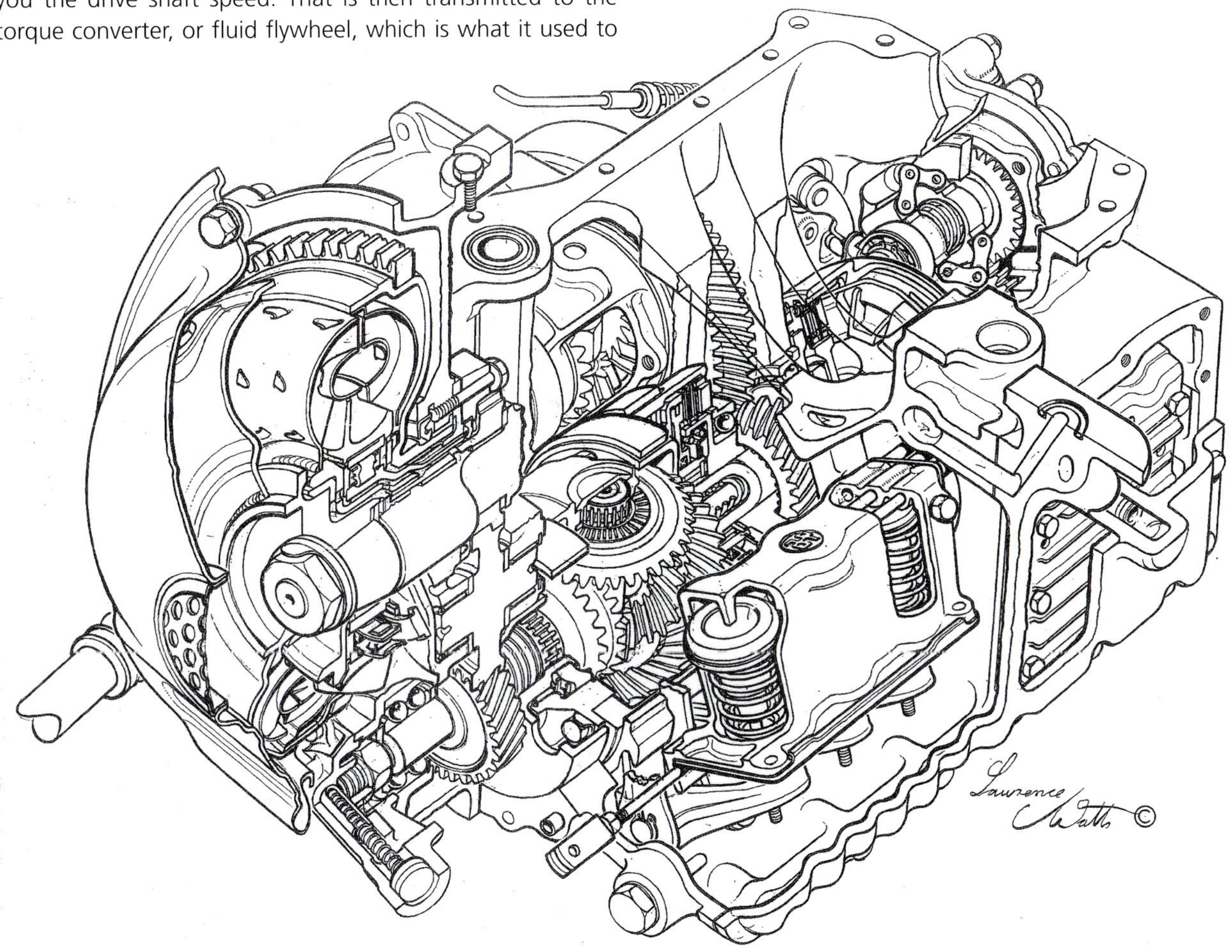

Wilson self-changing gearbox as fitted in the Armstrong Siddeley Sapphire and Hurricane models, later led to fully automatic gearboxes.

Six wheels on my Panther

From the simplicity and purity of line of the Bond Bug, to the complexity and over-the-top excesses of the six-wheel Panther, which 'Autocar's' Peter Garnier commissioned Lawrie to draw in 1977, provide a contrast in what an editorial artist was called to draw during what has been called the 'golden years' of the sixties and seventies. It seemed that anything and everything was the order of the day. 'I go along with that,' agreed Lawrie, 'because I never knew what I was going to be faced with next. Whether I liked it or not, I had to take a professional view and give it my best shot, and that's what happened with Bob Jankel's Panther six wheeler, made in Surrey. It was a weird mixture comprising an 8.2 litre turbocharged Cadillac engine in the rear, three seats abreast in the front, a central steering wheel for guiding the four small-diameter front wheels, an idea perhaps copied from Ken Tyrrell's Formula 1 cars of the early 70s, retractable headlamps and two spare wheels of different sizes to fit front or rear. A greater contrast than that and the Bond Bug would be hard to imagine, but I know that half the number of wheels would provide me with twice the fun to drive!'

Recognisable in the drawing of the Panther are typical Watts touches, such as the recess in the right hand door for an early mobile phone, and a rectangular panel with dialling numbers. Cutaways in the body show the chassis, which is a shame as the crude channel-section structure is the least attractive feature of the car. It would have been more at home beneath a Bedford truck. As Lawrie's drawing was completed before the car, he never got to drive it, nor did he recall anyone who drove it. 'How the same people who made the cracking little Kallista sports car conceived this car is beyond me,' concluded Lawrie, 'but it was a gamble which did not succeed as few were sold.'

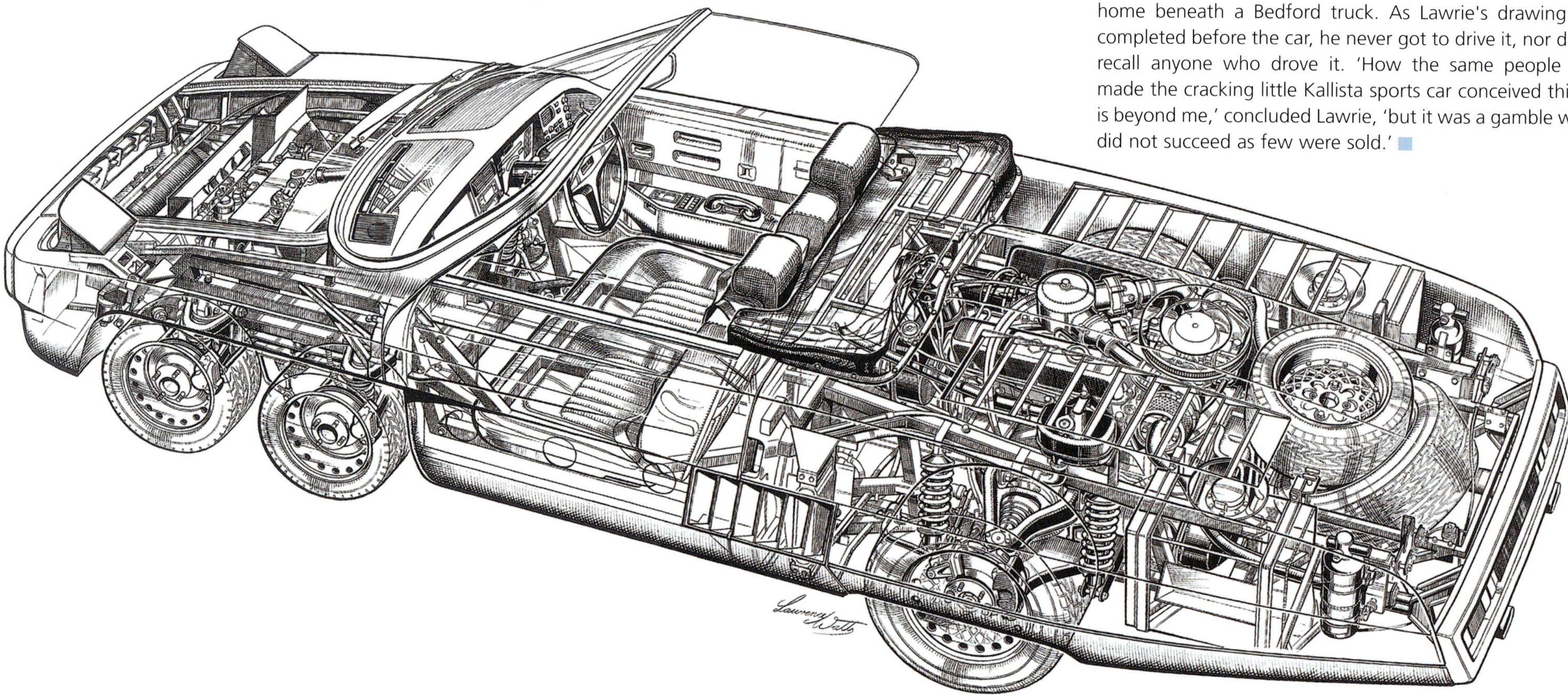

Bob Jankel's Panther six-wheeler was an exercise in 'Suck it and see' marketing of the 1970s. Lawrie still gave it his best shot, as can be seen in the plethora of detail, while reserving judgement on the viability of the project. It bombed. Only a couple were made.

Gilding the lily, in hindsight

Leafing through his drawings all these years later prompted a thoughtful comment from Lawrie. 'I've recently realised that what we're looking at now is progression in engineering. Take that beautiful 3.5 V8 MGB-GT, for example. A superb car to drive but, like the Austin-Healey, it suffered from being tail happy - though not to the same extent as the Healey. Hindsight suggests that if the MG designers had moved the weight distribution rearwards when they fitted the powerful 3.5 V8, the car could have done justice to its promised performance. Instead, it was difficult to keep it between the hedges on a wet B road, or on a fast A road for that matter. From a glance at the drawing, it's clear that without too much effort the chassis could have been re-engineered to position the engine farther back, re-locating the gear box around the rear transmission, and providing independent rear suspension. That would have improved the handling enormously and allowed the car to release its true potential. Thus revived, it could have extended the longevity of its production run. But due to lack of funding, lack of initiative, plain disinterest - or a combination of all three - nothing happened and MG, one of the most charismatic names in the history of the British car industry was allowed to wither. Worse still, it's re-badged successors had nothing in common with the original marque except the magic octagonal emblem. How very sad!'

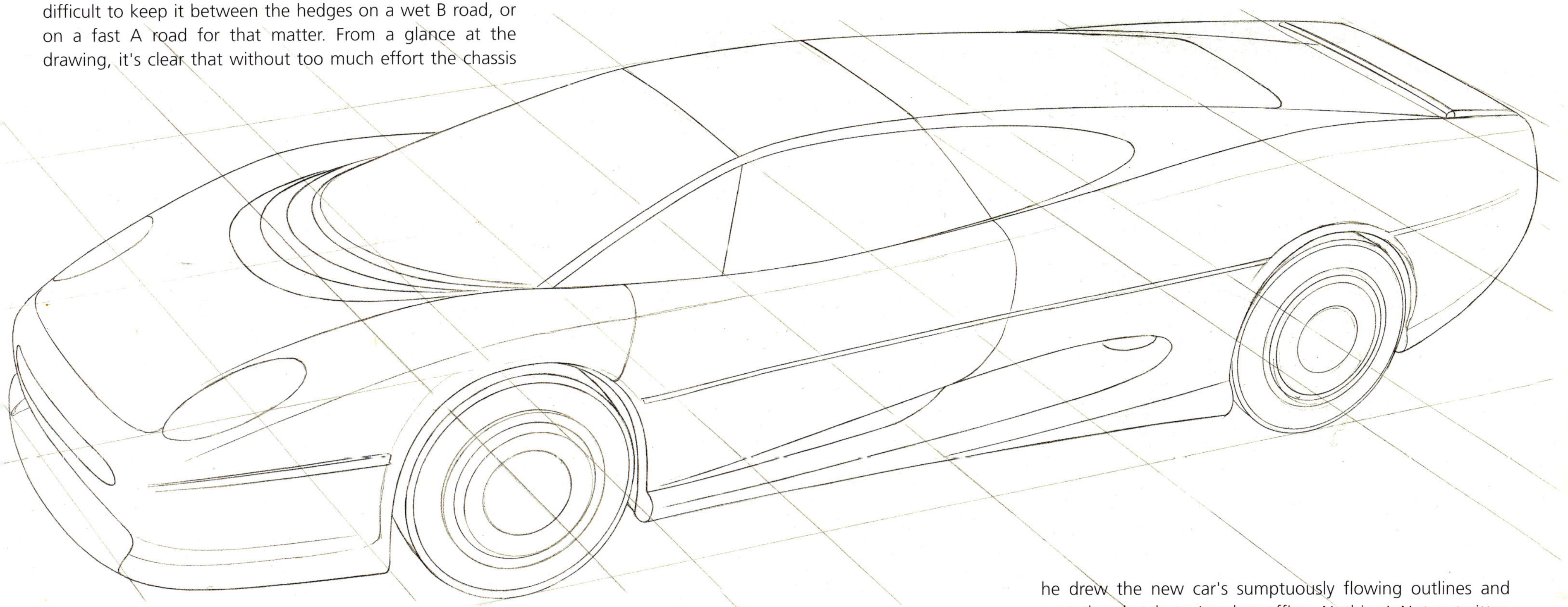

XJ220 outline

Given a tip-off about a secret new Jaguar being built in a body shop hidden in the back of beyond, Lawrie stumbled into one of the best-kept secrets of the early 80s, the stunning new Jaguar XJ220 supercar. On his own initiative, he drew the new car's sumptuously flowing outlines and sent the sketch to London office. Nothing! Not a twitter. Silence.... It was only when he saw a detailed drawing later in the 'Motor' that Lawrie realised there had been a tacit agreement between the two magazines that this was to be 'Motor's' scoop (they needed extra readership more than parent magazine 'Autocar'). 'Disappointed? You could say....!' he groaned.

Triumph TR7

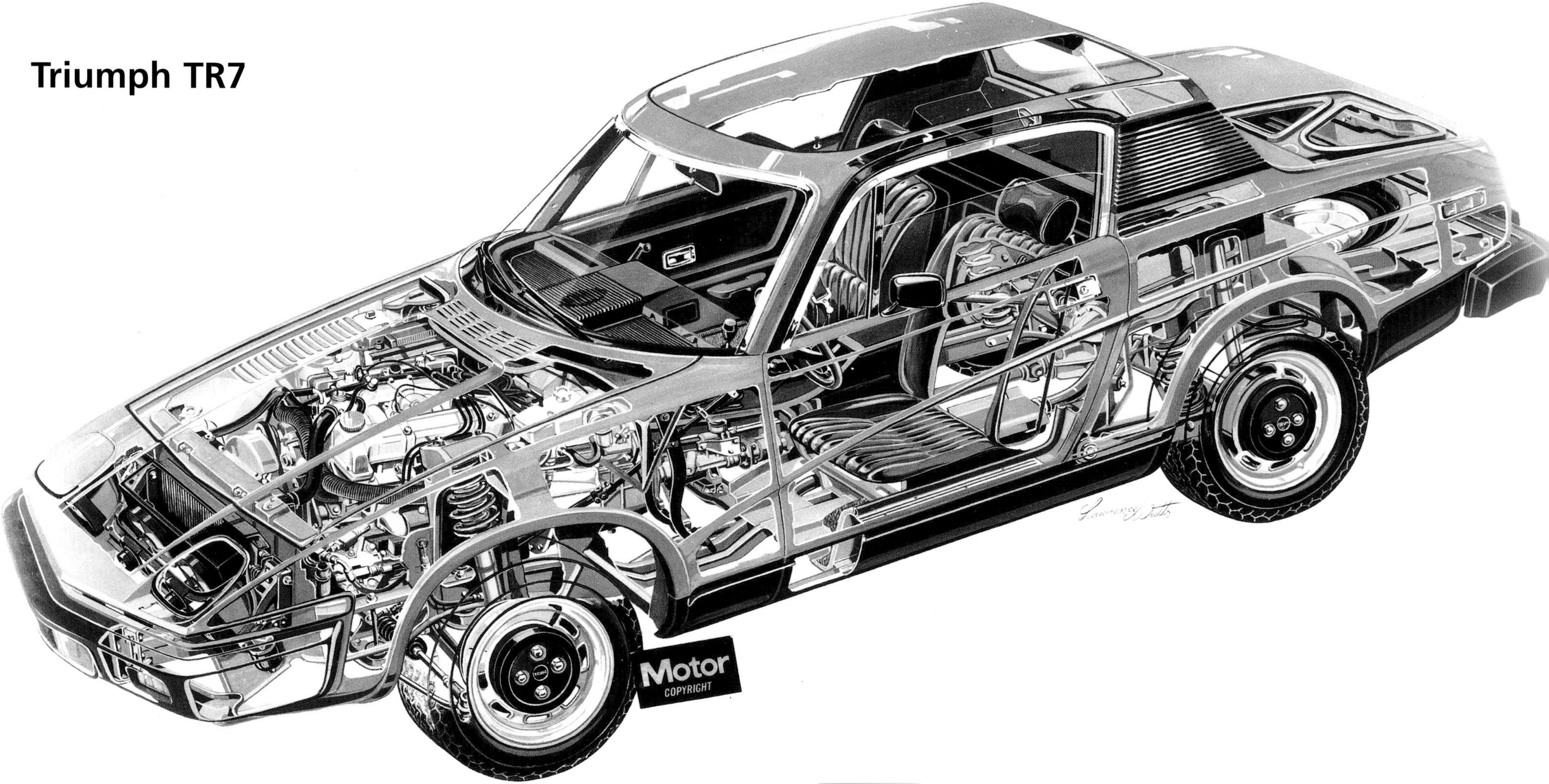

Last of an illustrious line, the TR7 Triumph was state of the art (as compared with its forerunners when Lawrie drew it). Through the cutaway in the body he shows the independent rear suspension, a long way from the solid axle and cart springs of the original TR2! At the front, pop-up headlamps followed the fashion of the day and blended well into the sleek Italian styling. This export model is shown with left-hand drive for the American market where the TR7 met a warm reception.

Gentleman's carriage

For years, Lawrie had an interest in what he calls 'gentlemen's coaches', nothing to do with Rolls-Royce, rather the horse driven variety. This interest was nurtured by annual visits to the Royal Stoneleigh Show, a few miles from his home in Coventry. 'I was sent there every year by one of the farming papers and, when ever I had a break, I'd wander around to the Royal Mews,' he recalled. Eventually he decided to draw one of the magnificent Royal coaches because ' it was such a wonderfully elegant, and functional, structure. It was just crying out 'draw me'. So I did, but first I had to get permission from the Royal Mews to take photographs of all the Queen's coaches. Officials were very helpful in making the coaches available when they were at the Royal Show, so I was able to get a complete set of colour photographs which not only saved me a trip to London but would have proved invaluable in helping me with my drawings. Incidentally, if I'd done them, nobody was interested in publishing them, it would have been for my own interest because I felt that the coaches are beautiful, irreplaceable, and a piece of our history. Coincidentally, about this time I was sent to Gloucester on an assignment for a publication unconnected with coaches. But when I arrived at the factory, standing there was a magnificent coach! Of course, I forgot all about my assignment and got talking to the man in charge, who turned out to be a coachman himself. He suggested that I bought the bible, a bound book called 'On The Box Seat'. That tells you all you ever needed to know, and a lot more, about coaches including the etiquette that goes with them. So I got absorbed in coaches for the remainder of that year, it soon became another hobby of mine. While I was in that factory in Gloucestershire, I started to sketch my new friend's coach and that picture is the final pencil sketch. There's no shading because I intended to paint it. However, that was a daunting prospect because it would have had to be done as accurately as possible but I ran out of time....... I never did get to paint the Royal coaches either! Their detail is breathtaking and the trim inside is gorgeous, it would have been a huge challenge for me to reproduce. I thought at the time that it would make a nice collection but didn't appreciate just how much work would be involved. Anyway, I'd no commission to do this work which would have taken quite some time. While this was going on, I met the Prince of Wales when he visited the Royal Show to open the new 'Farmers Weekly' stand. The first words he said when introduced to me was 'I suppose you're going to tell me you're overworked and underpaid!' I knew that I was talking to someone with a sense of humour, so in the course of conversation I told him of my plans to paint the Royal coaches. He was very interested and encouraged me to go ahead but, with so many other things piling up in my day job, and no commission to cover my time working on the project, nothing came of it. So this will be the first time that that drawing will have been published.'

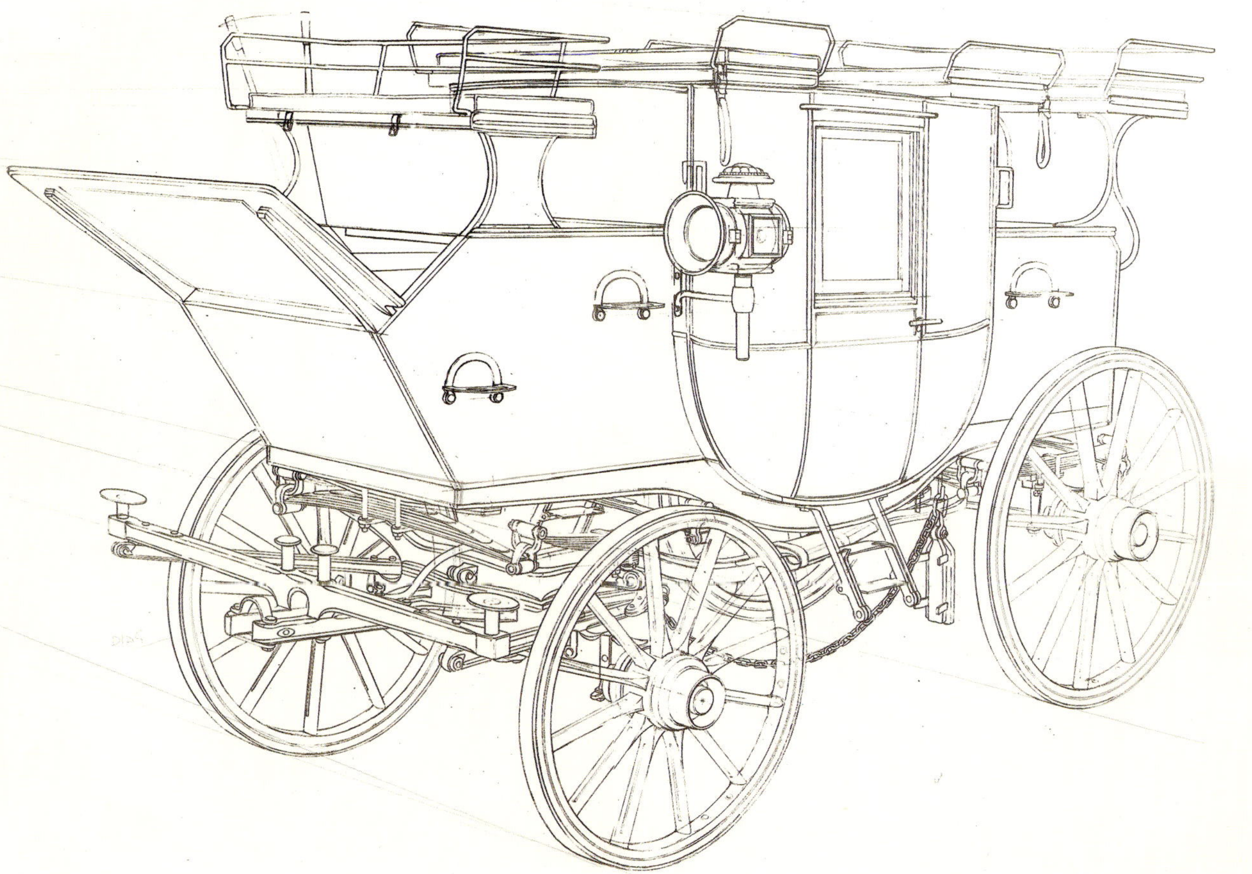

Mechanical Farming

In a long and varied career, Lawrence Watts was privileged to work alongside some of the foremost automotive and aeronautical engineers and designers in this country and abroad. The results of that work have been seen over a span of 40 years in prestigious technical magazines the world over. Yet when asked what gave him the most fun to draw his answer may seem surprising. 'Although most of my time was taken up with aircraft, cars and bikes, the project which provided the most fun in executing was something I did for the Potato Marketing Board! The drawing had so many diverse features that I had to work out how to show them all in action. For instance, a huge packaging factory with lots of processes in action, people all over the place, lorries delivering tons of potatoes to conveyor belts, girls hand grading potatoes into different boxes, potatoes on grading chutes, fork lift trucks collecting the spuds off the chutes, then loading their cargoes into crates before being packed onto waiting lorries at the other end of the factory. Look closely at the drawing and you'll see literally hundreds and hundreds of potatoes, all individually drawn. Furthermore, they're drawn to show perspective! Each of the personnel is busy doing something, so there's a real immediacy, something's happening no matter where the eye alights. To show all this, I had to 'lift' the roof off the building in order to give a bird's eye view. I also had to show the direction in which the production system was flowing. Yes, it was a challenge and great fun to do. Even more satisfying when eventually it appeared as a two page pull-out in the December 1981 edition of 'Power Farming'.

Down on the farm

I knew Lawrie through our work on 'The Motor Cycle'. It was not until we started writing this book that I realised how much wider was his sphere of influence on other publications, and I knew nothing at all about his extensive portfolio in the farming world, in which he moved freely and remained busy long after retirement. Most of the other artists I knew specialised in one or two fields, so I asked Lawrie how he made the transition from very specialised automotive and aeronautical worlds to mechanical farming. His answer was revealing. 'Shortly after I drew the Austin Gypsy, I found myself drawing farm buildings and how they worked. Before that, farm buildings were more or less sheds which had not changed in hundreds of years. Then they gradually started to evolve into something approaching factories, handling farm produce through the use of specialised mechanical equipment. And that was what interested me, the mechanical aspect. My previous experience came mighty useful when called upon to illustrate new methods of handling farm produce, not many - if any - artists were specialists illustrating farm machinery and new types of specialist buildings, simply because there were so few. Plus which, Iliffe (later IPC) publications such as 'Power Farming', 'Farmers Weekly', and 'Farm Mechanisation', to name only a few, needed an editorial artist with technical training to illustrate the 'rural factories' springing up all over the place. A new challenge for me, and I loved it! I came into direct contact with the owners, in many cases just a farmer and his son, who benefited from the new automated processes by being able to operate the system with minimum effort and no staff. It just happened that I got on very well with farmers, so that was a bonus.'

All of that led to one of the most celebrated of Lawrie's drawings, and his favourite, a potato packing factory which he drew for 'Power Farming' and the Potato Marketing Board. 'The new departure showed me things I'd never before experienced so I was learning all the time.'

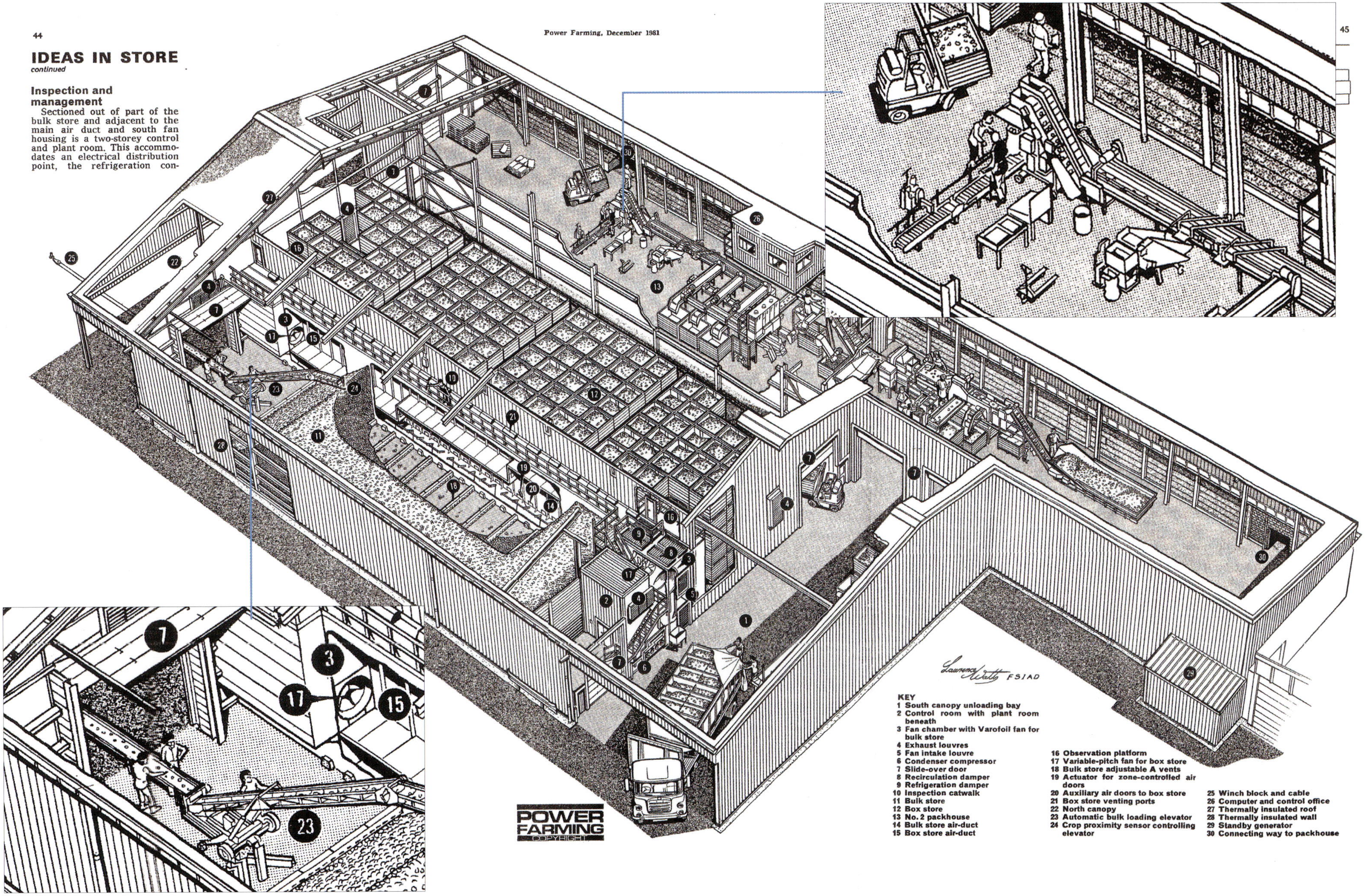

44

Power Farming, December 1981

IDEAS IN STORE

continued

Inspection and management

Sectioned out of part of the bulk store and adjacent to the main air duct and south fan housing is a two-storey control and plant room. This accommodates an electrical distribution point, the refrigeration con-

45

Laurence Watts FSIAD

POWER FARMING COPYRIGHT

KEY

1 South canopy unloading bay
2 Control room with plant room beneath
3 Fan chamber with Varofoil fan for bulk store
4 Exhaust louvres
5 Fan intake louvre
6 Condenser compressor
7 Slide-over door
8 Recirculation damper
9 Refrigeration damper
10 Inspection catwalk
11 Bulk store
12 Box store
13 No. 2 packhouse
14 Bulk store air-duct
15 Box store air-duct
16 Observation platform
17 Variable-pitch fan for box store
18 Bulk store adjustable A vents
19 Actuator for zone-controlled air doors
20 Auxiliary air doors to box store
21 Box store venting ports
22 North canopy
23 Automatic bulk loading elevator
24 Crop proximity sensor controlling elevator
25 Winch block and cable
26 Computer and control office
27 Thermally insulated roof
28 Thermally insulated wall
29 Standby generator
30 Connecting way to packhouse

Cowed!

'While talking to the farmer, I felt a slight nudge in the back. Absorbed in what the farmer was saying, I did nothing. A second, firmer nudge, grabbed my attention and there was this beautiful cow looking adoringly at me! Soon as I turned around, she lifted her head onto my shoulder and drooled all over me! At which the farmer remarked 'You love cows, don't you?' to which I replied 'I'd not thought about it until now. So, yes, I suppose I do like them, though this is taking it a bit far!' 'Oh, don't worry,' the farmer reassured me, 'she's just taken a fancy to you, but when she's ready to eat, she'll be off after the others.' Word got back to the Farmer and Stockbreeder office, which had commissioned the drawing, and Stephen Howe, who later became managing director of the farming group of journals, decreed that in future Watts would be allocated any assignments which involved cows! You can't get more involved with farming that that.'

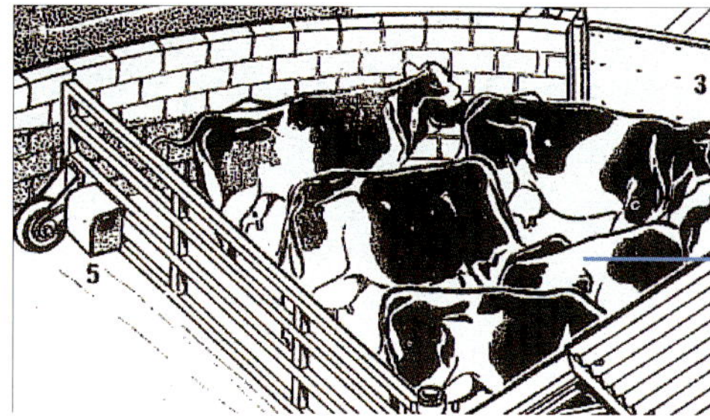

Musical cows on a merry-go-round

Another assignment accelerated his learning curve, as he describes: 'I was learning all the time, and discovering things I'd not known about myself. For instance, in 1969 I went to a farm near Oakhampton in Devon, to draw a machine called a Rotolactor, which comprised platforms in tandem which revolved six or seven times a minute, and upon which cows stood while being milked at the rate of 120 animals per hour. At milking time the farmer played music in the milking shed. Responding to this, the cows made their way to the collecting yard, filed into the milking area and onto the rotolactor platforms where each cow was allocated its own milking bay and there connected to the teat cups which linked each cow to the central milk extraction machine. An added incentive for each cow was three and a half lbs of cake in a trough in the milking bay, which the cow ate (with considerable relish) while being milked. Time had been calculated precisely for each cow to be milked dry, the animal was automatically disconnected from the machine and went out into another collecting area until the complete herd had been assembled and were led back out into the pasture. The entire operation had been done automatically only because of the willing co-operation of the cows!'

Look closely at Lawrie's drawing and you will see a masterpiece of detail. For instance, all the cows have different markings. As one cow exits the foot bath (key 9), another's head appears around the corner, already it's dribbling into the foot bath! Others are looking for their mates (7), or being attached to their milking stations (33) where they will stand obediently while milk is extracted. Although we can't see all of them within the building, we know they are there because Lawrie has shown their legs beneath a cutaway of the circular wall (53 and 54).

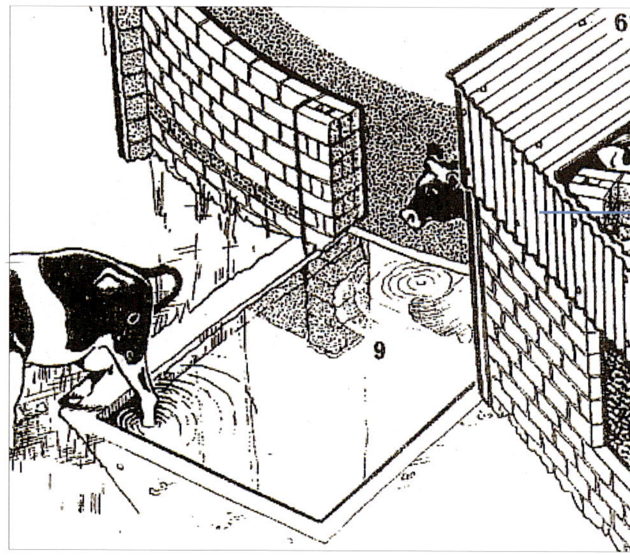

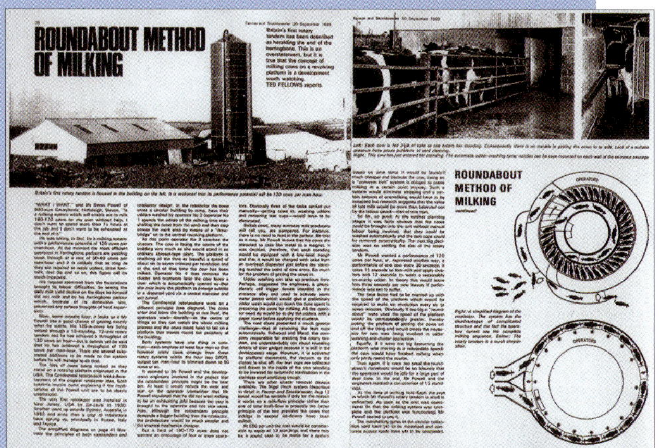
ROUNDABOUT METHOD OF MILKING

ROUNDABOUT METHOD OF MILKING

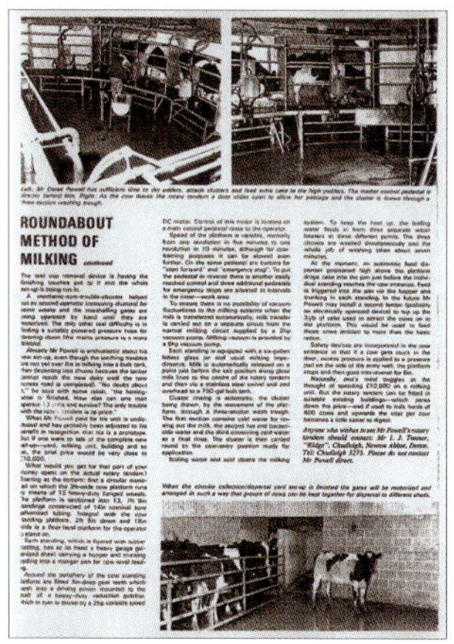
ROUNDABOUT METHOD OF MILKING

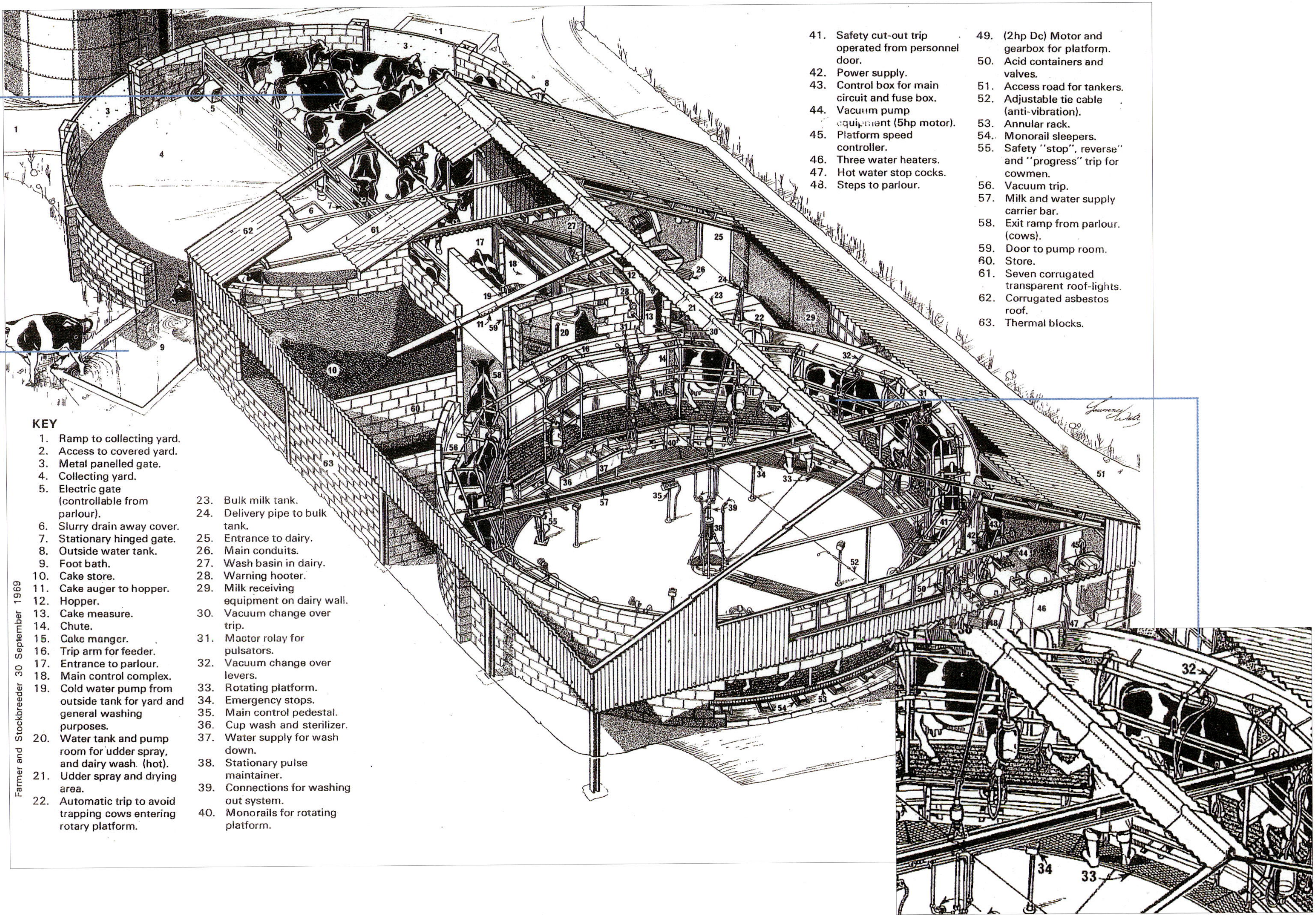

Farmer and Stockbreeder 30 September 1969

Turner era turned into a new life

His farming connections introduced Lawrie to a man who was to make an enormous impact on his life. 'I was sent to Turner Engineering in Alcester, Worcestershire, to see a new plough with a reversible blade. Designed by Tony Turner, this plough was a major innovation as it saved time by cutting out the need to turn, or to circle the field, at the end of each run. When you think how often that happens in small or large fields you can appreciate how much time was being saved. At the time, Turner Engineering was making commercial size grass cutting machines, used by councils throughout the UK. In the course of conversation, Tony Turner asked me if I'd be willing to draw, in my own time, a cutaway of his grass cutting machine for his own use. I saw no problem and duly did the drawing. That was the start of a friendship which exists to this day. A man of great charm, Tony's enthusiasm knows no bounds and, bubbling over with new ideas, he describes them to me and I interpret what he wants and provide the working drawings. I've been doing that since I 'retired' for the past 17 years, and he still calls me up. I've got a feeling that I'm now part of his team, in a non-executive capacity! Certainly, he's like a breath of fresh air, and I enjoy every moment of working with him. Anyway, Turner Engineering spawned Turner Developments to handle new machinery which we produced and it was marketed under the name of the holding company GreenMech. That's the name, and green is the colour, you will see on the shredding, chipping and stump grinding machines used by most councils throughout not only the UK, but also many overseas countries.'

Narrow shave

Flying, motorcycling, fast cars, they all involve an inbuilt risk factor. But it was none of these which caused Lawrie to think 'Is this the end?' Something much more unlikely, drowning in grain. As he recalled, 'There was this massive new bulk storage system which 'Power Farming' wanted me to draw near Scunthorpe. It held 3,400 tonnes of grain from four farms in the locality. Following my normal routine of understanding how a system worked, I planned to measure and photograph the inside of one of the giant storage bins while the facility was closed for lunch. So there I was, on the floor of this 25ft deep bin, happily working away, when I heard machinery starting up. Strange, because everybody had gone to lunch. Thinking no more of it, I continued working while vaguely aware that the conveyor which fed the 11 storage bins was running. Then the thought crossed my mind, what is fed to which bin? It could have been peas, corn, oil seed rape, anything. I soon had the answer as grain started rushing in. Frantically slinging my camera gear over my shoulder, I climbed up the access ladder, but I could only climb as fast as the bin was filling..... If you ever need an incentive to climb a ladder quicker than you ever thought possible, just imagine a sea of grain embracing your legs, as it eventually did mine, knowing that death by suffocation awaits you should you be overcome, or tempted to stand on an aerated crop of grain....The dust also created a massive problem as it increased in intensity, filling my lungs and eyes as I neared the top. Choking, and with tears pouring down my cheeks, I managed to haul myself over the top. Just as I did so, the motors switched off. The bin was full.... None too pleased with what had happened, I went in search of the guy who had off loaded the stuff. He nearly had a fit! Nobody had told him that I was in the bin and he was so shocked and upset that I felt sorry for him. When the foreman returned and heard what had happened he, too, was full of remorse. If I'd not escaped, he'd have been answering some awkward questions, such as which of the 11 bins had Mr Watts disappeared into? Remember each one held about 77 tonnes, and I'd have sunk to the bottom..... Anyway, they say when you've fallen off a horse, or a bike, the best remedy for shock is to get straight back on again, so I cracked on with my job until it was finished. Only then did I remember that Eileen, who had come along for the drive, was in the car outside, oblivious to what had happened...'

In spite of the shock, and by dint of using photographs, many measurements and copious detailed sketches, Lawrie managed to get sufficient information to enable him complete this superb drawing back in Coventry. To provide a measure of size, he drew a man up on the gantry looking into a bin, looking for all the world as if he's shouting 'Are you comfortable down there, Mr Watts?'

GIANT STORE FOR QUALITY GRAIN *continued*

THE BRICE-BAKER ECONOSTORE AT KEB FARM

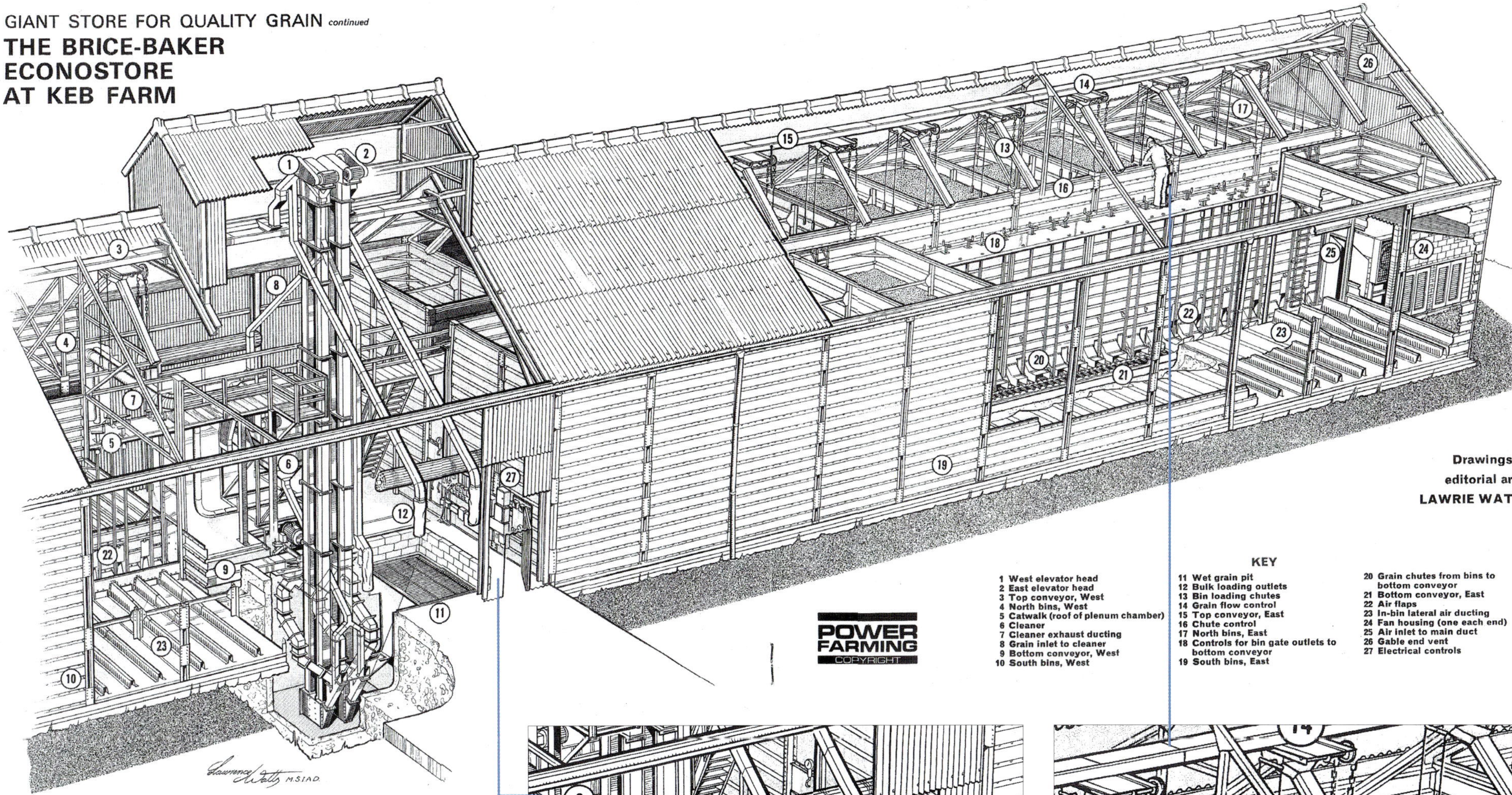

Drawings by editorial artist LAWRIE WATTS

KEY

1 West elevator head
2 East elevator head
3 Top conveyor, West
4 North bins, West
5 Catwalk (roof of plenum chamber)
6 Cleaner
7 Cleaner exhaust ducting
8 Grain inlet to cleaner
9 Bottom conveyor, West
10 South bins, West
11 Wet grain pit
12 Bulk loading outlets
13 Bin loading chutes
14 Grain flow control
15 Top conveyor, East
16 Chute control
17 North bins, East
18 Controls for bin gate outlets to bottom conveyor
19 South bins, East
20 Grain chutes from bins to bottom conveyor
21 Bottom conveyor, East
22 Air flaps
23 In-bin lateral air ducting
24 Fan housing (one each end)
25 Air inlet to main duct
26 Gable end vent
27 Electrical controls

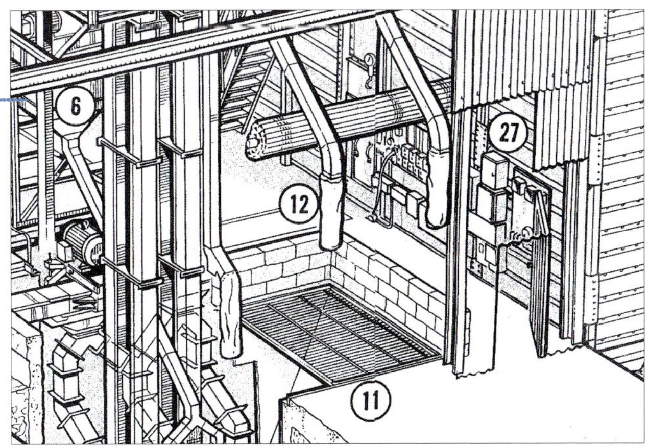

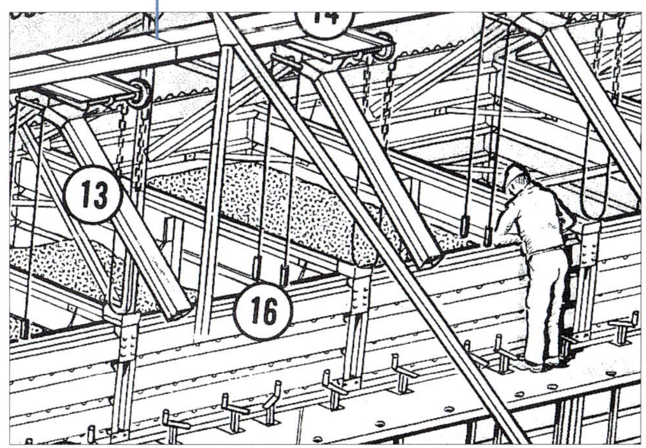

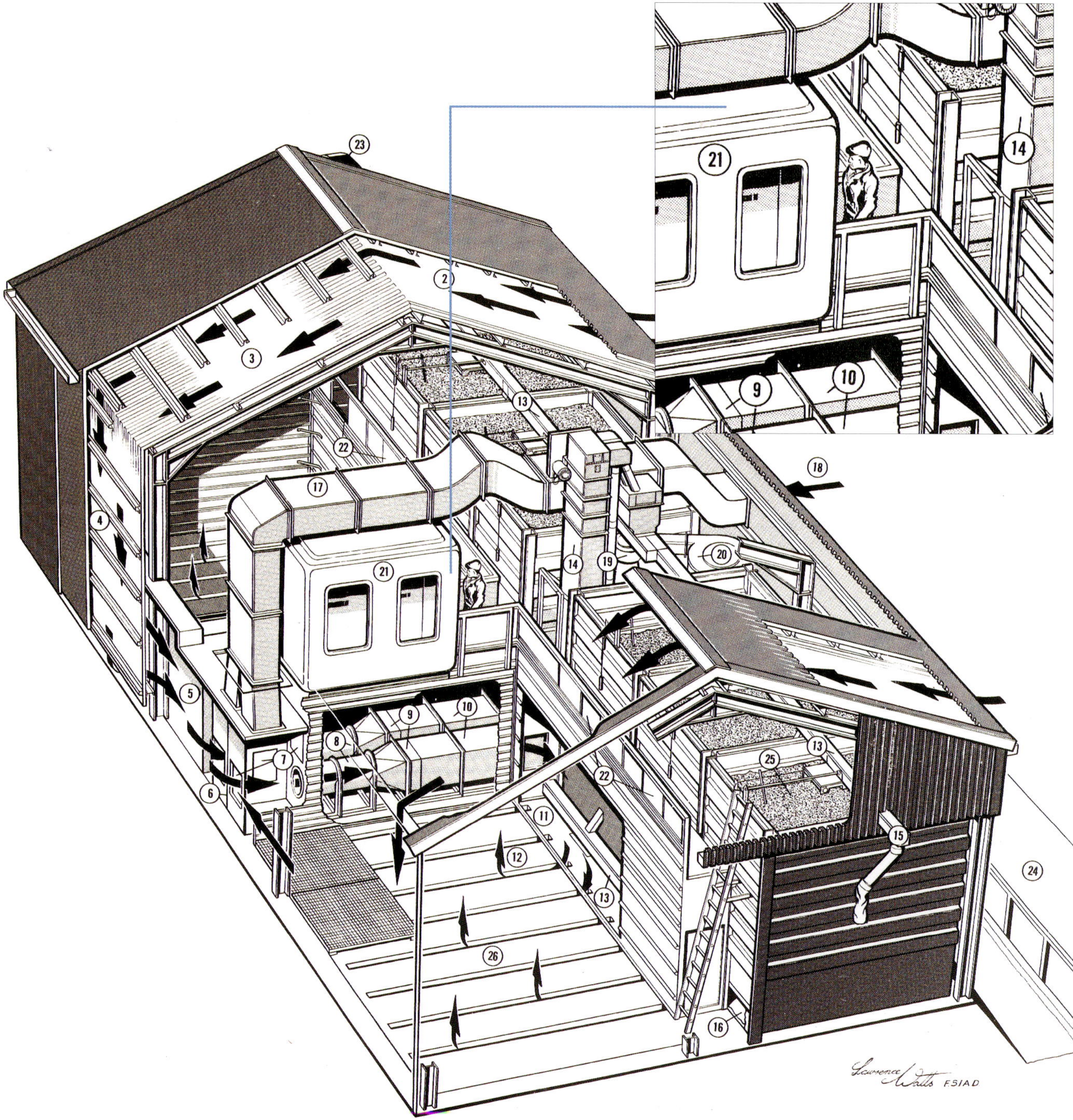

The ideal farm store

With a brief supplied by prize-winning Scottish students, Lawrie in 1985 drew what was called 'the ideal farm storage facility'. 'It was a cracking design,' he recalls. 'My brief was to interpret what they supplied in notes and sketches as their answer to the traditional old farm building, unchanged over hundreds of years. The design incorporated a complete thermostatically-controlled ventilation system including under floor, storage bins, and an integral office, all of this within a compact space. The building had a double skin, with air drawn in under the eaves on one side, circulated between the skins down the other, underfloor, and then around the storage bins before being expelled by means of machinery installed in the centre of the building, beneath an office which I installed on a second floor. I also drew a delivery system to the storage bins, so that a lorry could draw up outside beneath a pipe with a flexible pick up and the grain, or whatever, could then be drawn up the pipe and directed to the appropriate bin. The reverse operation could be done for delivery when the need arose. For economy, most of the building used readily available materials for bolting together on a secure base and which could be changed if they became damaged, or modifications were required. It was fun to draw and I hope that my interpretation of the students' brief proved beneficial to farmers.' Characteristically, he drew a suitably clothed farm worker entering the office, 'just to provide scale'. Likely story; he did it because that's his signature!

Irrigating the Fenland

A massive machine for irrigating arable fields in the Norfolk fens was of interest to Lawrie not only because of the ingenious nature of the device but also because it was designed by an ex-motor cycle engineer from the defunct BSA factory in Birmingham. 'It was interesting to see his interpretation of the brief,' recalled Lawrie 30 years later. 'The fields which he was required to irrigate were huge and 12 feet below sea level, and were too big for any normal irrigation system. So he designed a mobile one which relied on a stationary engine and winch positioned in the middle of each field to drive pulleys which allowed the machine to traverse the field while a laser beam kept it on course. The machine comprised a large drum which was mounted transversely on a wheeled chassis. Around the drum was coiled a water pipe which picked up water from collection points at 200/300 foot intervals around the field. To avoid missing a pick up, a laser beam kept the pick up points in formation. Having been treated with fertiliser after being picked up, the water was then fed through valves into a vertical steel pipe which, at its upper end, was bent at right angles into a perforated pipe which extended outwards, and parallel to the area to be irrigated. Attached to either end of the chassis were cables which ran through a pulley at one end and a winch driven by a stationery engine at the other end. This pulled the machine across the field in a straight line while valves fed water into the irrigation pipe. It was an ingenious labour-saving solution to a considerable headache.'

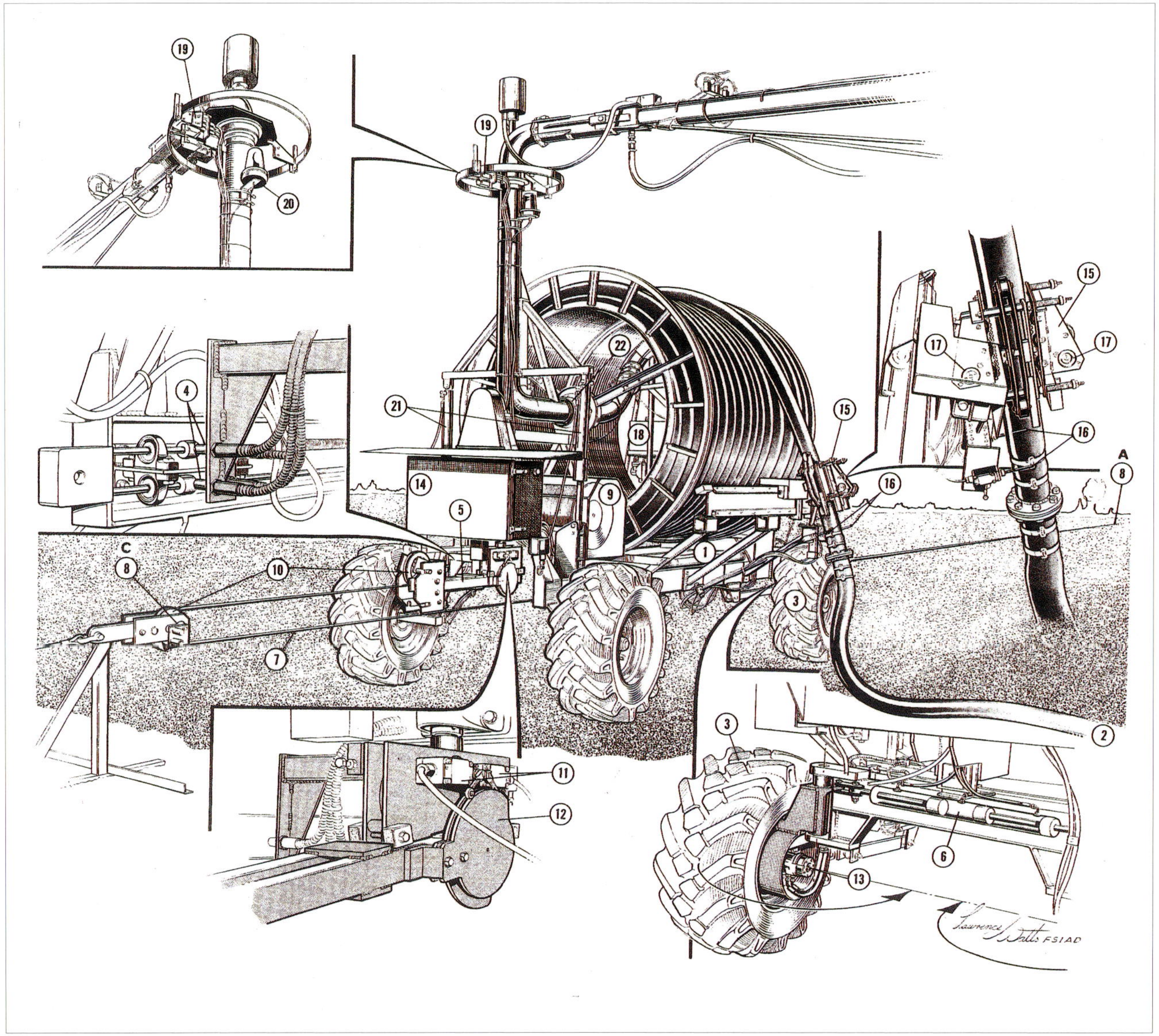

Dowler Gantry System

What puzzles me when driving around the countryside during the growing season is the frequency of 'tramlines' through growing crops, large swathes flattened by tractor tyres when spraying crops (as a lifelong organic food consumer, I'd ban crop spraying!). Farmers don't much like 'tramlines' either but accept it as a price to pay for the perceived benefits of spraying. An ingenious Cotswold farmer, David Dowler, decided to cut the waste and also the time taken to irrigate growing crops. A keen motorcyclist, he was handy with tools, had an inventive mind and, most importantly, thought laterally. Hence his invention of the Dowler Irrigation Gantry. To cut down on the number of wheel tracks, and the attendant compacting of the ground - not once, but every time spraying took place - he placed the wheels as far apart as possible and joined them by means of a gantry which was adjustable for height, it could be inches or up to eight feet, above and parallel to the ground. The gantry was hollow to carry hydraulic pipes and controls to a pair of wheels at either end, one large and one small wheel. The large one was hydraulically driven, thus providing infinitely variable power, while a smaller wheel behind it acted as a caster. At a stroke, he'd reduced waste caused by tractor tracks and soil compacting by a third. In so doing, he'd also gained the means of irrigating, or seeding, a wider patch of soil at one pass of the tractor via folding and extending arms through which the seeds, fertiliser, or water was fed through metered jets from a large tank. On some models the tank was centrally located, on others it counter-balanced the engine and cab at the opposite end, depending on weight distribution. Another benefit was the time saved; a field could be covered in about two thirds of the time, a significant consideration on a large farm. Technically, the gantry was more efficient because the span ensured level distribution due to cutting out tilting of the tractor during spraying, which was also the cause of spray drifting in windy conditions. 'It was a very ingenious system,' remarked Lawrie, 'and when I was introduced to David by Tony Turner, we got on like a house on fire, a real gentleman. Soon after we met David asked me to help refine his original design, that's how well we got on.'

David Dowler's first gantry was just a simple collection of angle iron bolted together; from this it gradually evolved into a highly sophisticated piece of machinery. As Lawrie explains, 'Imagine this huge structure across the field, how is it going to turn around? In fact, you don't have to turn it around. You merely turn the wheels through 90 degrees, then drive the machine sideways along the edge of the field until the wheel you're sitting over is now in line with the wheel track made by the wheel farthest away from you. In other words, you're running back in your own track with one wheel and starting a new track with the wheel which is farthest from you. The big (driven) wheel, which was leading on the downward trip, is now at the front and the smaller wheel is leading. As the driver's seat swivels, it can be turned to face

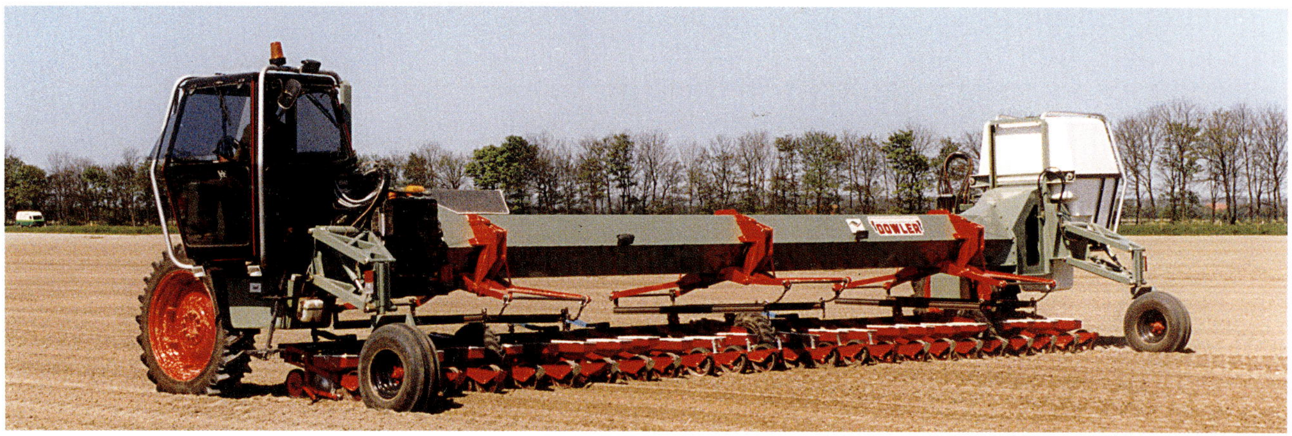

In action farrowing a field (breaking up the surface before planting). Note how the accessories are clamped to the beam.

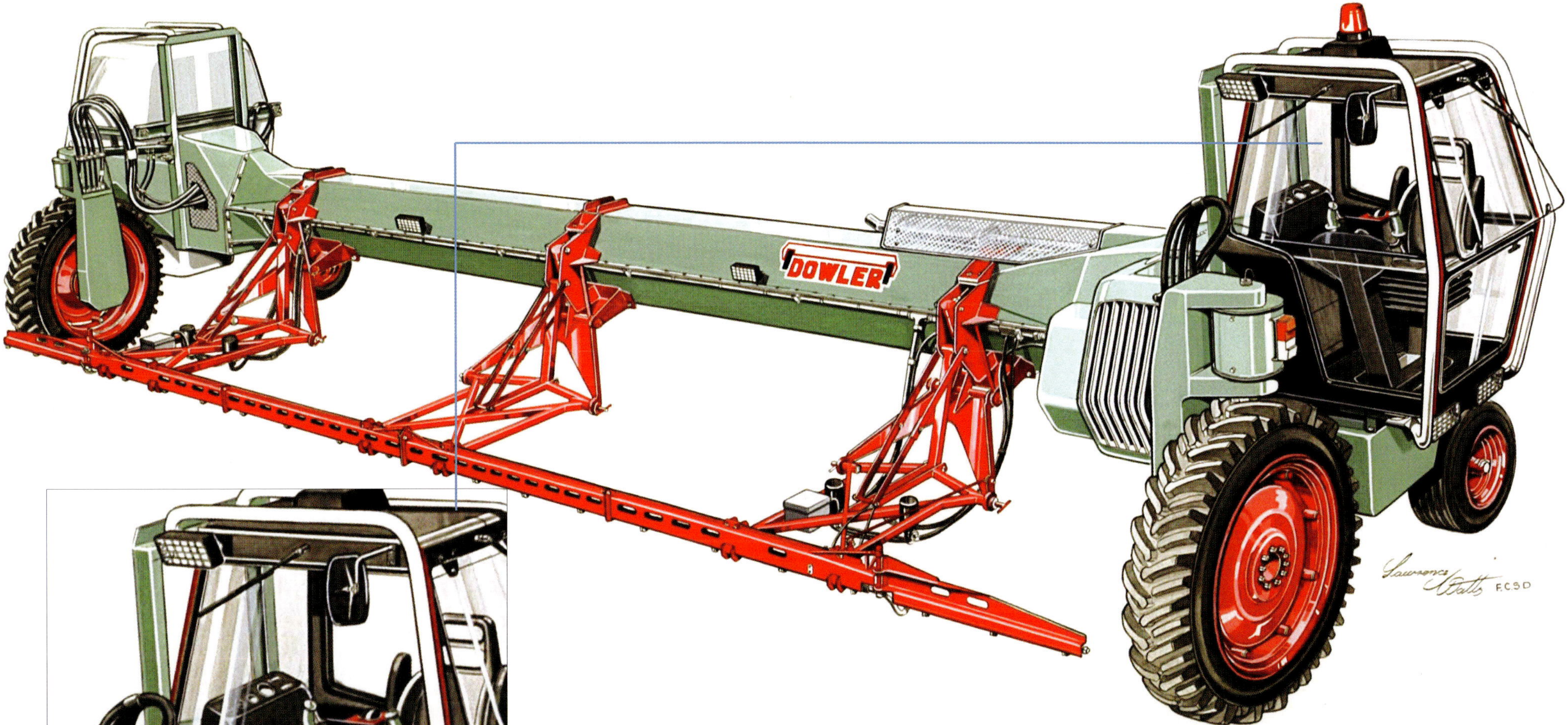

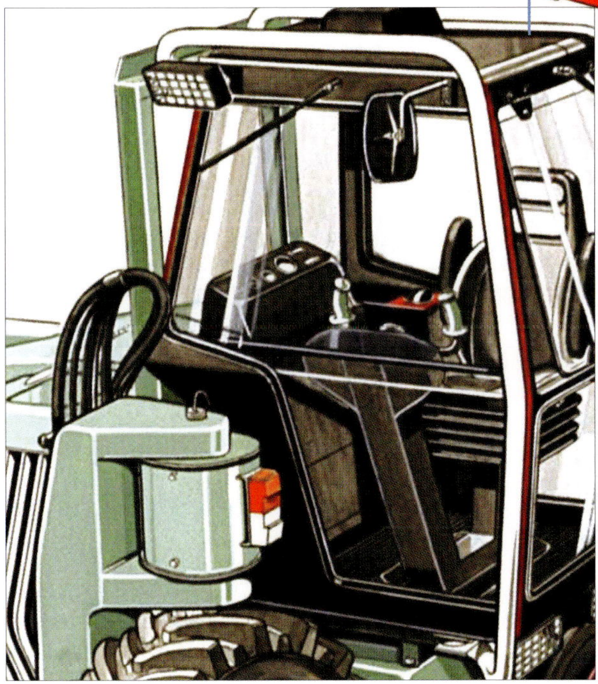

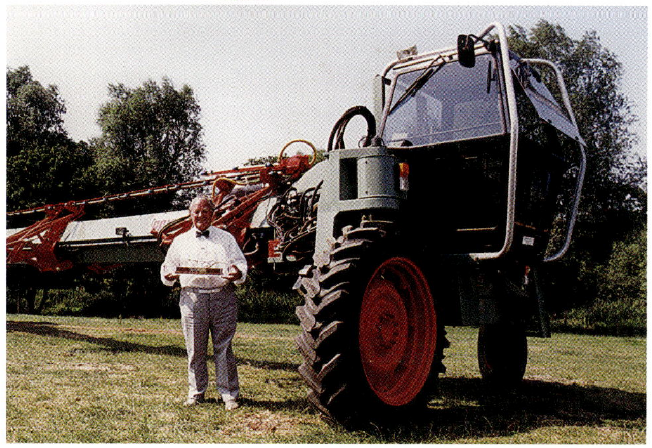

Lawrie with the prize, 'Best in Show' presented by the Queen at the Royal Show at Stoneleigh.

the new direction of travel. As the horizontal beam feeds downwards, it is unaffected by the reversal of direction. So when the machine is in position, the taps are turned on and normal service is resumed. The cumulative effect of this machine's attributes is that working time is reduced by one third, a massive saving in a day's work for a farmer, regardless of the size of field being covered, and the area wasted and compacted by trundling to and fro through growing crops is reduced by at least one third. Another benefit is that the sooner seeds or fertiliser is spread, the sooner growth will appear, and it will be more evenly grown than previously. So it requires little imagination to see the benefits the system provides to farmers with large fields trying to make the most of good weather for sowing or fertilising…'

Although Dowler's prototype machine was made of unsophisticated angle iron, he worked on it so that all the essentials were right, using it successfully for 18 months on his undulating farm in the Cotswolds. At this stage he approached Lawrie and asked his advice about finding a manufacturer, which Lawrie did. About six months later, Dowler invited Lawrie to see the first pre-production machine. Lawrie walked into the barn, took one look at the new machine and asked 'What the hell's that?' As Lawrie recalled, 'David's machine stood about 4 or 5 feet off the deck, a lightweight construction, but the new contraption alongside it was like the Forth Bridge! It was massive. Made without any regard to lightness or manoeuvrability, it was the antithesis of David's design. Incensed, I told him that this production prototype was a load of rubbish. He agreed! David's response was, 'Think you could do any better?' 'Sure!' I replied. Bad move! Before I knew it, I'd talked myself into a job! So I got stuck into a re-design of his original, 12 were made and sent afield as far as Australia and America. One day in the early '80s I turned up at the Royal Show at Stoneleigh. David rushed over and asked if I'd got a coat. 'No, why should I?' I replied. 'You'll need it,' he said, 'because you're going to meet the Queen!'. So I rushed home, got a coat and tore back to the Show, in time for David and me to be presented with the top award in the Show, the Silver Plough, for the 'Most Innovative Machine In The Show'. The major manufacturers were there - Massey Ferguson, John Deere, the lot - and we'd beaten them! I was very pleased because I'd designed every inch of that machine, the special hub-centre steering, the hydraulic motive system, the two-wheel drive (a 4WD came later), the air-conditioned cab, spraying system, everything. A feature of the design which appealed to crop sprayers were 20ft folding arms which, when extended on each side, doubled the overall width to 80ft. You don't have to imagine how much time this saved in the tedious business of crop spraying or irrigation, especially in the enormous fields of North America and Australia. The stuff which was to be sprayed was carried in a large hopper on the opposite end of the

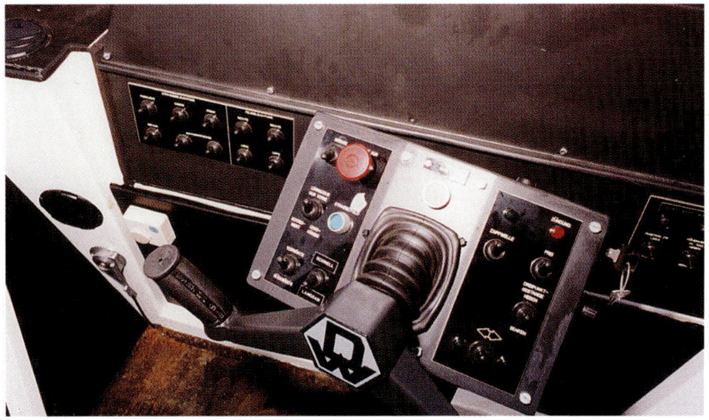

Streering wheel owes its shape to aircraft design, driver's seat rotates and steering column and instrument panel turn with the seat.

machine, acting as a counterbalance to the cab and engine, which was essential to provide equal weight distribution for the hydraulic drive to the big wheels of this 12 ton vehicle. A change of equipment allowed the machine to distribute seeds, with the facility for planting those seeds at pre-determined intervals. After a time we realised that the machine which started life to facilitate irrigation was going to be far more versatile by being used for fertilisation and seeding. Then we met a need for secondary cultivation by designing attachments for harrowing, turning over soil before planting. Stuttgart University Agricultural College asked us to supply a machine with a span reduced from the normal 12 metres (40ft) to 8 metres for strip cropping in their narrow fields. Of course we complied, they still had most of the advantages inherent in the standard-width machine but with the addition of a traversing three-point linkage which was unique at that time. Every tractor has a three-point linkage but I designed one to traverse the length of the beam via a rectangular 'rail'. It was moved by a heavy-duty chain and allowed the attached kit to be positioned precisely wherever the operator wanted.'

I was fascinated by the cab because it looked like no other tractor cab I'd seen, more like a chunk off a luxury car. 'Yes, that cab was air conditioned and had 360 degree vision,' Lawrie confirmed. 'The seat rotated through 180 degrees, with the controls matching the seat. This enabled the operator to manoeuvre the vehicle without the need of a second party on the ground. The W shaped steering 'wheel' owed its shape to aircraft design, with an embossed DW (Dowler Watts) in the centre. With an eight-cylinder diesel engine alongside the cab, it was necessary to insulate the cab against noise.'

One day David Dowler asked Lawrie how he felt about a four-wheel drive version of the machine. The advantages were better grip while at the same time four-wheel steering could be designed into the package. Although he agreed with David's theoretical advantages, Lawrie was unhappy about some of the engineering required. 'With the single wheel steering fitted, no problem, the back wheel casters, following the front wheel faithfully. But with four wheel drive, different kettle of fish, because all four wheels have got to steer as well as drive. Colossal problem! And I wasn't very happy with the leg which carried the steering on the single wheel drive. Anyway, I got stuck into the design sorting out the problems as they occurred but then disaster struck. Without warning David Dowler, a fit active man in his 70s, died. I was bereft. I'd lost a great friend in his prime and, for a time, the bottom fell out of my world, as it did for Mrs Dowler and her family. With David's passing, all existing work and future plans, including the revolutionary 4WD drive Dowler, were shelved. The end of an era for me.'

Although activities ceased following David Dowler's death, it is worth recording Lawrie's progress with the four ▶

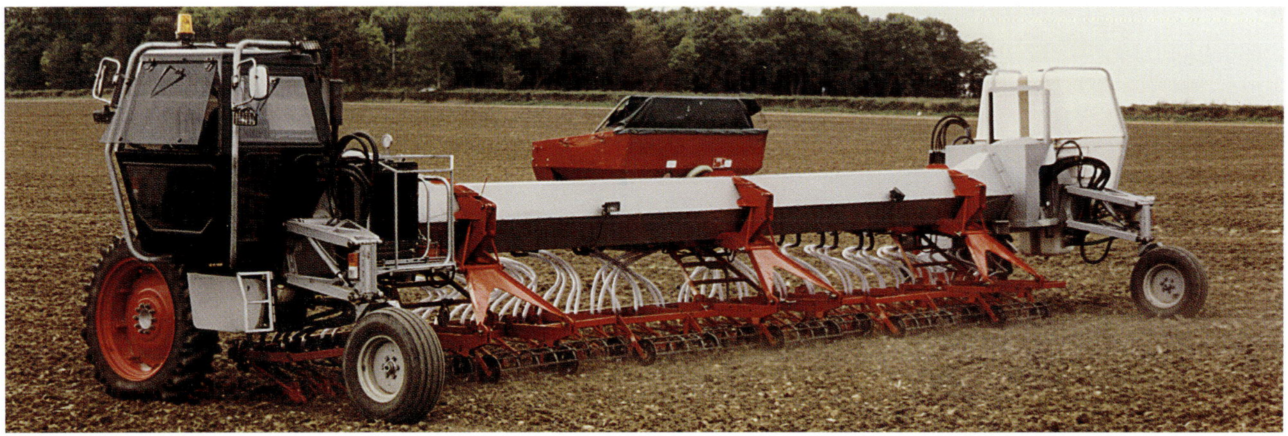

Planting seeds, fed from the central hopper.

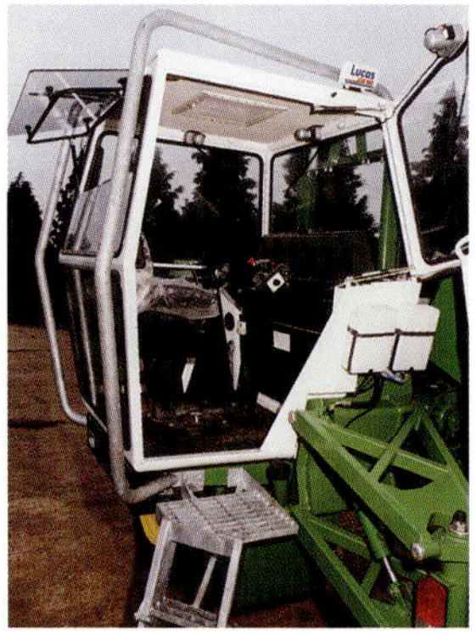

Entry to the air-conditioned cab. Large expanse of glass provides all-round vision.

Drawing of the 4wd prototype which was never made.

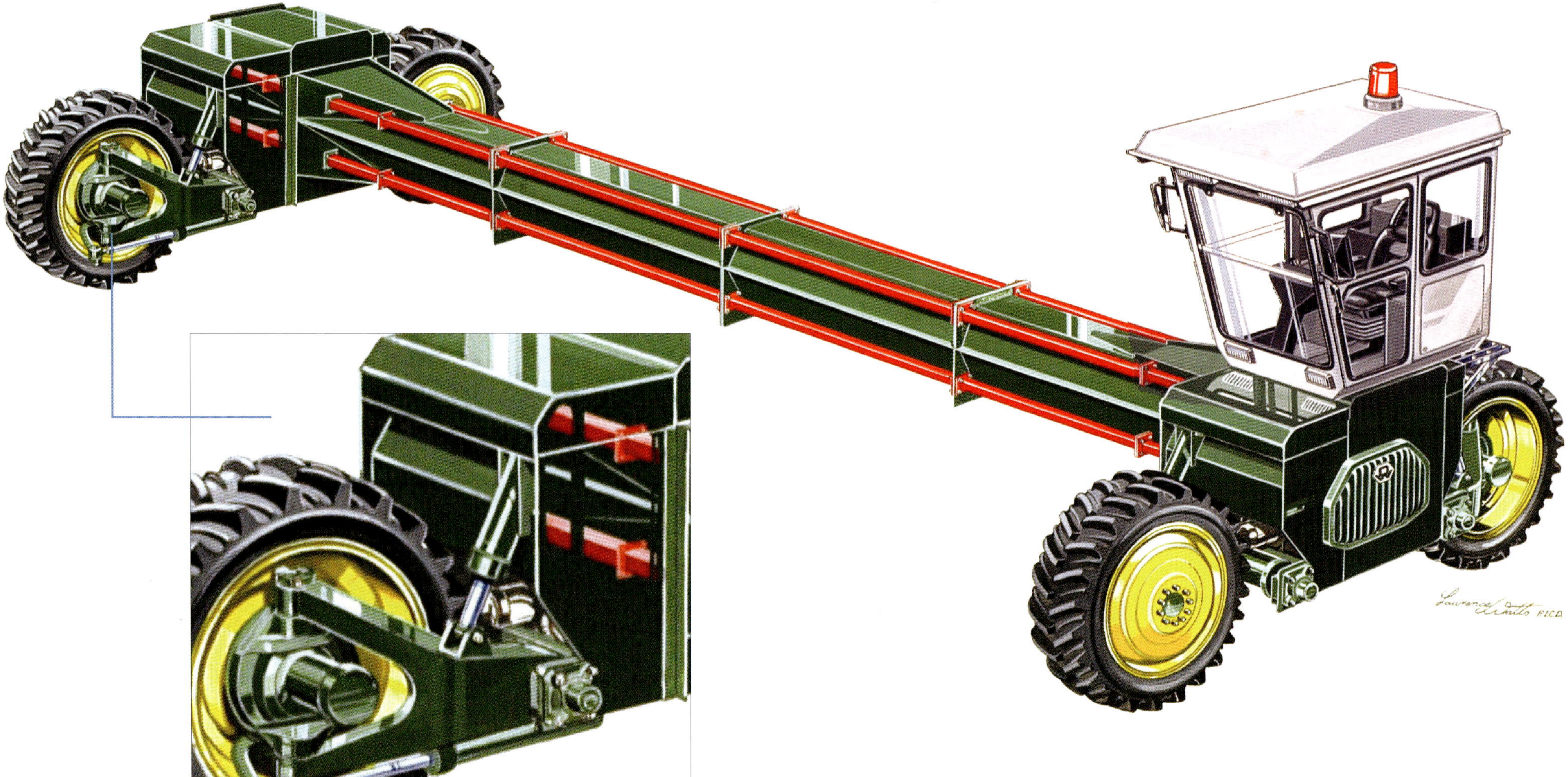

Hub-centre streering and hydraulically driven wheels all round.

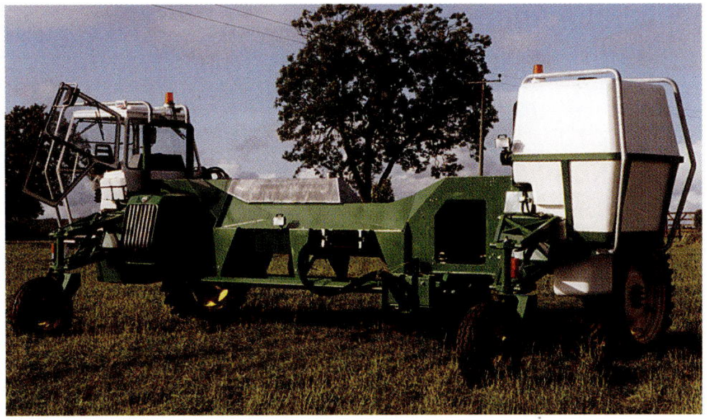

White hopper contains the stuff to be sprayed.

wheel drive model. It incorporated hub-centre steering, a re-designed six-sided central beam and it was this which provoked a heated argument between Lawrie and his old mate Vic Willoughby. Lawrie recalls, 'I was in the middle of some experiments here when Vic arrived one day. Although he'd no interest in farming, nor in the Dowler machines, it did not deter him from having his penny's worth. Straight away he waded in with unasked for comments, along the lines of 'That's no good, you want a tube!' Patiently, I explained that a tube is not inherently stable. It will torque until it collapses, folded on its own. But if you put corners on the beam, it becomes stiff and resists twisting under stress. Vic remained unconvinced. So I made a hexagonal paper beam for him. Out of the same material, I made a similar sized tube. I then invited him to hang something out of the middle of each. The tube collapsed. The hexagonal beam didn't. In spite of this, Vic remained unconvinced, so we agreed to differ. On the Dowler production machine, the 12 metre long beam was constructed of continuous steel plate which had been folded for us by a ship building firm, one of the few used to handling steel plates this big. It was an expensive operation, and the finished product required a big trailer to transport it from the shipbuilder to our factory. So, for ease of manufacturing, the new machine used 8ft standard-sized steel sheets bolted together to form four sections with internal bulkheads for strength. This structure could be handled by an assembly team in the factory; it also had the advantage that removable plates provided access to the interior for installing and servicing pipes and other equipment, which had been a bit of a pain on the earlier machine. Outside the main beam were four separate smaller beams which took the load of whatever implements were attached to the machine. This relieved the main beam of any load carrying stresses. The new construction method made a world of difference; instead of the huge crane needed for lifting the one-piece beam of the original we now had 8ft sections which could be manhandled around the factory. Installing the hydraulic piping and other equipment inside the beam was simplified and it was accessible for servicing or replacement. From an economic perspective, there was a massive saving in construction time and cost because I also took the opportunity to simplify some of the original design and manufacturing methods. For instance, alloy castings replaced the former steel fabrications for the wheel and steering assemblies. By so doing we gained strength, saved weight and drastically saved on cost to such an extent that although the new machine was so much more sophisticated it was a third lighter and not too much more expensive than the machine it replaced. Although all the castings had been done, the prototype had not been completed by the time David died. This was a great pity in more ways than one because in addition to its farming activities I felt that the machine offered a lot to civilian governments, and military, for their specialised requirements. Just before he died, I'd been urging David to consider other applications to exploit the versatility of this new machine. Fitted with flails, it could have been adapted as a minesweeper for military use, especially in the desert. Because the machine could be driven on the road it was also ideal for airport authorities such as clearing ice on runways or, with different fittings, detecting runway debris such as the stuff which allegedly caused the ill-fated Concorde disaster in Paris. In truth, we were on the brink of an exciting new venture with almost unlimited potential! I'm sure that David also realised this before he died…'

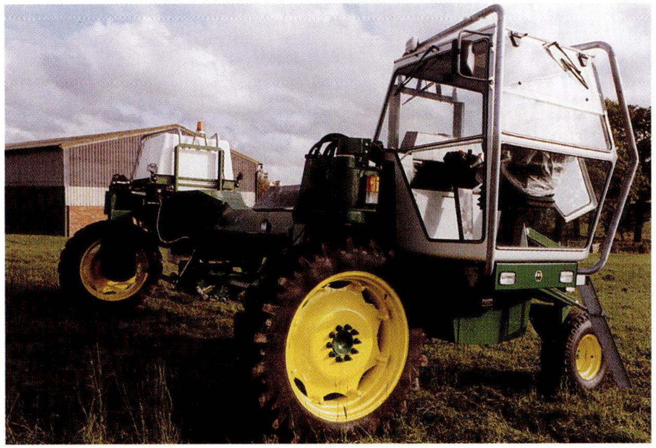

Driver's cab and engine at one end, hopper at the other. Large wheels hydraulically driven, rear caster wheel turns. In this picture, the driver's seat is facing the gantry, preparing to turn.

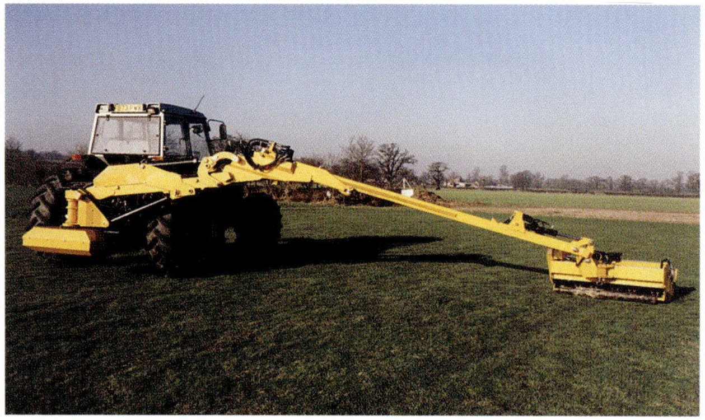

Turner long-reach flail cutter, for embankments, here fully extended to eighteen feet.

Chipper family

Life went on and Tony Turner kept Lawrie busy as he, too, was an innovative character in the Dowler mould. 'He had been in grass mowing machines,' Lawrie explained, 'then he developed a long reach flail mower for inaccessible motorway embankments, or in our Fens or along Dutch canal banks, followed by a family of chipper machines. The main benefit of this new design was safety, it dispensed with the need for tractors and machines on slopes, cutting out rollover accidents which took a toll of lives every year. The main feature of the design was an extendable metal arm, which opened out to no less than 18 feet and which had an hydraulically-driven flail on the end. The device was bolted onto a standard three-point mounting linkage on the tractor. An important operational feature is that when the operator was sitting in his cab, he didn't have to look backwards to see what was happening because the arm operated almost at right angles, and slightly forward, to the cab, thus keeping it within the driver's range of vision. I'd drawn this design from a conversation with Tony, gradually working up the idea in my mind - as I'd done on many other occasions with car and motorcycle designers - until I had a clear mental picture of what Tony wanted. Then I went off, drew it, and presented it to him. Nine times out of ten it was exactly what he wanted. So then we had a working conference which refined the idea from which I made proper working drawings. And that's how this new piece of kit evolved, and similar ones thereafter. Incidentally, although I would have done anything to help Tony expand his range of machinery, I drew the line at what are commonly called 'technical illustrations', such as for catalogues or spare parts lists. This was not my scene, rather not draw than be forced to do them! My job was editorial drawing, visualising how things worked, or would work when made by interpreting the designer's intentions. So I frequently drew things which had not yet been made but of which I already had formed a clear picture in my mind. That's the major difference between ▶

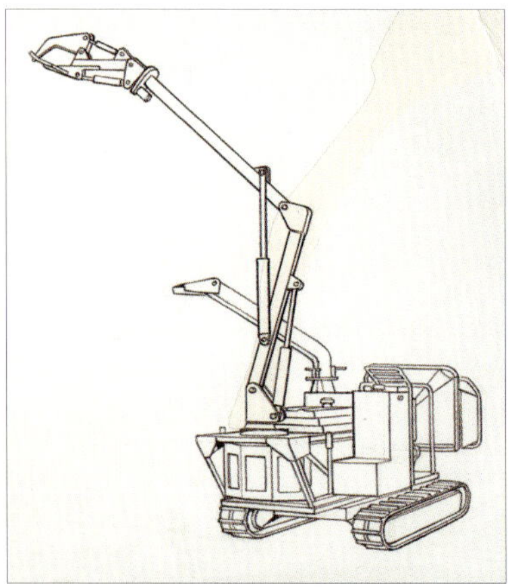

Long reach with jaws attachment for gripping tree branches while lopping - intergral saw cuts simultaneously.

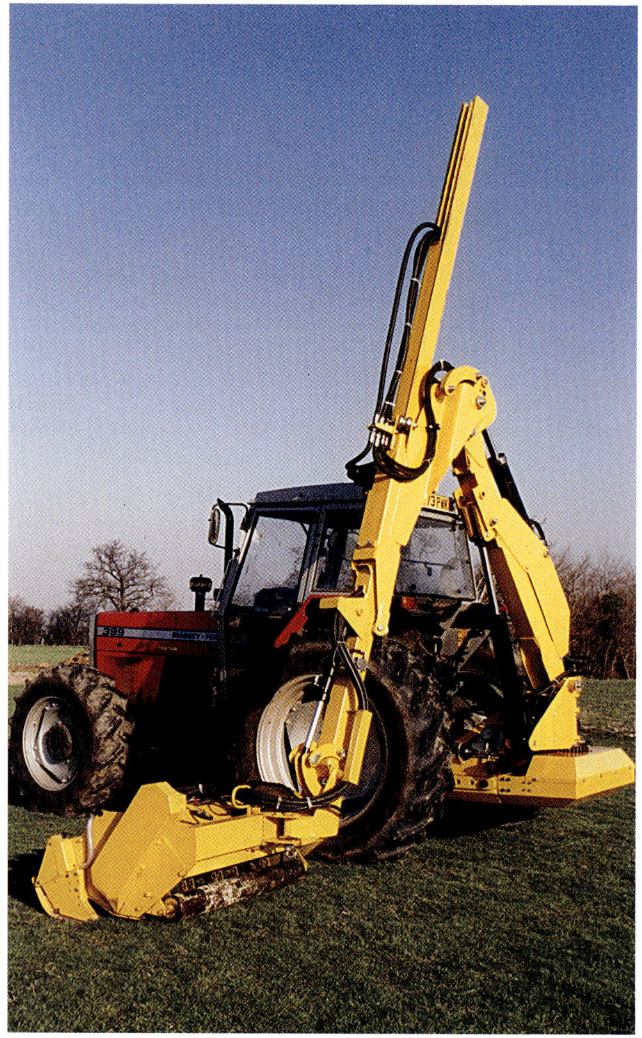

Turner long-reach flail retracted for close-up work.

what I do and what a technical illustrator who draws the object in front of him, does. That's why I so enjoyed working with David Dowler, and continued to work with Tony Turner. They brought out the best in me and I believe that some of my most innovative work has been with them. Prior to that, I'd been working with designers who were usually employed by large firms which had their own agenda, but David and Tony being their own bosses had no such constraints. So they could afford to give me a free rein, sit back and see what materialised. This enabled me to express myself as a designer, then transform the concept into the form which David or Tony wanted, and they in turn went into the factory and transformed it into material form. That really encouraged me, knowing that before I'd even reached home that evening the guys were at work transforming my drawings into metal. Yes, it was one of the happiest times of my life and I'm pleased to say that my efforts benefited both David and Tony. Everybody won!' So if you see a green machine, with GreenMech writ large on the side, mowing your local greens, chipping wood bark from tree felling or pruning branches of roadside trees, you'll know that they are part of Tony Turner's chipper family and almost certainly started life on Lawrie's drawing pad.

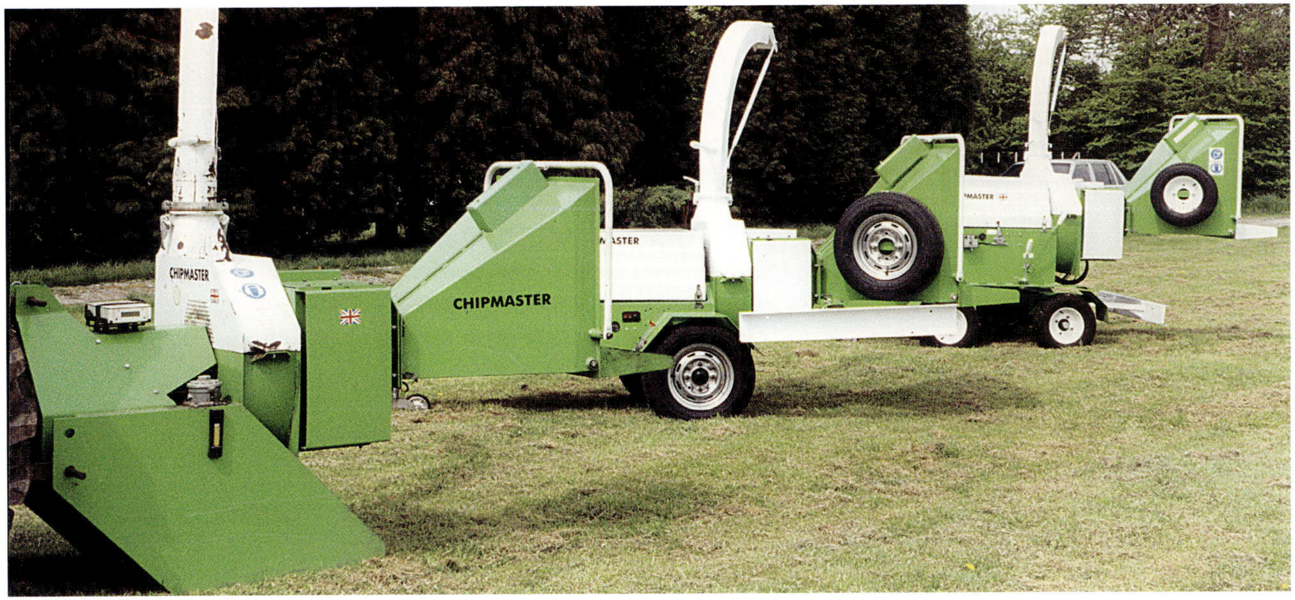

Some of Turner's Chipmaster range.

Once bitten...

After field testing one of Turner's grass cutting machines on a farm in the Cotswolds, Lawrie returned the machine to the farm yard. While passing a parked trailer, he felt a blow on his upper arm. Thinking he'd misjudged the corner of the trailer, he went back and had a look, nothing was sticking out. Mystified, he walked on, did his business and got into his car to drive home. After a while he felt so tired he pulled over and had what turned out to be a deep sleep. He resumed his journey but still felt drowsy. In bed that night, his upper arm felt increasingly uncomfortable and by morning had swollen alarmingly. So he popped off to the doc. It didn't take long to discover that Lawrie was suffering from snake bite! The 'bump' he'd felt as he passed the trailer in the farm yard had been a snake striking him. Fortified by some antidote, he immediately phoned the farm and asked them to search the trailer parked in the yard for a snake. They called back to say they'd found, and killed, a 2ft long black adder. 'The moral is,' says Lawrie,' if you see a trailer parked in a farm yard, give it a wide berth, you never know what may be lying in wait for you....'

Busman's holiday

While on a gliding holiday in Devon, Lawrie took a telephone call from Stephen Howe, Editor of 'Farming Weekly'; would Lawrie please do an urgent job for him just down the road in Cullompton? And send in the completed drawing asap? 'Just what I needed in the middle of my hols,' groaned Lawrie. 'Anyway, I did what he asked and set off to find the farm and see the size of the problem. Blow me, it was ginormous! A series of nine silo-type storage containers ranging from large to enormous were being used for storing and drying grain from farms between Exeter and Taunton; the drying and storage processes were controlled from a giant tower. Such a vast complex; the only hope I had of assessing the layout was from the air, no, not by hanging from a crane but from flying over it and taking photographs. Luckily, the farmer and his two assistants - it took just the three of them to run this giant outfit - agreed with my low flying plan. So I borrowed the club's glider tug, the high-wing aeroplane which towed our gliders up to flying height before releasing them, and flew over the site, taking as many photographs as I could through the door aperture (I'd removed the door to facilitate a clear view for photography). The size of the site was such that it would fill my camera's viewfinder only if I flew at an altitude of 600 or 700 feet. I photographed it from every angle, providing me with a bird's eye view from which I could get a good perspective for drawing it back at base. A ground level visit provided the essential detail information. To show the co-operative nature of the project, and to provide scale, I drew a lorry full of grain being received into the depot; the lorry driver is hanging out of his cab telling the other man in the picture details of the load, in return receiving unloading instructions from the foreman - you know he's a foreman because he's wearing a

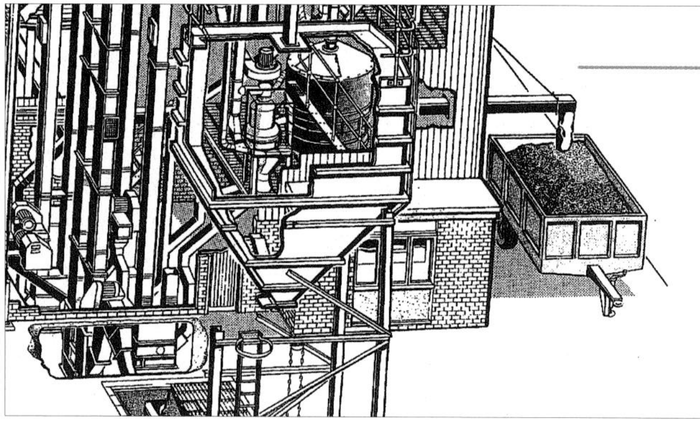

regulation warehouseman's long coat. The lorry is to pull into the unloading bay, a few yards in front, gradually raise the cargo on its hydraulic rams and release the grain into the hole in the ground. You'll note that the grid has been pulled back, ready to receive the grain…'.

This cameo, which represents probably less than 10% of the whole, is a perfect example of Lawrie's style; telling a story through an eye for detail. The raison d'etre for this collection of huge containers is in the back of the lorry, unseen under a protective tarpaulin, it's there because we imagine it to be there. How subtle! So he leaves us to ponder this until we ask ourselves, what next? That's the lead in to the main part of the story. Assuming the lorry driver does as he's told and unloads the grain into the sump hole, machinery will be activated according to which switch the foreman engages in the control panel in the little office (window visible in the right hand corner of the control tower). He will already know how much grain is to be unloaded and his machinery will be

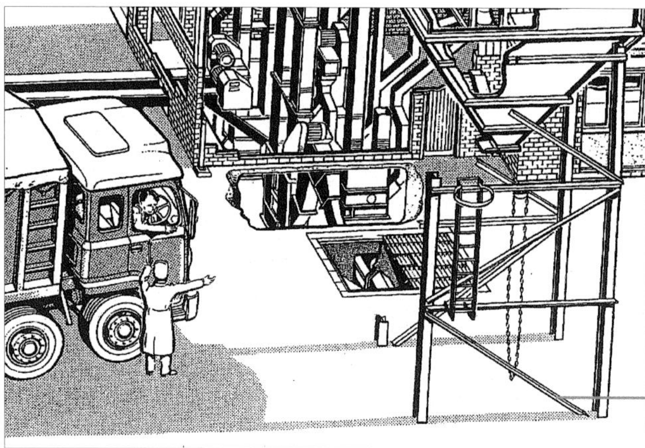

programmed to log this so that each co-op member will receive back the exact amount they booked in, less weight of the moisture/water removed in the drying process. Grain is usually damp, or even wet, when harvested; the optimum moisture for grain which is to be sold is 12 %, so that's what farmers aim at. Grain is sold on its condition, the better the condition, the more money the farmer receives.

Unfortunately, few farmers are sufficiently affluent, or produce enough grain, to have their own drying facilities and fewer still can afford the time to supervise drying and storing at the optimum temperature. Hence the need for this huge drying store. It meets the farmers' requirements and all it needs in return is to be fed lots and lots of grain, so it's ideal for a farmer's co-operative. Plus, it helps farmers gain the true market price. Responsibility for matching the incoming grain to the right silo rests with our friendly foreman, who has already got the information he needs from the driver of the lorry. This info is matched to a silo containing similar grain and moisture content. The incoming load is then directed through a series of pipes and fed into the silo where it will be processed with grain from other members of the co-operative.

You may ask why the silos are in three different sizes? Lawrie explains, 'The three biggest, on the left of my drawing, cater for different types of incoming grain. Because of the high moisture content of new grain, this will take the longest to dry, that's why extra large capacity is needed. When the moisture content gets closer to the ideal 12%, usually by the warm air vents in the floor, that grain will be re-directed via ground floor pipes to a smaller silo alongside. The final conditioning before going to market takes place in the three smallest silos, partly obscured behind the control tower. As the loading bay is hidden by the tower, I hit on the idea of showing just one pipe unloading into a farmer's trailer just around the corner. Next stop, the market! I so enjoyed doing this drawing that I actually thanked Stephen Howe for wrecking my holiday. He thought that I was extracting the mickey!'

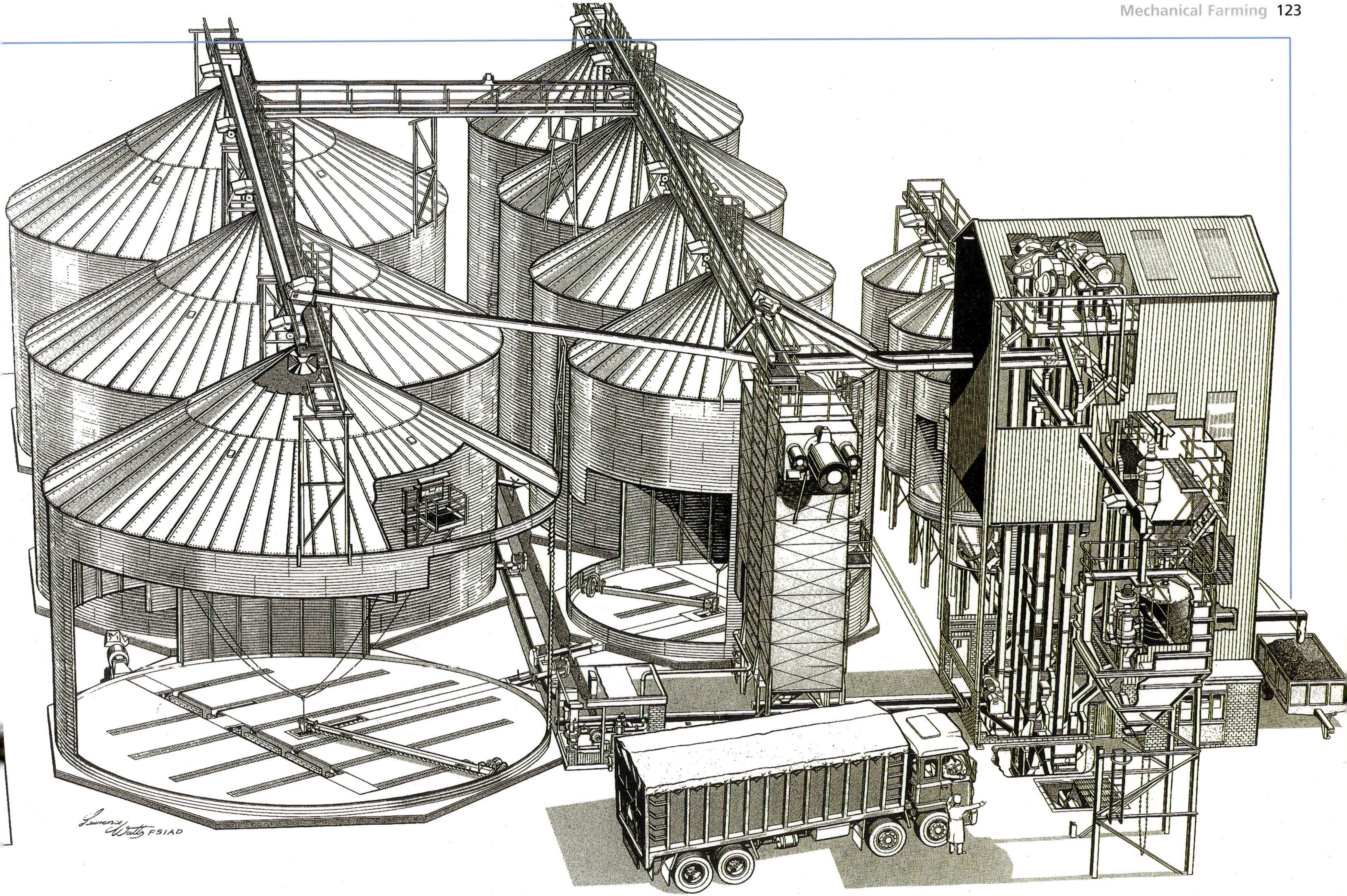

Grain from Devon co-operatives is stored and dried in these silos. Ready for market, grain is loaded into trailers at the side of the building.

Freddy Flail

An exception to Lawrie's not doing stuff for a parts handbook or similar was a booklet he did in conjunction with Tony Turner. Called 'Safer Grasscutting with Freddy Flail', the booklet was intended by Turner for operators of his machines and dispensed safety tips through a cartoon character called Freddy Flail, created by Lawrie. So successful was it that it became the industry's official safety code.

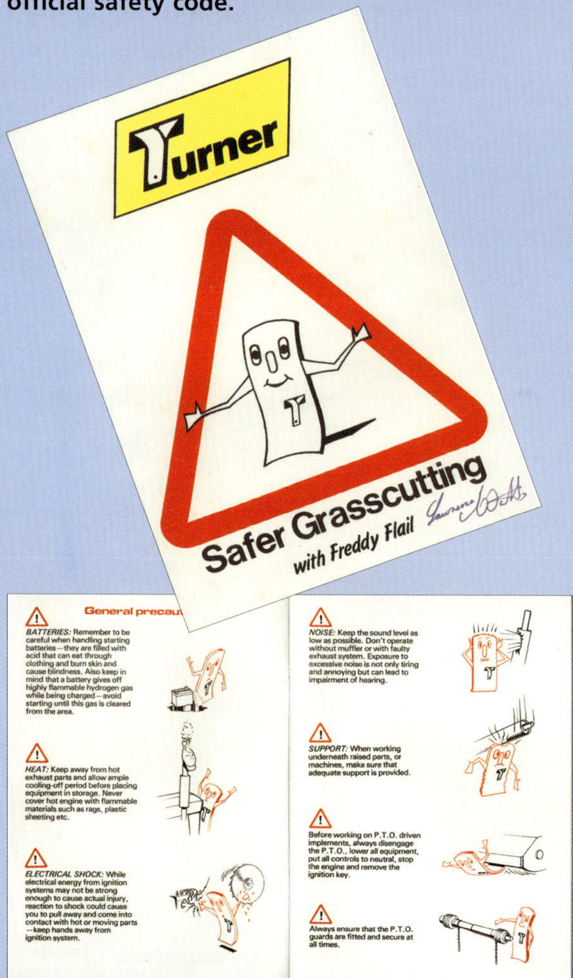

Economy of scale

Back in the early '80s, the buzz word in farming was 'economy', making your farm more economical to run, maximising your returns, minimising staff levels, investing in the future. All of these were incorporated in Felix Farm at Binfield, Berkshire, where Lawrie drew a purpose-built dairy for 'Power Farming' in May 1985. A new design of dairy, first of its type in this country, was to accommodate the merging of two herds of Friesians under one roof to make a 150-cow, one-man unit. 'It succeeded brilliantly in meeting its target,' reported Lawrie, 'and I was impressed with the ingenious design. Fundamental to it was the use of electronics for milking, feeding, and monitoring the condition of the 150 cattle, all under the control of one herdsman.'

To show how the dairy worked, Lawrie resorted to his tried and tested technique of lifting most of the roof off, but to get the right perspective he had to see it from a position on an imaginary ladder about 300 feet above and diagonal to the building. This was the only way he could get everything he wanted to show into his drawing. So much of this was in detail that he used a numbered key to identify everything. Aligned north-south, the 57 metre-long main building had as its main feature 75 cubicles in two banks separated by a 2.2 metre wide passage, with another line of cubicles along the eastern wall. The wide passageway allowed a herdsman to feed the cattle from a forage box, and also provided room for a tractor and scraper for cleaning purposes. Cattle coming into the building at the southern end had to pass a console (3) in the office by the door, where the herdsman entered each cow's identity number into a data logger (4). This held details of each animal, including how much feed it should receive; the food would then be metered by a micro processor to an overhead hopper which fed the cow while it was being milked (11). Only a magnifying glass can do justice to the amazing detail in the milking parlour (1). For instance, in the confusing array of pipes, rails and feeders of the 16 milking points, Lawrie's shown two sets of teats awaiting udders! In the feed concentrate hopper store (11) he's drawn a shovel propped next to the open door, just as the herdsman left it. Down at the other end (17/18) you'll find that the washing up troughs need emptying as does a pail on the floor near the sink. While on the subject of detail, just take a look at the brickwork on the outer walls of the parlour; each brick has been individually drawn, matching other walls in the building which are also to scale. Not just drawn, but in perspective and from a great height as well! 'Yes, it took some time,' sighed Lawrie, 'but it was worth it, as were the vertical lines of the corrugated space boarding (34). The asbestos sheeting (33) and cladding (35) would not be tolerated today, but at that time the health hazards were not fully understood, so I was quite happy drawing them.'

Lawrie is proud of this drawing because it encompasses everything that he's set out to do; telling a story about a building which houses cattle, but with a major difference, making it interesting. First, he set out to grab the onlooker's attention, then he's had to retain it, which is why he's included such an enormous amount of detail. To liven it up, he's added some action, this time through his affinity with cattle. Of the 70 cows visible in the drawing, no two are alike, nor are they just standing there looking bored. Most of them are doing something, whether it's flicking a tail, eating the fodder the herdsman has just distributed, moooing, looking out for a friend (it's been established that cattle do make friends), taking a snooze, having a pee, taking a look outside, drinking, or...I could go on. A farmer would immediately identify the cows as Friesians, because Friesians all have different markings. So that's what 'farmer' Lawrie has drawn...70 different cows! And at the end of the day I wouldn't be surprised if he went in there to wish each of them 'good night old pal....'.

Although the project achieved its object in improving the efficiency and conditions at Felix Farm, six months after it was completed new milk quotas were introduced which reduced by 20% the farm's output. The reward for all that work - bureaucrats deny this marvellous new building the opportunity to show its potential!

MANAGEMENT MATTERS

THE 150-HEAD, ONE-MAN DAIRY UNIT AT FELIX FARM, BINFIELD, BERKSHIRE

continued

"The degree of automation in the parlour also gives the cowman more time to observe the animals, while ensuring that he has to concentrate on each one as it enters, in order to key its number into the computer."

Another advantage is that controlled feeding has enabled the calculation of rations to be more precise, thereby eliminating wastage.

As a further management aid, the office contains a large window through which the herdsman can observe the whole of the interior of the building.

The new dairy unit was funded partly from the sale of an off-lying farm, and from bank borrowings. A fact which causes Mr Trowell to comment: "From the herd management point of view, the building is a dream, but the circumstances surrounding it could have been better."

Those 'circumstances' were the imposition of quotas, after only six months of the new unit coming into operation; and the loss, on appeal, of their secondary allocation.

"On our original quota of 1·25 million litres," explained Mr Trowell, "we would have been looking for a yield of 5,700 litres from the Friesians at Felix Farm and 4,000 litres from the Guernseys.

"Even so, with restrictions on what was fed out of the bag in order to improve margins, we would have had to be on the ball to achieve our target."

But with a final allocation of only one million litres, Mr Trowell is having to re-consider his approach to milk production.

"We thought," he says, "that having invested a considerable amount of capital, on a green field site, we would have been given a workable quota; we are now in the position of having to pay for the new unit out of a reduced income forced upon us.

"The emphasis now is on producing quality milk at the lowest possible cost, not quantity. That means making far better use of our grassland and virtually cutting out concentrates completely."

The current winter feeding regime for the two herds is zero-grazed forage maize followed by silage (3,500 tonnes in five clamps around the estate), brewers' grains and concentrates.

"In order to reduce the concentrate bill further," explains Mr Trowell, "we are considering lengthening the grazing season by increasing the maize area and re-introducing kale. And, although the land is not suitable for cereal production, we may put in 80 ha-or-so of barley, purely to put through the cows. That is unless the market price falls to such an extent that it will pay us to buy it in."

Grassland management is also being re-assessed.

"As we are now aimin cost quality milk," says l ell, "we are going to hav our approach to gra: conservation. The G which were on establisl kept their condition f than the Friesians on nε

"I think we are going t revert to a system of p grazing pastures around with the outlying fields top-quality leys, wh receive extra nitrogen cropped as hard as po hay and silage."

Factors beyond his co caused Mr Trowell to ı evaluate his approach production, but he still the new dairy unit as a l investment.

"There is no quest concludes, "that we are have to manage the cows ciently in the future an we have the environ enable us to do that.

"Had there not l restrictions on what could produce, we wo been home and dry, and that circumstances do us to do the building the deserves."

Drawings by ε artist Lawrence

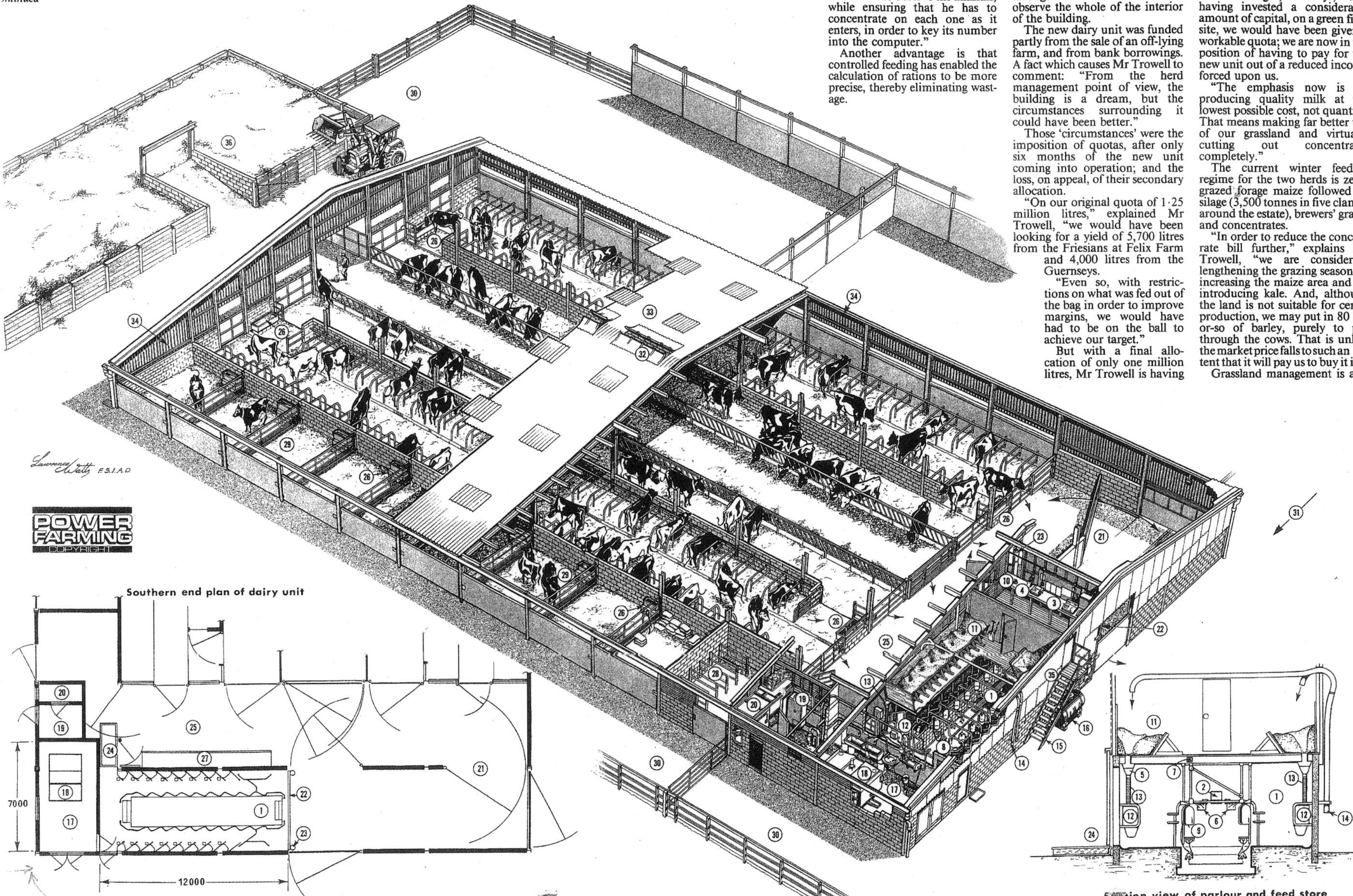

Southern end plan of dairy unit

Section view of parlour and feed store

KEY TO DRAWINGS

1 16-point mid-level Isolir
2 Console at parlour entr
3 Office console
4 Datalogger and print-ou
5 Feed dispenser
6 Parlour automatic contr
7 Overhead wiring
8 Parlour controls
9 Mid-level recorder jars
10 Office with observation
11 Feed concentrates hop
12 Isoline mangers
13 Feed delivery from over
14 Feed intake point
15 Steps to overhead offic
16 Pressurized water supp
17 Milk pump assembly ur
18 3,455-litre bulk tank
19 Power room
20 W/c and shower
21 Collecting yard
22 Guillotine entrance doo
23 Parlour door counte weights
24 Foot-bath exit from par
25 Dispersal yard
26 Water troughs
27 Cattle race
28 AI stalls
29 Isolation and calving cu
30 Closed yards gates to
31 Main entrance
32 Open ridge
33 Asbestos sheeting
34 Space boarding
35 Asbestos cladding
36 Slurry store with loadin

INFORMATION NOT ON YOUR FIRST PRINT

SOME COPY MISSING FROM THIS PRINT.

Nautical

'Queen of the Red Sea'

So what does a successful businessman like Tony Turner do in his leisure time, assuming he has any time for enjoying himself? Well, Tony organises his life so that he has time to pursue other interests, that's essential to keeping him fresh and on top of the day job. How does the idea of the Lady Jenny V, a 96ft motor yacht, grab you? Based in the Red Sea? Why there? Tony Turner's business acumen suggested that his hobby could be turned into a business, and as the Red Sea is a diver's paradise, why not make the Lady Jenny a base for diving and exploring? With 30 cruises a year in the 1980s, the Lady Jenny offered on-board accommodation for passengers who wished to explore and photograph wrecks of all ages from Egypt to the Yemen. And it was available whenever Tony took time off to unwind....

So successful did it become, it earned for itself the title 'Queen of the Red Sea'. A handsome vessel, the Lady Jane was drawn by Lawrie to illustrate a promotional brochure for Lady Jenny Charters Ltd. This switch from agricultural machinery was another challenge for Lawrie but one which he gladly accepted. 'There are many interesting features on that ship and I was able to harness my previous experience and bring out those features by means of cutaways in the hull to illustrate what went on below decks, so to speak. To show how the ship sat in the water, I drew lots of little wavelets, fore and aft, by judicious work with pen and ink. Glance at the drawing and you'll notice the sleeping quarters are in the lowest deck, marked out by a row of portholes, the next deck up is the living quarters and a classroom for lectures and meetings while the upper deck and superstructure is the control area. As on a mini liner, there are guest cabins, but rather fewer; four with twin Pullman berths, one with six berths (no, I don't know what went on there…), one single, and two more twin berths which I've tried to show by means of cutaways in the hull. The stern of the second and upper decks are clear of permanent structures apart from a small crane used for lifting the diving gear and oxygen bottles, so I visualise them as meeting or recreational areas. I first got to see the ship when Tony had it taken to Hamble for outfitting, that's where I drew it. During the 10 years or so which the ship operated in the Red Sea, incidentally it was the biggest one there, the Lady Jenny appeared regularly on television when they used to film undersea. It almost came to an end when it was confiscated by the Ethiopians during their war. Tony hot-footed it down there and, with a friend called Roger Mansell, snatched the Lady Jenny back from under the noses of the Ethiopians. With a couple of seamen, Tony steamed into safer waters. That was typical of him! No time wasted in jaw jawing, just direct action! I also did a large painting of the Lady Jane for promoting the Red Sea trips at a 1970s London Boat Show. About seven feet long, the painting had a rotating antenna mounted on top of the mast which flashed a light towards a mirror on the opposite side of the stand. The mirror was directed towards the sea in the picture which, being covered in crumpled and painted baco foil, caught the reflected light which provided a life-like impression of sunlight on dappled sea.'

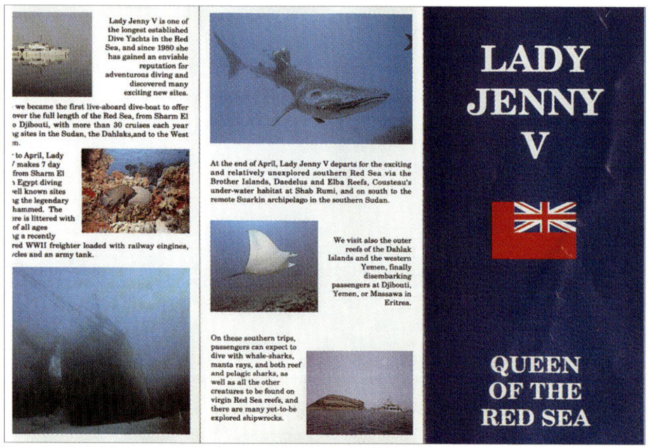
Lady Jenny V is one of the longest established Dive Yachts in the Red Sea, and since 1980 she has gained an enviable reputation for adventurous diving and discovered many exciting new sites.

we became the first live-aboard dive-boat to offer over the full length of the Red Sea, from Sharm El o Djibouti, with more than 30 cruises each year ıg sites in the Sudan, the Dahlaks,and to the West n.

to April, Lady makes 7 day from Sharm El Egypt diving ell known sites ıg the legendary hammed. The re is littered with of all ages ıg a recently red WWII freighter loaded with railway eingines, cles and an army tank.

At the end of April, Lady Jenny V departs for the exciting and relatively unexplored southern Red Sea via the Brother Islands, Daedelus and Elba Reefs, Cousteau's under-water habitat at Shab Rumi, and on south to the remote Suarkin archipelago in the southern Sudan.

We visit also the outer reefs of the Dahlak Islands and the western Yemen, finally disembarking passengers at Djibouti, Yemen, or Massawa in Eritrea.

On these southern trips, passengers can expect to dive with whale-sharks, manta rays, and both reef and pelagic sharks, as well as all the other creatures to be found on virgin Red Sea reefs, and there are many yet-to-be explored shipwrecks.

LADY JENNY V

QUEEN OF THE RED SEA

Seascape

'Unlike previous drawings, I had to create an impression of the ship sitting in the sea, so I needed to use a subtle variation of line to give the impression of wavelets around the ship.
This was achieved by initial gentle pressure on the nib - the start of the wave - then gradually increasing the pressure to widen the nib - for the broadest part of the wave - then easing the pressure for a fine line at the tail of the wave. By keeping them close I created shadow, such as under the stern, and reducing the size as the waves faded into the distance created perspective. A simple technique, but only if you practised it!'

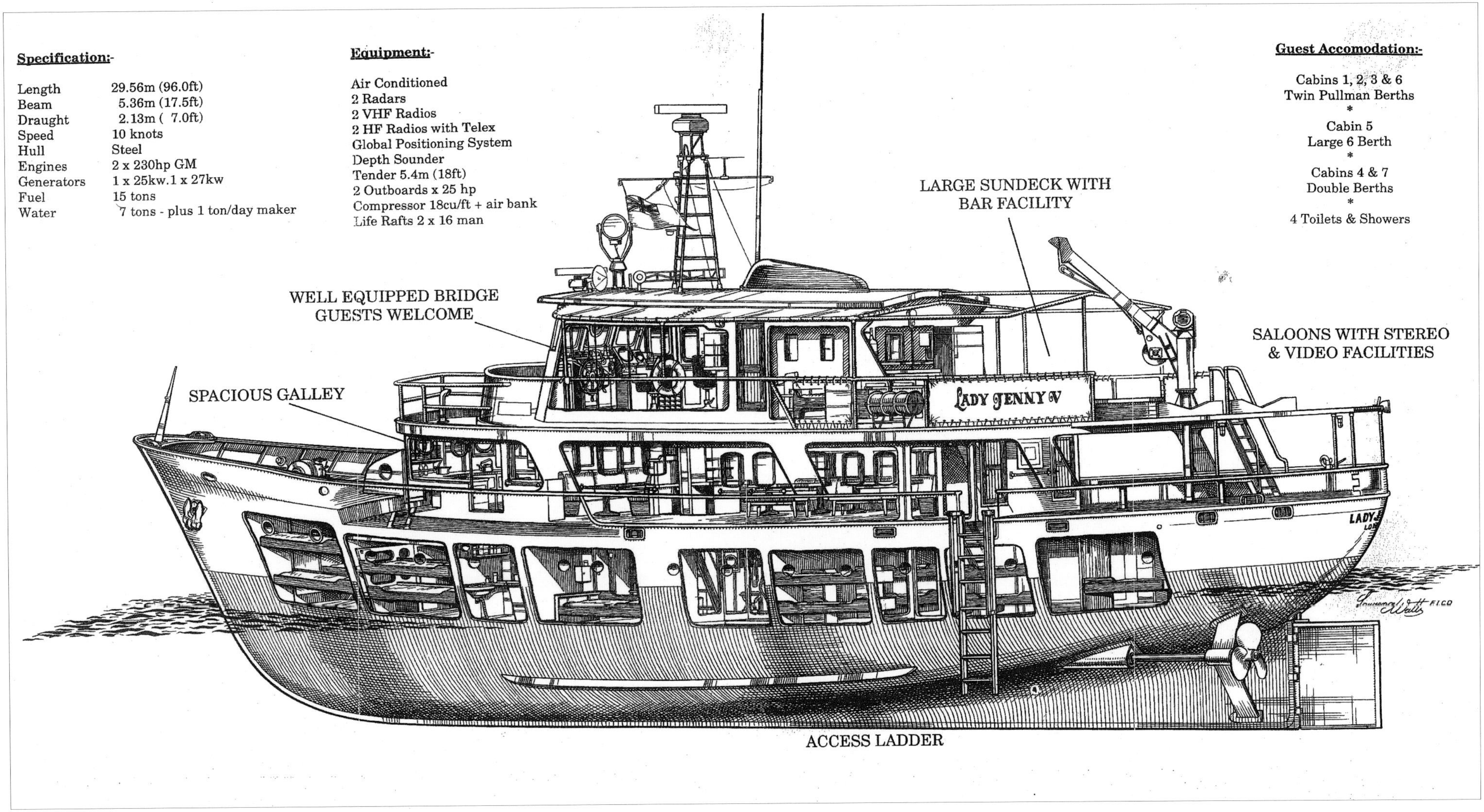

The Lady Jenny was Tony Turner's hobby, but it paid its way by taking parties of divers through the Red Sea for 10 years. Lawrie drew the ship while it was being refitted in Southampton Water at Hamble.

Blue Riband challenger

Lady Jenny V was not the only boat to be drawn by Lawrie, he also drew a potential challenger for the Blue Riband, the award for the fastest crossing of the Atlantic. Sponsored by Newcastle Breweries, Richard Noble's 'Northeast Spirit' looked more like an aquatic paper arrow than a boat, so long and tapered was its profile. Powered by a very large single gas turbine aircraft engine, this provided drive via six shafts connected to 'cauliflower' propellers, so named because that's how they were shaped to reduce drag. Some additional thrust was provided by exhaust gases from the engine. The craft carried a crew of six; three drivers abreast, three seated behind who monitored the radio, instruments and engine, or if not doing that were resting off shift. It was intended that the record attempt would be done in one controlled dash across the Atlantic to New York. A number of hatches on deck provided access to the massive fuel tanks which occupied most of the hull. But the project never came to fruition because, on the quiet, the Aga Khan was also having a go. He got his boat in the water first and claimed the record. So Noble's attempt was abandoned and this drawing is all that remains of a gallant attempt that never happened.

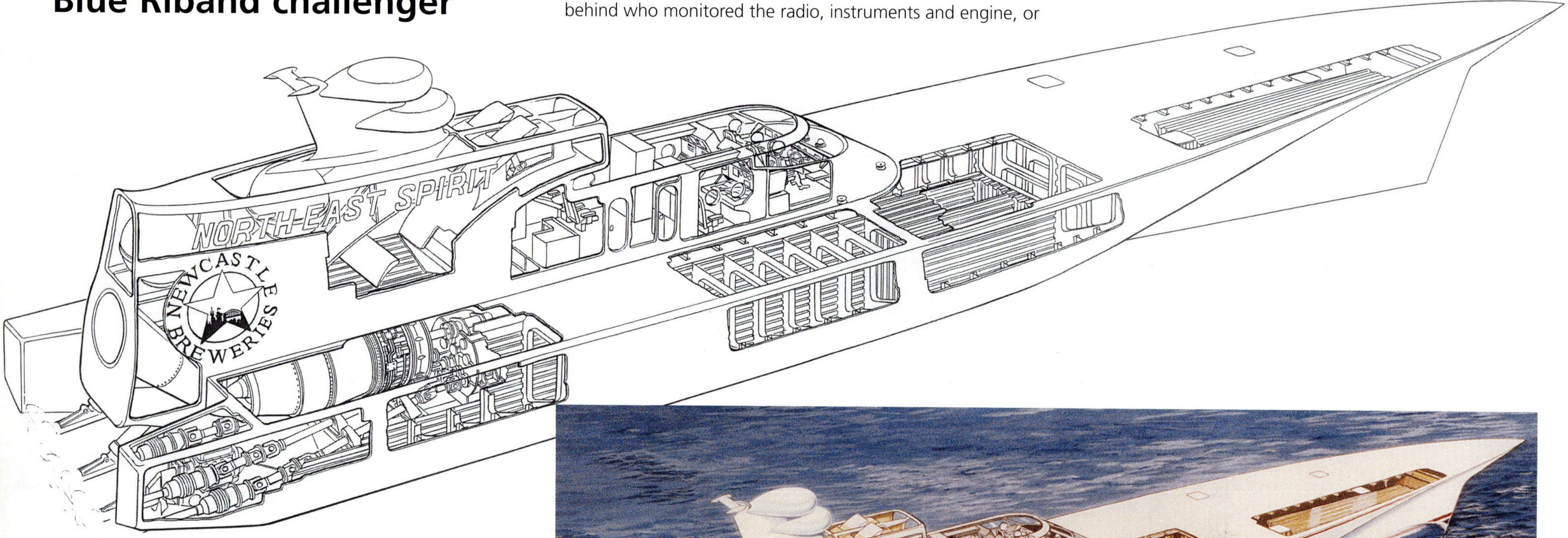

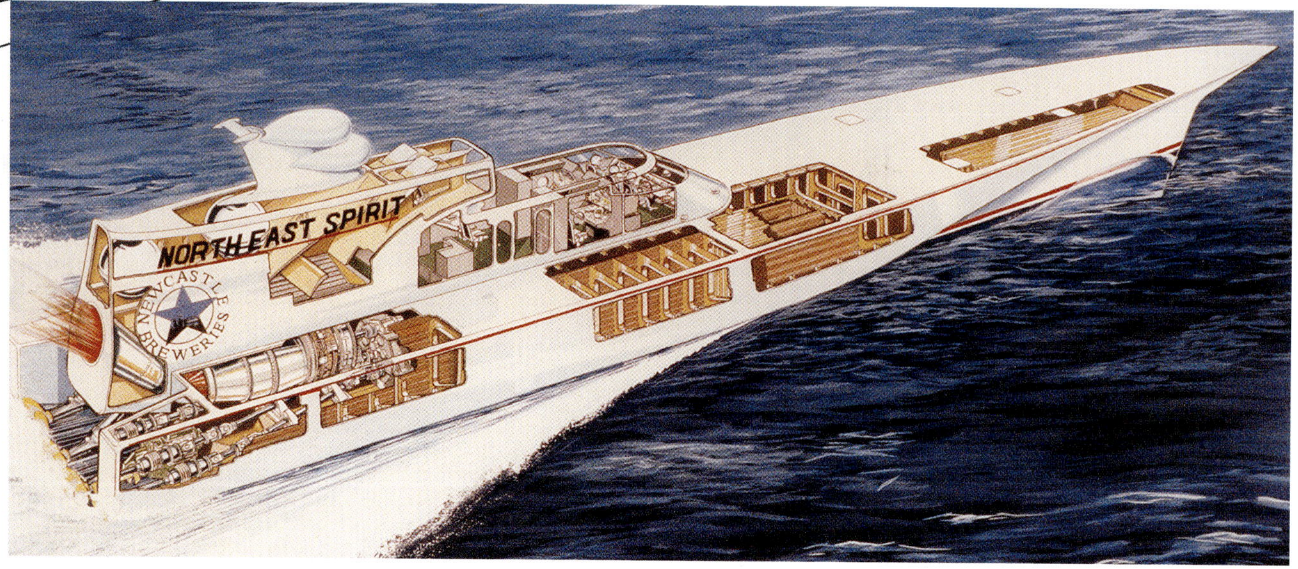

Hovercraft, a British innovation

'Before hovercraft had even been thought of I saw a glimpse of the future when in the factory of Alfred Herbert Tools one day in 1949,' recounted Lawrie. We were discussing a drawing he'd done of a Vickers VA-2 hovercraft which appeared in a BP promotional brochure of 1963. 'For transporting machine tools around the factory floor, Alfred Herbert used a large metal tray with a skirt which trapped the air beneath the tray. Plugged into the factory's compressed air system, this energy source provided a cushion of air beneath the tray when required, such as delivering a newly completed heavy tool to another part of the factory. An operator switched the compressed air to the tray which he then guided to where the load was required. It demonstrated quite well the principle upon which the hovercraft was subsequently founded, utilising a cushion of air to support a moving platform. This stimulated my early interest in hovercraft when they eventually appeared years later.'

In the '50s and early '60s, the British-invented hovercraft was the great hope for the future of rapid transit on land and water. By utilising internal combustion propulsion for forward motion over a cushion of air, and aircraft principles of lift with flight, combined with an ability to travel over land or water, the hovercraft encapsulated many of Lawrie's interests. Recognising this, 'Flight' magazine sent him to Vickers' Swindon factory in 1963 to draw the VA-2 hovercraft. 'I had been paying particular attention to the development since the early days,' continued Lawrie,' in fact, I had to return to Swindon a couple of times because they were building two types of hovercraft, one of which was intended for the military as a sort of landing craft. It seated about 20 persons and although I was extremely interested in it, there was a secondary, personal, interest for me. The man in charge of the electrical development was an old friend of mine, George Turner, who now lived near the factory. I'd known him years before when he was the electronics guru at Armstrong-Whitworth. Apart from his day job on the hovercraft, he was also running a gliding club at Swindon and, as the chief flying instructor, George now offered me the choice of the club's gliders. So, instead of moping in an hotel room in the three evenings I spent in Swindon, I was off gliding with George and the local enthusiasts. An unforeseen bonus and another example of the kinship which followed me through life between those interested in flying, or my other love, motorcycling.

'As the hovercraft was new territory for me, I became so absorbed that I made Vickers promise that when they were testing the VA-2 they would invite me. Good as their word, they contacted my Coventry office and I was on site the next morning when tests started. Shortly after, I was invited aboard to experience the strange feeling of hovering, while watching everything that was not nailed down on the ground being blown away in the slipstream from under the craft. Unlike an aircraft slipstream, which came in a straight line from behind the engine outlets, the air which formed the cushion beneath the hovercraft came from 360 degrees because of the downward thrust from the fore and aft fans. When power was applied for 'take off', the craft was lifted onto the cushion of air generated and surplus air was expelled from behind the curtain upon which the craft rested on the ground. This air, expelled at high velocity, blew away everything which was not firmly anchored.

'With forward thrust applied, we began to move slowly along the runway for the first time. It was unlike any flying experience I'd had before, though you had to have a pilot's licence to fly the hovercraft. That whetted my appetite, and I followed the development of hovercraft thereafter, especially the homebuilt ones. Home for the British Hovercraft Championships was Stanford Hall, scene of Percy Pilcher's flying exploits at the beginning of the 1900s. The various categories of hovercraft competed over courses laid out on the grass, then around or across the lake and back again. One of the most popular engines for these home made craft was the Velocette 200cc two stroke twin, as used in the Viceroy scooter. Being light, compact and horizontally opposed, it was perfect for the job as a propeller could be driven directly from the end of the crankshaft. In fact, it was such an appealing engine that I 'persuaded' Velocette's managing director, Bertie Goodman, to part with two engines which I've still got in my garage. The projects for which they were intended are still uncompleted... but that's another story.'

The versatility of the hovercraft perhaps encouraged the initial ambivalence about the uses to which it would be put. Ferrying people over land and water was an obvious civilian one, but in what sizes? The VA-2 which Lawrie drew had a length of 30ft 4in and beam of 15ft. Eight years later, I took my Renault 16 across the Channel on a scheduled flight from Dover to Calais on a huge twin-engined Hover Lloyd. We drove on board, parked in our allotted space, climbed up into the passenger cabin, strapped ourselves into aircraft-type seats and prepared for take off - or more correctly, lift off. When the engines for hovering (activating the giant fans) started, the noise was similar to being in a 'proper' propeller-driven aircraft. As the pilot engaged the twin propulsion engines, one either side of the passenger cabin, their propellers reassured us that we were about to fly. As Lawrie had found, it was an experience with a difference. ▶

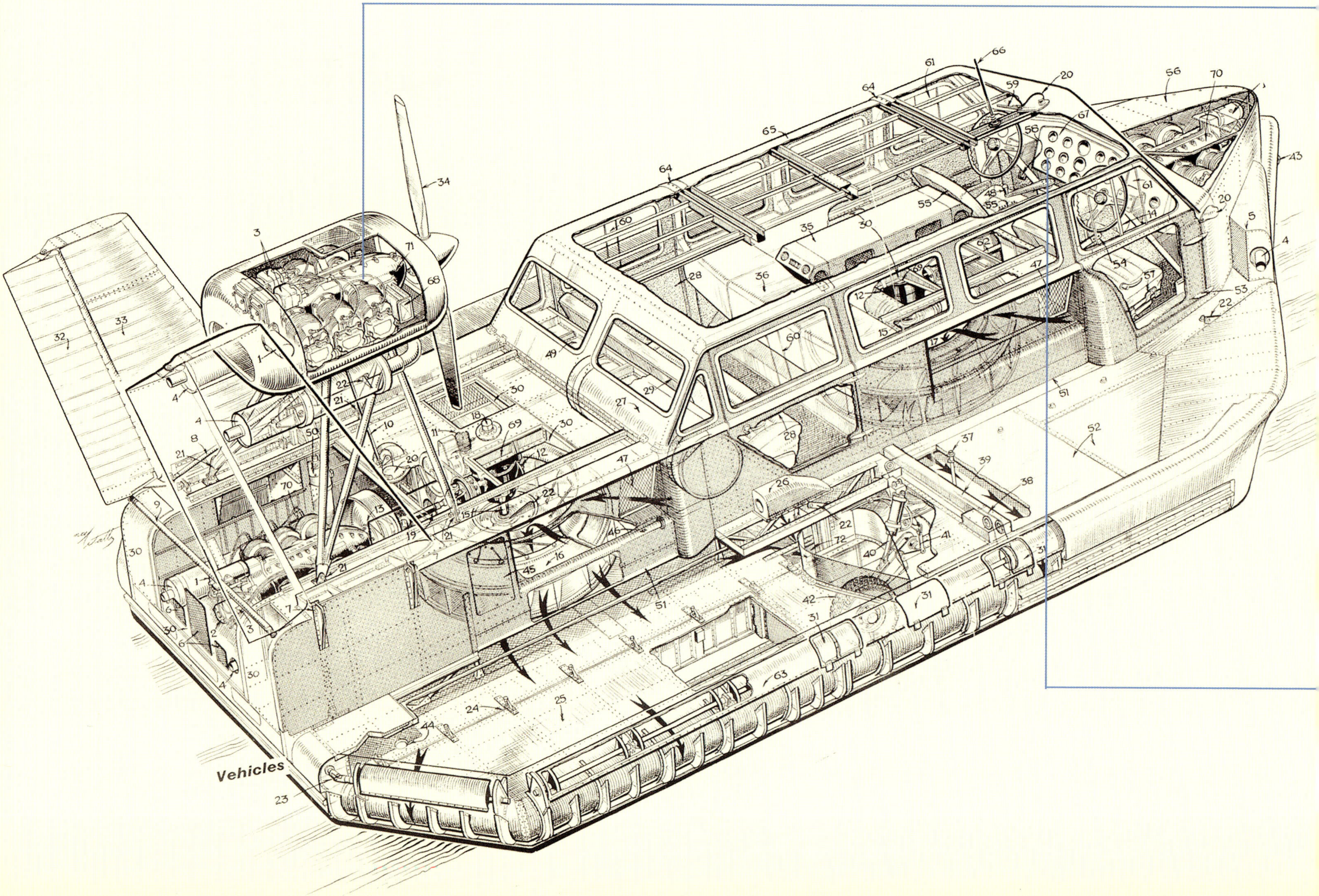
Vehicles

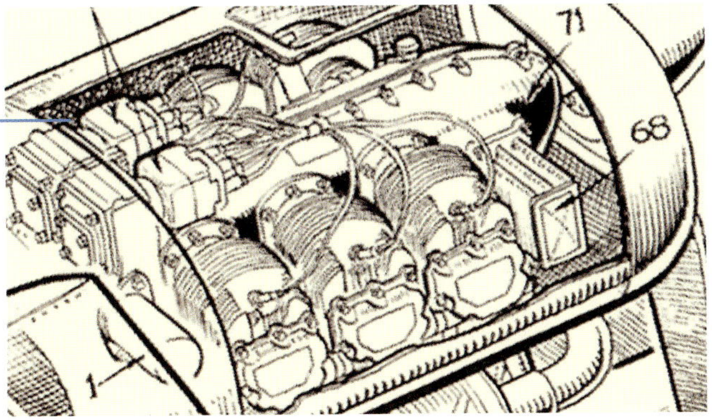

Firstly, turning about our own axis when the pilot lined up the craft to descend down the slope into the sea was novel, especially when accompanied by a high level of engine noise. That increased substantially when he opened the throttles as the craft reached the water and taxied for the gap between Dover's harbour walls. Once through there, the pilot unleashed the twin engines and the craft lurched forward and headed for France. The motorcyclist in me found the mixture of engine noise, some vibration, the occasional 'cruuump' as the craft hit a wave hard, quite intoxicating. Mentally urging on the craft to greater speeds, it literally flew across the wave tops of the relatively calm sea at speeds which I estimated to be around 50+mph. At that time, and judging by my first experience, the hovercraft made the lumbering old cross-channel ferries seem redundant. I'd been firmly convinced that I'd just sampled the way ahead in cross-channel travel so that I was sorry when Calais, and the end of our journey, appeared. We swooped towards the shore at speeds which would have been suicidal in an orthodox ship - assuming it could reach them! - before gradually losing speed and floating up the concrete landing ramp in a whirlwind of sand and dust. The craft then turned around with a perceptible curtsy before gently sinking to the ground as the motors were switched off. It was an experience to be relished, perhaps more by the young than senior citizens....and that may have been a factor which contributed to its ultimate lack of commercial success, allied to its inability to perform in high seas.

As various bodies expressed opinions on the craft's best uses, British, American and Russian military saw tactical uses for the craft around the world and jumped straight in with their requirements, either buying or building under licence. It was the Americans who adopted the hovercraft as a strategic and civilian model and developed the potential which had been promised from the outset. In the UK, the country which had invented it, redundant hovercraft were later stored at the Royal Naval Station at Lee on Solent, opposite Cowes, where they are on view to the general public, a monument to British inventiveness, and political indecision.

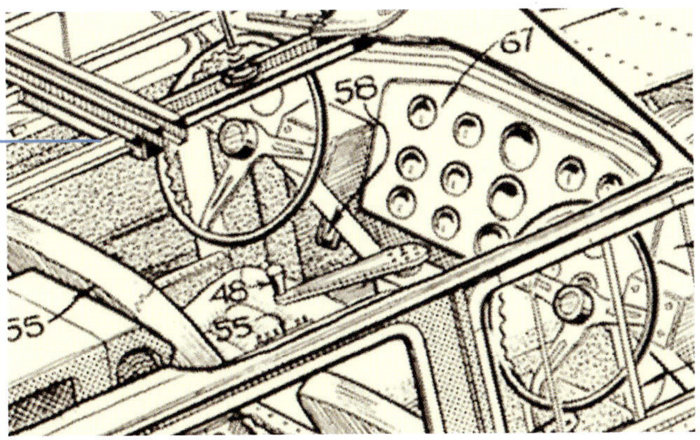

Flymo model hover

Anyone who has operated a Flymo lawnmower will be familiar with the lift from the Flymo's rotary blade. Designed for cutting grass, the blade doubles as a propeller which, as in the hovercraft, generates lift by trapping the air against the internal sides of the Flymo's familiar rectangular body. The trapped air provides a cushion upon which the craft hovers, allowing it to be guided in any direction by long handles. The principle is the same with the hovercraft, the main difference being that instead of the Flymo's single blade a number of blades attached to a central rotor form a fan which creates an exponentially stronger downward thrust as the speed of the fan increases. Two powerful fans, fore and aft, driven by a pair of Rolls-Royce Continental piston engines producing a total of 266shp at 2,450rpm, lifted the 6,000lb VA-2 seven or eight inches above a surface of hard standing or still water. Unlike the manually operated Flymo, the VA-2's forward motion was provided by a single Rolls-Royce Continental engine, mounted on struts above the stern of the craft. That engine developed 230shp at 2,600rpm and drove a 7ft 6in diameter propeller to a cruising speed of 40 kph and a maximum range of 75 miles. Later versions used more powerful engines which provided greater range.

Aeronautical

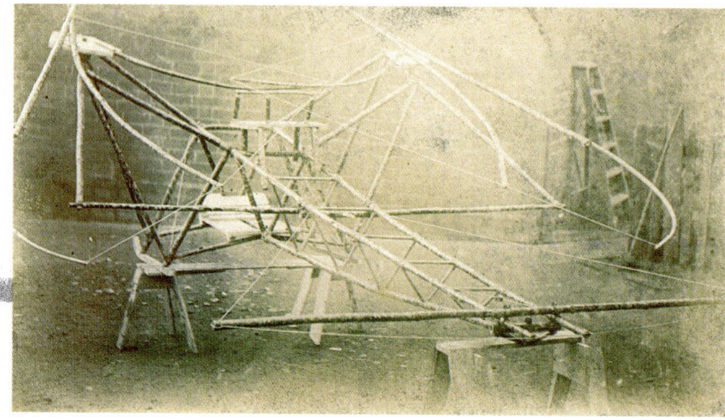

Early days

Within minutes of first meeting Lawrie, after I joined 'The Motor Cycle', he was talking about aircraft. When reminded of this he offered no defence. 'Aircraft have been in my blood since the day I was born,' he explained. 'What I did not learn until much later was how closely connected my grandfather was with aircraft. When I started to build model aircraft, I used to get my dad to hold the bits while I glued them together, always helpful was my dad, then he'd take me up on the common flying. For some reason which I've forgotten, I had this grand idea of building my own flapping wing aeroplane. It was when I was well into the design work that my dad told me about my grandfather's involvement in such aircraft. In fact, my father used to hold the bits while his father assembled them, all three of them in fact. And now my dad was doing the same thing for me! After some success with model aircraft, I became secretary of the Coventry Model Aeroplane Club, which gradually grew to more than 100 members, and we flew our models right through the war period. I built a model Dornier, not quite three feet wingspan, complete with all its German markings. I was so proud of this that I photographed it hanging from a piece of cotton in our back garden, and rushed off with the film to the chemist to be developed. Daft move! The police arrived on my doorstep asking who'd taken the photograph? Where was it taken? When I told them it was only a model, and the photograph was taken in my garden, they were unimpressed saying that they could find no record of a low flying German aircraft in our neighbourhood. It was only when I showed them the model that they handed back my photographs. But I was still in dead trouble! Apparently it was an offence to photograph military aircraft in wartime, especially as my box Brownie photograph was fuzzy, slightly out of focus, making it all the more realistic. Of course, you couldn't see the cotton thread holding up the plane, the only thing which gave an indication of scale was the top of the garden fence.' So I asked Lawrie the obvious question, 'You did that on purpose, didn't you?' which provoked a swift, 'No, I didn't'!' followed by a discreet move on to other matters. 'I remember entering a competition shortly after starting work in Daimler's drawing office but got diphtheria which, the doctors believed, I'd contracted from my habit of sucking the rubberised top of my pencil while I was drawing. They found that the germ was on all the pencils, so the entire supply in the drawing office was confiscated which, needless to say, caused consternation. It was not the Germans who stopped Daimler's drawing office from doing their bit for Britain, it was young Watts! I usually got the blame, this was no exception. Apart from the diphtheria symptoms, I was not too ill but had to stay home awhile as I was now judged to be a carrier of the disease. This gave me the incentive to enter three aeroplanes in each of the five sections in the model exhibition, a total of 15 models. They varied in size from tiny scale models up to full scale flying models, one of which was a Meteor. Obviously, I wanted to

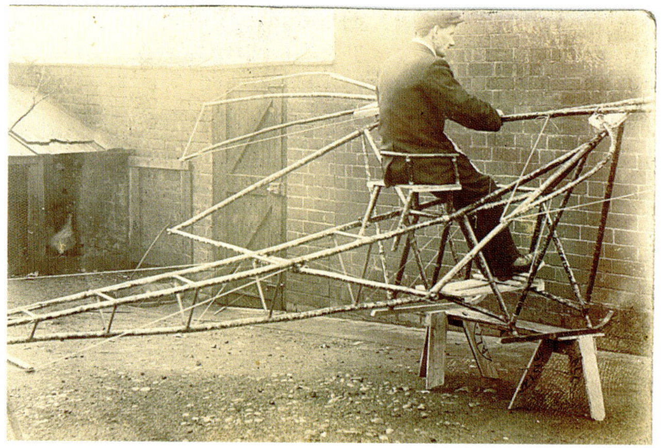

Lawrie's grandfather was one of the early pioneers of flight, building this flying machine in 1911. Seated in the machine he tries the motive power, flapping 'wings' each side (seen here without covering) which were operated by muscle power, rather like oars on a skiff.

Lawrie's original concept of a flying wing glider which he designed, built and flew himself.

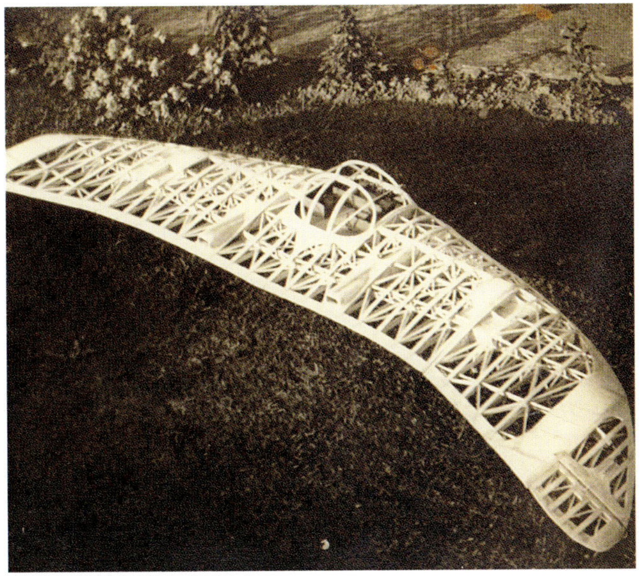

go to the exhibition but was told I couldn't because of my diphtheria. So I hatched a plot with my dad whereby I'd hide in the boot of his Opel Olympia and when he reached the exhibition, which was held in a large field, he'd pass me the word and I'd clamber out and discreetly wander into the marquee where all the models were displayed. To say I was flabbergasted, to the point of shame, would be the understatement of the year because I found that my models had won first, second and third in all five categories! For good measure, I'd also won a special prize for the Best Model in the Exhibition!! It was unbelievable....

'Encouraged by my success, I entered the Handley Page Flying Wing competition for up and coming designers. This was restricted to gliders in free flight and as there was no radio control then, the model had to have pre-set controls according to the speed of the aircraft operated by little drag flaps which would regulate the ailerons. The aim was to fly within a prescribed area, usually a big circle, which had to be ▶

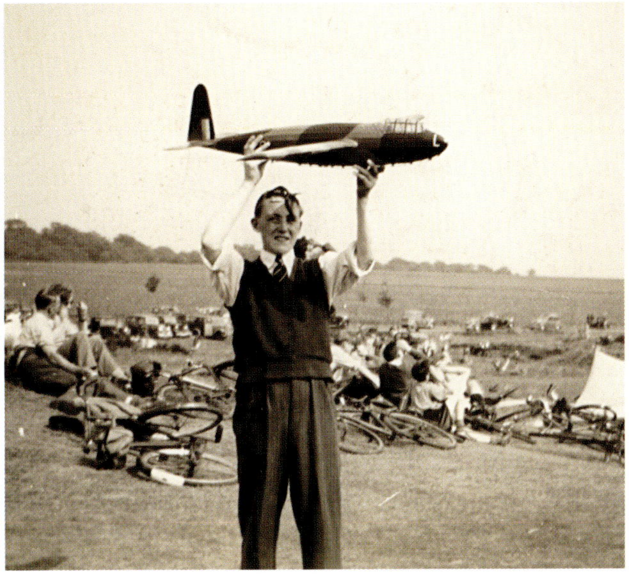

Teenage Watts with one of his models. All were original designs, first drawn to scale then built, usually in balsa wood. No kits in those days!

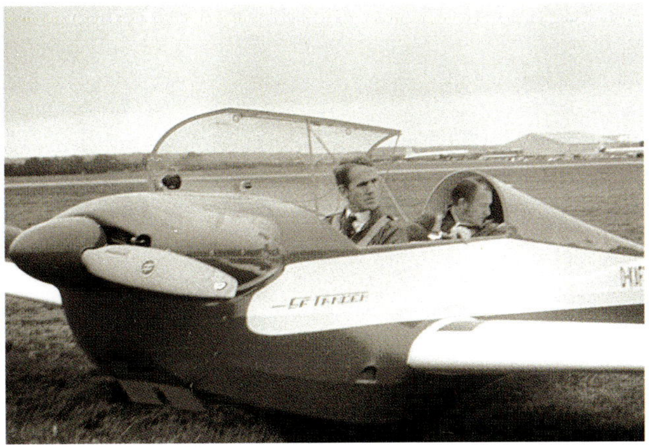

The day Lawrie received his licence to fly motor gliders, at Lasham Airfield.

completed a number of times before landing. I was interested in flying wings and had already done a crescent-shaped flying wing with drooping wing tips, the technology for which I'd been working on at that time for my planned flapping wing machine. The wing tips formed the outer supports for the wheels and a nose wheel was also fitted, as on a normal glider. I entered, and won, the Midland area competition which qualified me to go to the national competition at Radlett to Sir Frederic Handley Page's factory for the flying competition. I made a box for the model to take it by train, as it was too big to go in my dad's car. We got to London OK, then found a Tube line to Radlett, but in the station, we were told my box was too big to take on a train! Dismayed, we retraced our steps. While we tried to fathom a way to get to Radlett outside the station, two chaps approached us, 'What's up?' When we told them the problem with our too-big box, what was inside, and where we were going, they said without hesitation, 'We'll take you.' Strangers, just like that! They dropped us off at Radlett airfield, we expressed our heartfelt gratitude and exchanged our goodbyes. Then Dad and I looked at one another and chorused 'Fancy that!' Our two benefactors were Flanagan and Allan, the best known comedy act of the day! To cap it all, we were just in time to fly my wing in the national competition. Better still, we won! With my mum and dad, we went down to the Waldorf Hotel in London a month later to collect my £60 prize. That was big money then. I'd already won £15 at the area competition, so for once in my embryo career, I was ahead of the game. Shortly afterwards, I received a personal letter from Sir Frederic inviting me to join their drawing office team at Handley-Page. After long and hard consideration about this unexpected offer, I turned it down in favour of an earlier option at Armstrong-Whitworth at Baginton. What decided me? They were already developing a flying wing! Thus early in my career fate had been on my side, setting a pattern which was to follow me throughout my working life, in the right place at the right time - from my first job, aged 17 years, working on aircraft construction at Armstrong Siddeley cars, then to Daimler Cars, loaned to Rover Cars for the new Land Rover project. Now, aged about 17 years, fate decreed I should work on aircraft again at Armstrong Whitworth, in a job which could not have suited me better. As a bonus, I'd been taught to fly in a Tiger Moth by the chief test pilot, Eric Franklin, and he sent me on my first flight. I couldn't have been happier!'

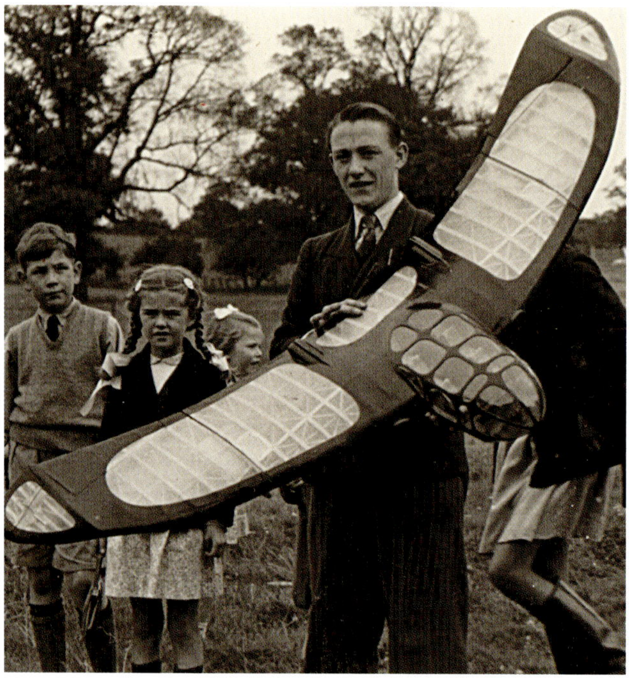

In its finished format, the flying wing glider won the Handley Page competition and Lawrie was invited to join the drawing office at Handley Page.

Upon the Flying Field One Day
by Lawrence Watts

Upon the flying field one day
An aero-mod young, grim and gay
Though slightly cracked upon the crust
Decided to fly his model through the sky, tail fust

He quickly altered rig and trim
Which wasn't very wise of him,
And then he wound up to the bung
And for a streak into the blue he flung, tail fust

He watched his model scrape the cloud
He quickly stumbled through the crowd
Alas to see his model 'Stuker' into the drink, tail fust

He spotted where his model had hit
And quickly dived in after it
But mud gripped him in glue like bond
He really should have looked beyond

This sorry tale, I now relate
His feet still stick up there in state
And may thermals bear him up aloft,
Tail fust.

Eric Franklin

'I'm still indebted to Eric Franklin for the start he gave me as a budding pilot. Remember, I was only 16 years old at the time which, even today, is mighty young to be going flying. So I hung on his every word. One of his aphorisms, which came over the intercom as I was close to flying solo for the first time, and which I remember to this day, was 'Do you want to fly like a bird? Then start to think like one!' Such good advice that I've passed it on to hundreds of embryo pilots I instructed during my 61 years of flying"

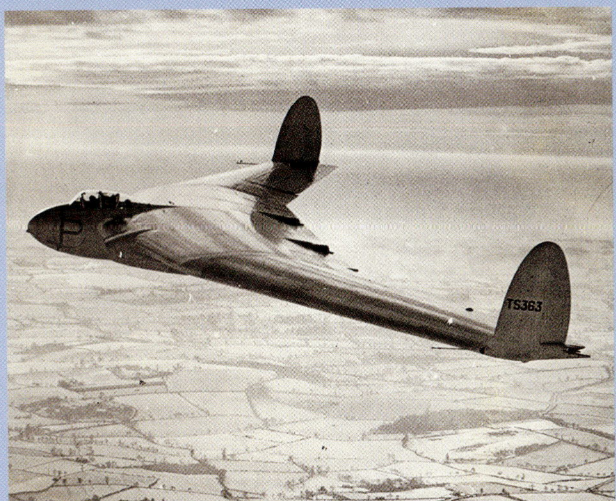

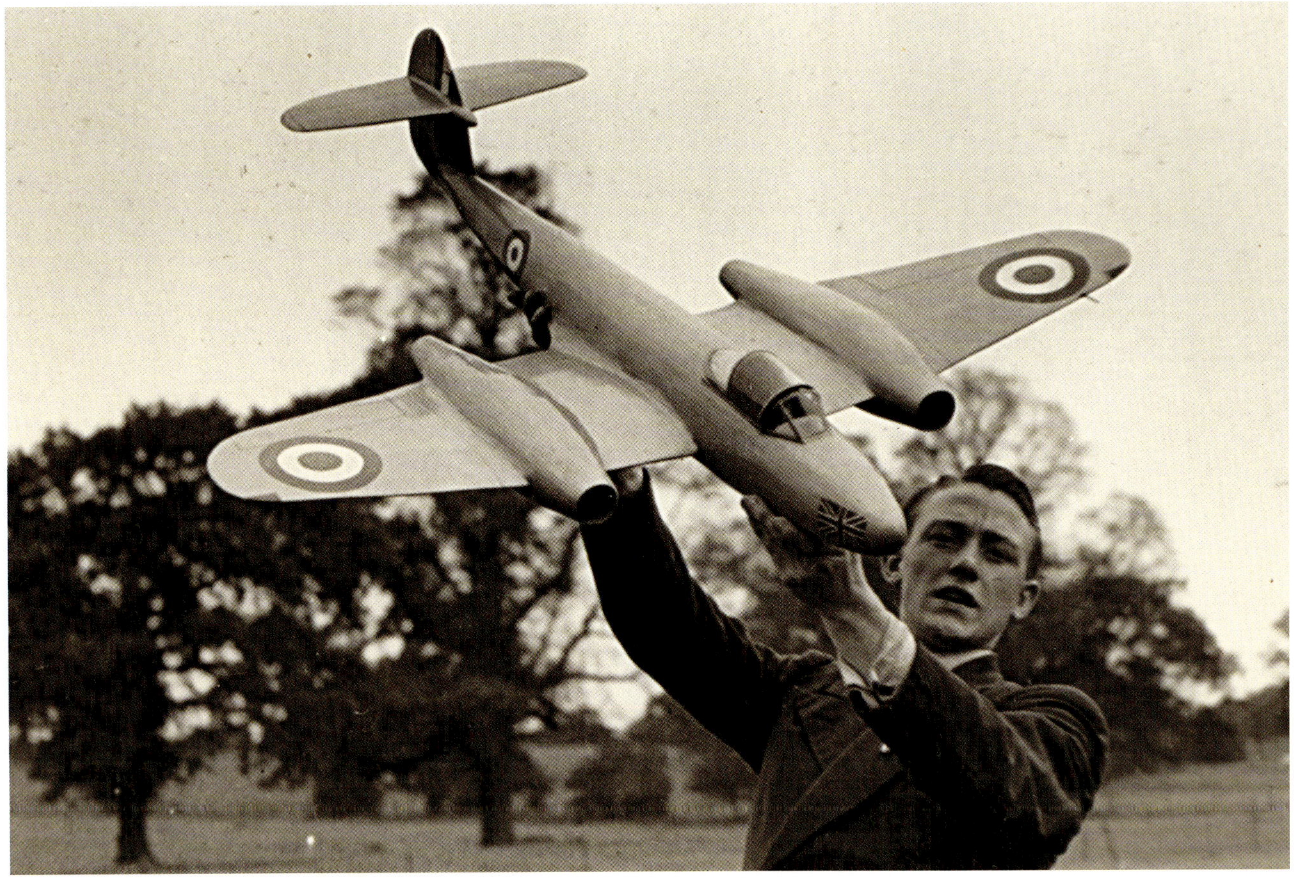

This stunning model of a Metor jet was tricky to fly - and to start. Once, Lawrie was a shade too liberal with the startup mixture and the engine ignited with such alacrity it torched his father's garden shed!

Chief test pilot of Armstrong Whitworth aircraft, Eric Franklin, taught 16-year-old Lawrie Watts to fly. Franklin is seen here in a twin-jet AW52, only two of which were built. The second crashed, causing pilot Joe Lancaster to become the first man to use a Martin-Baker ejector seat.

Apollo

Designed and built by Armstrong Whitworth, the Apollo aircraft was drawn by young Mr Watts, the first commercial sectioned drawing done by him and an obvious milestone in his burgeoning career. Lawrie told me that his copy of the drawing is believed to be the only surviving one, the original was used as the frontispiece on the cover of the Apollo's service handbook. 'At the time I was working in Armstrong Whitworth's drawing office and because I had introduced from Rover my technique of doing perspective drawings of the current project, colleagues who worked on that section of the Apollo could see from the sectioned drawing how everything fitted. That's when the managing director, Mr Woodhams, decided that after I'd helped in the design of the Apollo's tail plane, and done a sectioned drawing of it showing how everything worked, I'd be more valuable to the company if I were to establish and run a new department staffed by six artists. They would produce technical illustrations for the aircraft handbooks for use by the service and maintenance engineers. That all happened in March 1949, a crucial time for me because it was when I did my first 'serious' sectioned drawing and it was a highly significant one, though I was not to know it at the time. Shortly afterwards, Max Miller appeared at the factory to draw the Apollo for 'Flight' magazine. He was given my drawing by Mr Woodhams to 'help' him draw the Apollo, as I mentioned earlier. Well, it must have helped him better than I hoped because a week later Mr Woodhams had a letter from 'Flight' asking if they could interview me for a job with them! That's how, on the 9th of August 1949, just 12 days short of my 22nd birthday, I was employed by Max's firm, Associated Iliffe Press, as an editorial artist. In the right place at the right time and now, right in the job I'd set my heart on. Fate had dealt me a great hand. My hobby, aircraft, had led me to Armstrong Whitworth, then to 'Flight' magazine and a whole new world. I never looked back...'

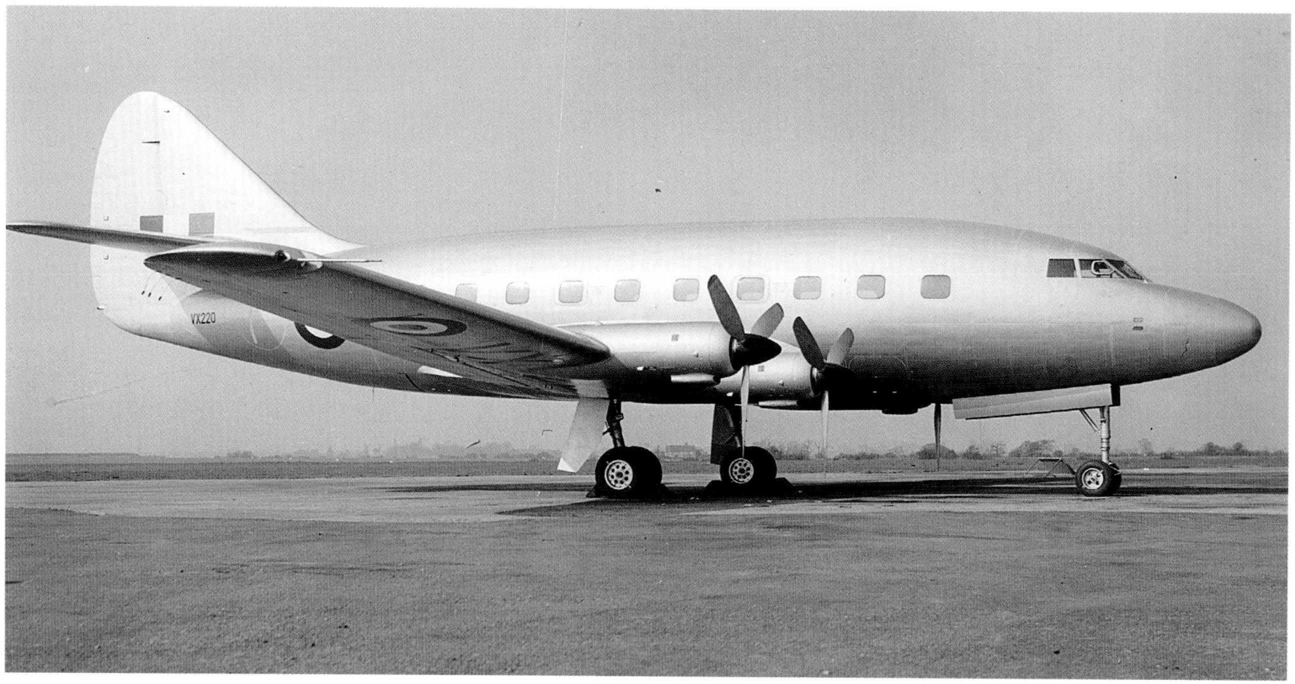

16/46	APOLLO	FRONTISPIECE

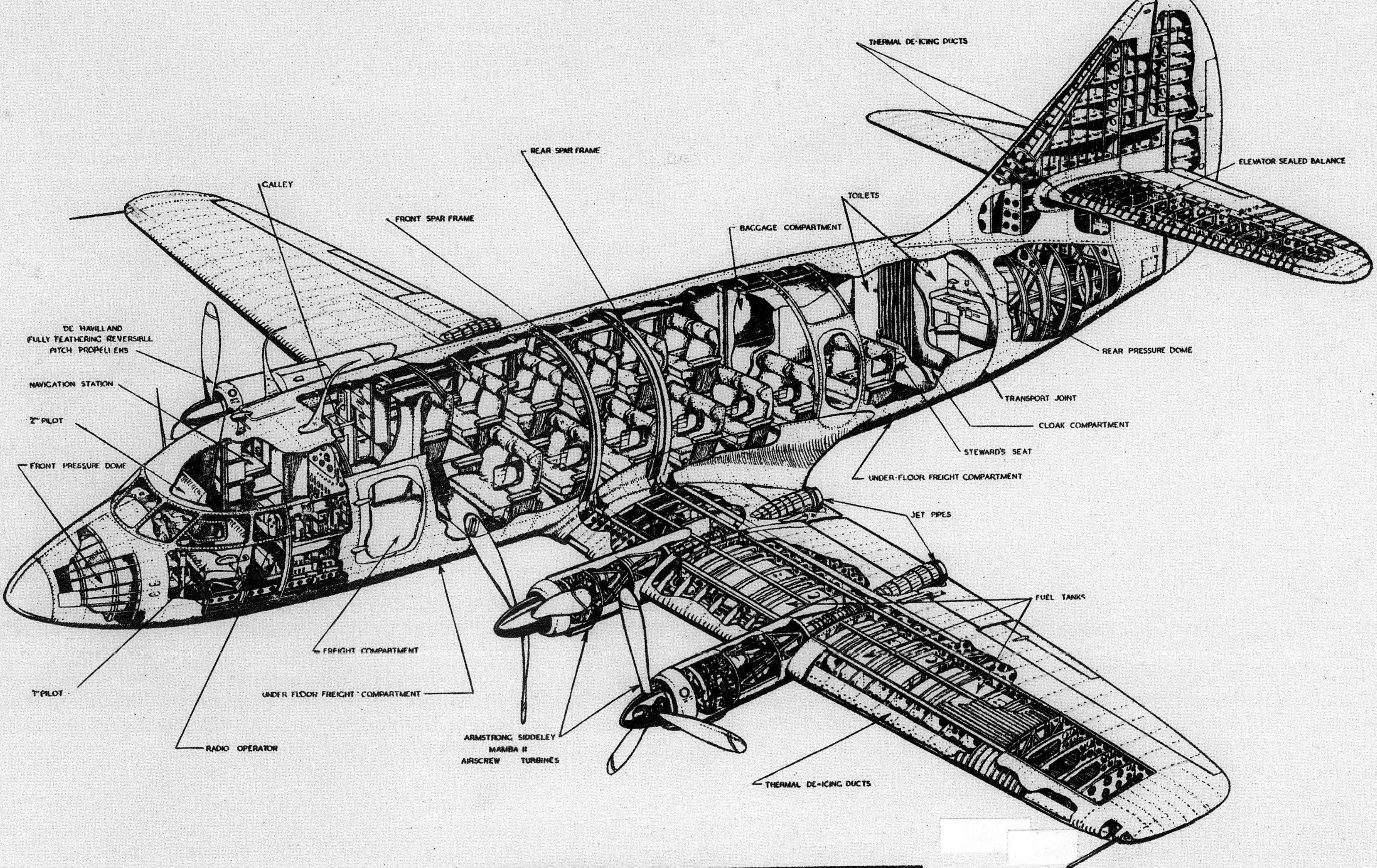

APOLLO AIRCRAFT

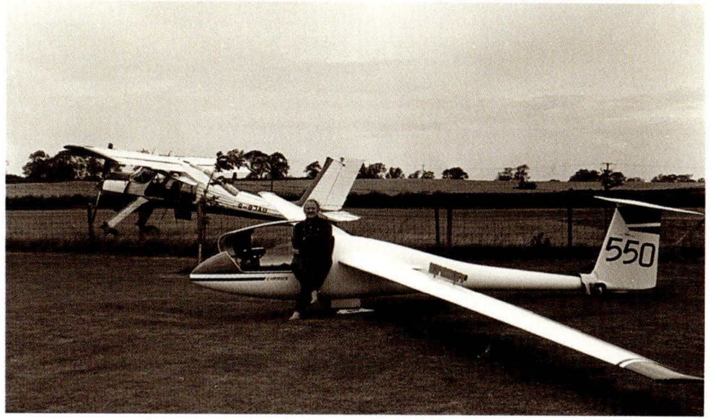

Lawrie beside his Cirrus 550 glider. Behind is his favourite tug aircraft.

Wings passing through air

'To understand where I'm coming from regarding my interest in flight,' Lawrie remarked one day, 'you need to know a bit about flight itself. For instance, a glider is an aeroplane without an engine; an aeroplane is a glider with an engine. Regardless, the principle of flight is the same. The four principles of flying are weight, lift, drag and thrust. The weight refers to the take off weight of the aircraft. To overcome this you need lift, which is provided by the passage of air over the aerofoil section of the wings through the air. On a powered aeroplane, the engine will pull the craft through the air to overcome the resistance of the air against the aircraft, which is drag. Once sufficient speed has generated to counteract drag, the lift achieved will overcome the weight. We now have two factors; thrust, which is forward motion, and lift, which is upward motion; combining the effects of these provides flight. So how do you compensate for the absence of an engine in a glider? There are two methods, both involve towing the glider forwards; either on a rubber rope, which has the effect of catapulting it forwards and upwards, but not to any great height or, more commonly, towing the glider forward on a stationary powered winch. Both aim to create sufficient ground speed for the passage of air over the wings to provide enough lift for it to become airborne. A third method uses a powered aircraft, a 'tug', which is attached to the glider by a rope. This tows the glider into the air and as the aircraft increases speed, so does the air rushing over the wing, thus creating lift. The rope remains attached until both aircraft reach an agreed height, usually 2,000ft, when the glider pilot casts off the towing rope, leaving the glider to seek out thermals (rising hot air) which will take it even higher into the sky. Some of the lift generated by the wing passing through the air is used for that very thing, to pull the aircraft through the air. Called a rotary couple, the centre of lift, which is some way between the leading edge of the wing and the trailing edge of the wing, the centre of gravity ▶

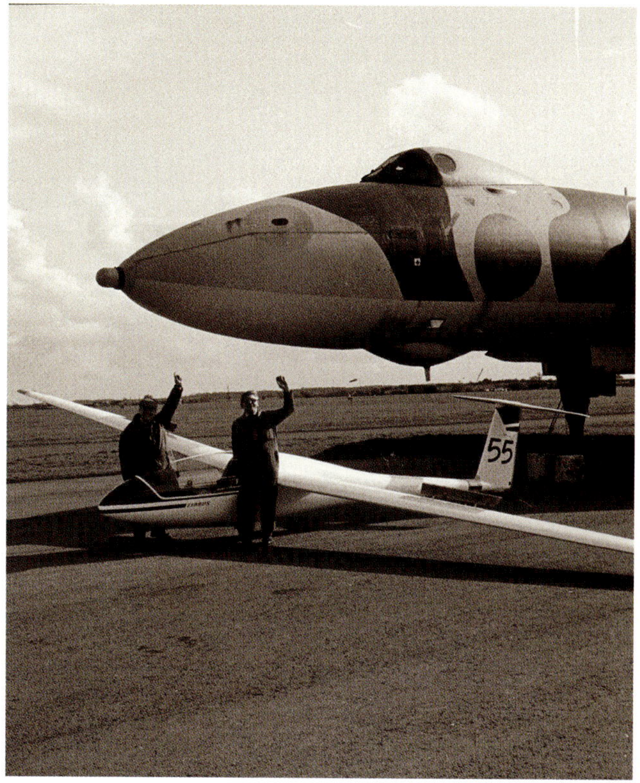

Lawrie and Cirrus dwarfed by the nose of a visiting Vulcan bomber.

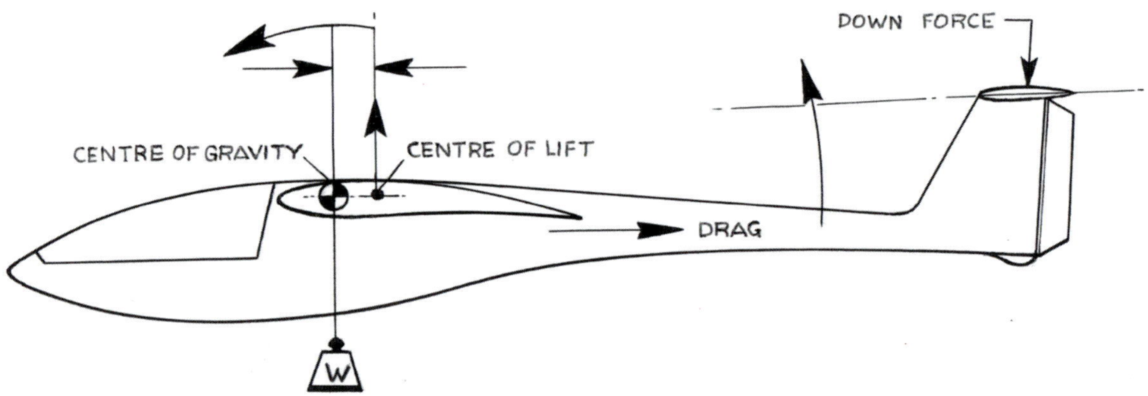

of the plane has to be ahead of that point. Thus, the centre of lift is behind the centre of gravity, causing the aircraft wanting to rotate about itself. Which is what we pilots call a rotary couple (ie: toppling over itself). Contrary to what people think, the tail plane is not holding up the rear of the plane. It's actually holding it down, preventing it rolling over. The net result of those three forces, is that the little bit where the rotary couple comes from, the weight and the lift, makes the aeroplane go forward. That percentage is greater than the drag, and that's where you get the thrust from with a glider.

'With an aeroplane, when there's power to spare, it's still a glider because the principle is still the same. If the engine stops, it can still glide for miles (depending on the height when the engine stopped), while remaining under control as it gradually loses height. The engine produces thrust which makes the aircraft go faster through the air. The trick is to control the thrust. If this becomes greater than is necessary, the aircraft will climb, not ad infinitum because eventually the drag will overcome the effects of thrust, slowing down the rate of climb until it levels off. That's why aircraft such as Concorde, with ultra slim shapes can go so quickly, they produce very little drag. Even without an engine, but a very slim shape, such as on my glider, you can go very quickly because of low drag. To illustrate this, my 60ft wing span glider on the end of the tug's tow rope at 70mph produces only 35lb of drag! It's all down to aerodynamics. The glider's been designed to slip though the air with minimum resistance to such an extent that it will attain 100mph with ease, which enables it to travel for many miles without a problem. Many's the time I've flown from the Midlands to the west coast, got there early, turned around and flown to the east cost before returning to Coventry. It's possible to traverse the country from one 2,000ft launch, but obviously you need lift from thermals en route and to balance your height to maintain what we call the best speed to fly.'

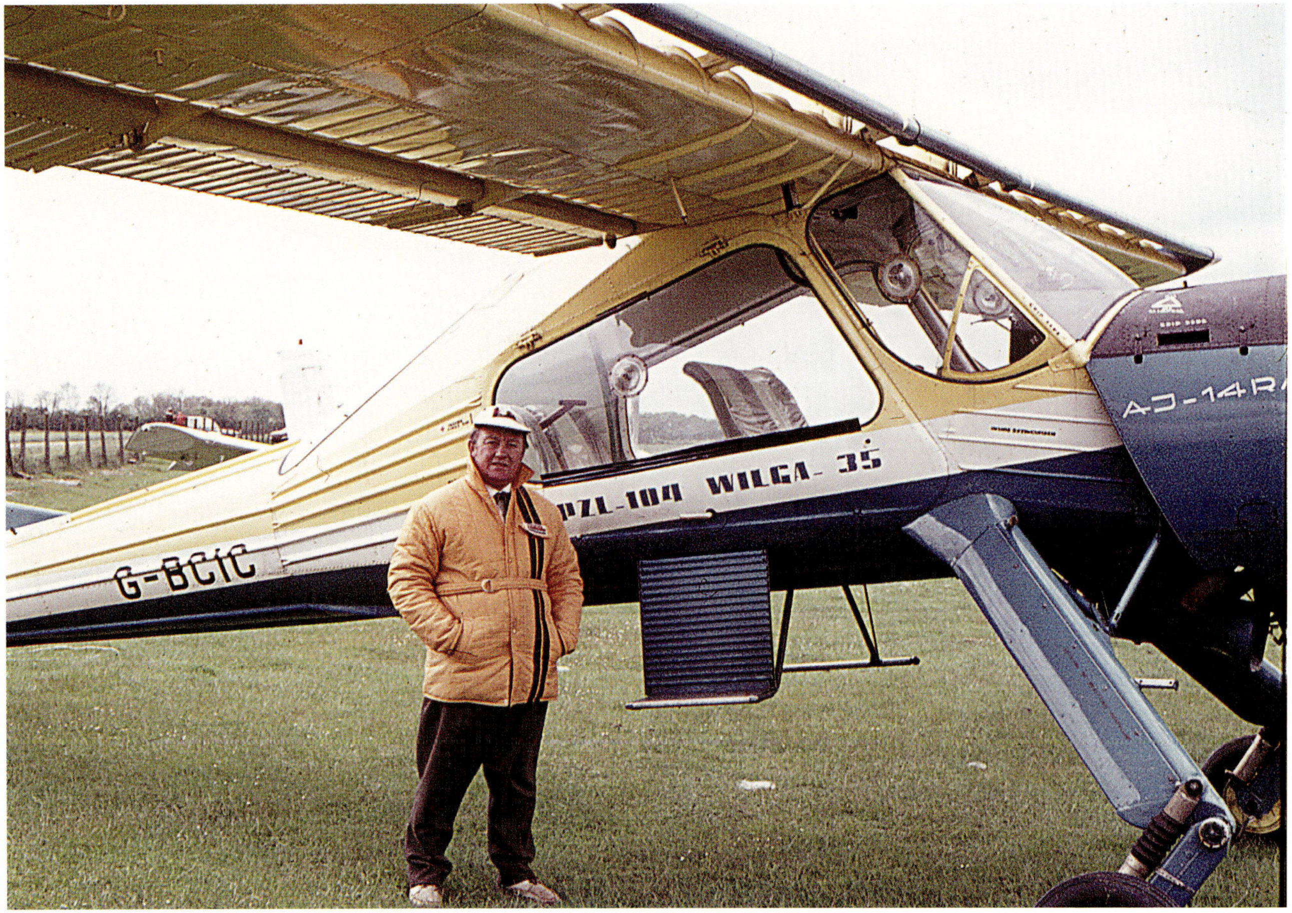

Gliders are towed into the air behind a powered aircraft to a pre-determined height, then the glider pilot casts off the tow rope and flies freely, relying on the wind and thermals. Here Lawrie poses beside his favourite tug, the Wilga.

Powered flight easy after gliding

Gliding is a great school for learning the essentials before graduating to powered flight. The Germans realised this when, after the First World War, they were forbidden under the Treaty of Versailles from having an air force. So if you wanted to go flying, you were by law restricted to gliding. When eventually the Germans formed their own air force, there was a huge pool of accomplished flyers eager to switch from gliding to powered flight in the Luftwaffe.

Ashes, er, to Ashes

As part of his flying duties, Lawrie was occasionally asked to deposit the ashes of recently departed club members over airfields and other places of choice. 'Of course, I always tried to interpret the wishes of the deceased or his relatives with due reverence, but it didn't always run according to plan,' he recalled. 'It seemed perfectly natural that an old mate who was killed in a flying accident should want his ashes to be spread from the air over Edge Hill, a place where he and I, and other club members, had spent many happy airborne hours. So we used a two seater open cockpit T21 glider and I flew it to a pre-determined spot where his ashes were to be dropped over the side of the cockpit by the co-pilot. At the appointed moment, I shouted 'release ashes' which he did. Seconds later, the slipstream blew the contents of the urn back over both of us so that I was flying virtually blind. My colleague was a passable imitation of Father Christmas, covered from head to foot in ashes. From somewhere beneath the covering I could hear him muttering 'Just like him, he'd never jump without a parachute.'

Antics and tragedy in the air

'I was a power pilot before I started gliding and friends I made while flying prop-driven aircraft also became friends in gliding; we enjoyed our gliding just as much as we did flying prop aircraft. One of the most memorable WW2 pictures was that of a Spitfire pilot who, having run out of ammunition while defending London, put his wing tip under an incoming German V1 flying bomb and tipped it over so it flew into the ground away from built up areas. I recall something similar when flying with two of my pals when we were returning in formation from a flying display in Walsall, Bill Woodhams and me in Tiger Moths and another pilot in an Auster. Suddenly, I felt my right wing coming up, so I instinctively glanced at my left wing. Nothing! So I looked across at the pilot, my old mate Bill, who was son of Mr Woodhams, managing director of Armstrong Whitworth when I worked there, and Bill was grinning from ear to ear. As a wakey-wake call he'd put his wing underneath mine and gently lifted it up! Crazy thing to do, but he was such a brilliant pilot, and he trusted me not to react badly, that he thought he'd enliven our jaunt. Bill had become a test pilot for the Popular Flying Association and, years afterwards, while test flying a home-built aircraft before certifying it, he was killed in a freak accident. It was a two seater which, instead of a passenger, had to carry ballast for the test. This being a farmer's aircraft, he'd used a sack of beans for ballast which he'd strapped in the passenger's seat. Unfortunately, sometime during the test flight the bag burst open and beans jammed the controls. When Bill tried to pull out of a dive, the controls were jammed causing Bill to nose dive into the ground. Definitely not Bill's fault, but it shows how the simplest error on the part of someone else can have the most horrendous consequences.'

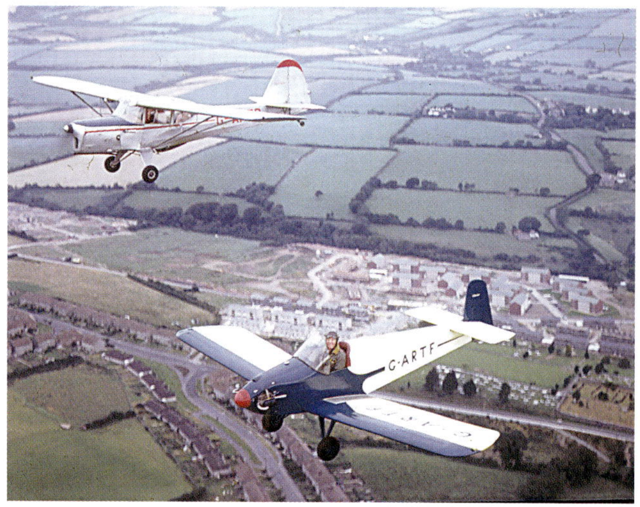

Test flying a home-built aircraft (lower) over South Wales, before granting it a certificate of airworthiness.

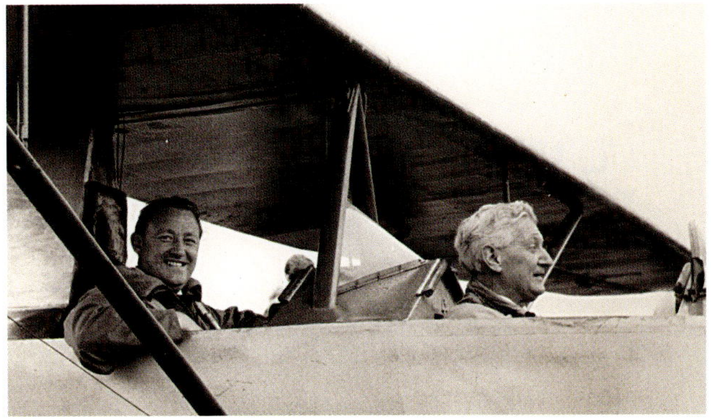

Instructor Watts! Lawrie takes pupil Doug Findon for a flight in a two-seater T31 Barge.

A load of hot air

When flying a glider, you log on to nature because you're relying on what nature provides in the way of winds, finding them, directing them, interpreting what they are doing, then harnessing them to your needs. It's a combination of factors which, mainly, comes with experience. Harnessing thermals, which is warm, or hot, air rising is at the heart of gliding.

Finding thermals is an acquired art form which, when mastered, decides how successful a glider pilot you will become. At first, you rely on instruments to show you outside temperatures, because you know that the warmer air rises and you aim to go with it. After a time you learn that a corn field gives off warm air, so you head for that. You also find that cool air above a lake is bad news so you avoid it. But a Motorway is absolutely fantastic, especially if it's really busy. The heat given off by that traffic is amazing and, I'm still getting lift even up to 3,000ft. No wonder there's a problem with global warming!'

Warming to his task, Lawrie then went on to explain how finding thermal clouds and harnessing their energy was essential to gliding. As he explained, 'The atmosphere of a cubic metre of air can hold more water invisibly when it's warm. If it suddenly chills as it rises, the water particles in that cubic metre of air will start to manifest themselves as droplets, which will take the form of mist. This is actually cloud. When the lift is exhausted then the droplets will disperse as mist, or if a big cloud is forming, such as a cumulonimbus, it will rain. Clouds are constantly changing, dissipating, while new ones are forming. All this movement means that air is being displaced, to be replaced with new air, according to the temperature. So there's constant movement of air, which is called wind. Cold air coming across the earth meets warm air, which is chilled as it rises into the sky. This is called a cold front. A warm front is the opposite; the warm air will rise over the cold air to create a weather front. As heat rises with the warm air, that's what we aim to use. If you see what's called a 'cloud street', that's a cloud forming. It's an indication that warm air, a thermal source, is thereabouts. If you can get under that street and into the warm air, you can probably fly for hundreds of miles as long as the warm air is still rising under the cloud formation. To maintain station in the thermals, you need to monitor your instruments frequently, especially the temperature gauge to ensure that you stay in the warm air, and also the altimeter to maintain height. Interpreting what the 'street' is doing provides a visual confirmation of your instrument readings to enable you maintain your place in the thermal. Keeping an eye out for other aircraft is not a bad thing either! So those people on the ground who look up and see a glider wafting through the air and think 'Lazy sod, just sitting there doing nothing, being blown along by the wind' don't know the half of it. We're working our little socks off to ensure that we remain aloft, and keep to our chosen route. Of course, we're enjoying every minute of it!'

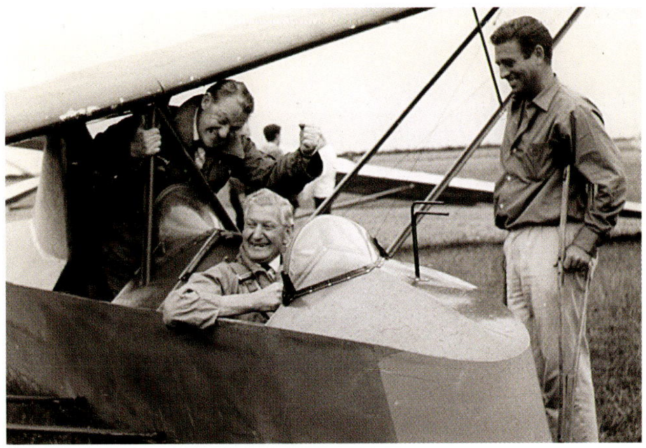

Sometimes it's hard work teaching! Instructor Watts about to hammer home a point to pupil Doug Findon, whose day job was importing cars.

Quick glider

'When I was talking about the modern sail plane and how it has improved in a very short time, especially when you remember that it's of no commercial value, it's purely for fun. My glider can be made to fly very quickly if I wanted it to. One day I was returning from a competition in Devon to Husband's Bosworth, home of the Coventry Gliding Club now known as the Soaring Centre. Ahead of me I could see an Auster plane being piloted by a good friend of mine, Mike Costin, one of the co-founders of Cosworth Engineering, with passenger Ben Rude, then chief engineer at Cosworth. They were also returning to Husband's Bosworth. I was about 500ft above them, so I thought I'd have a bit of fun, which I knew Mike, as a fellow glider racer, would appreciate. I gently pushed the nose of my glider down and started to pick up speed until I was going like the clappers and closing on them rapidly. Diving past at about 150mph, I waved and radioed an invite for a cup of tea when they reached the airfield. Then I pulled the stick back and climbed to my original flying height and resumed the journey. I'd arrived back at Husband's Bosworth, taken the wings off my glider and stowed it in the trailer before Mike and Ben landed! Their Auster, while being a good little aircraft, cruised at about 70/80mph while my glider in good conditions, cruised at around 80/90mph purely because of its superior aerodynamics. Proof, if need be, of the value of good aerodynamics.'

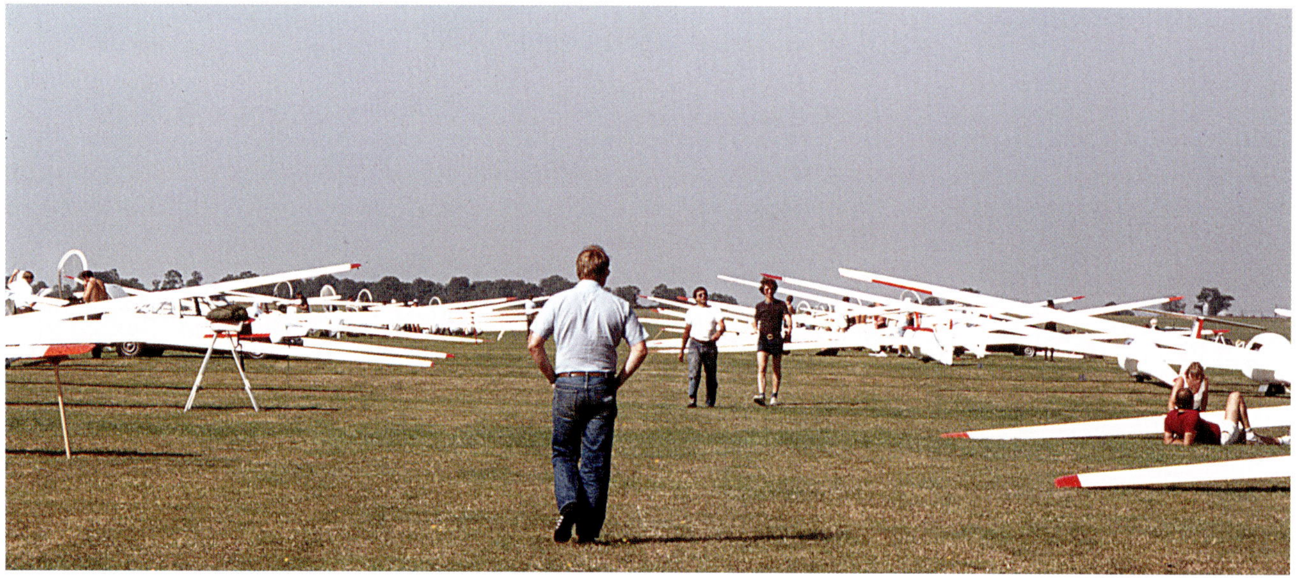

Gliders assembled for a competition at Husband's Bosworth Airfield. Families come too, sometimes to help...

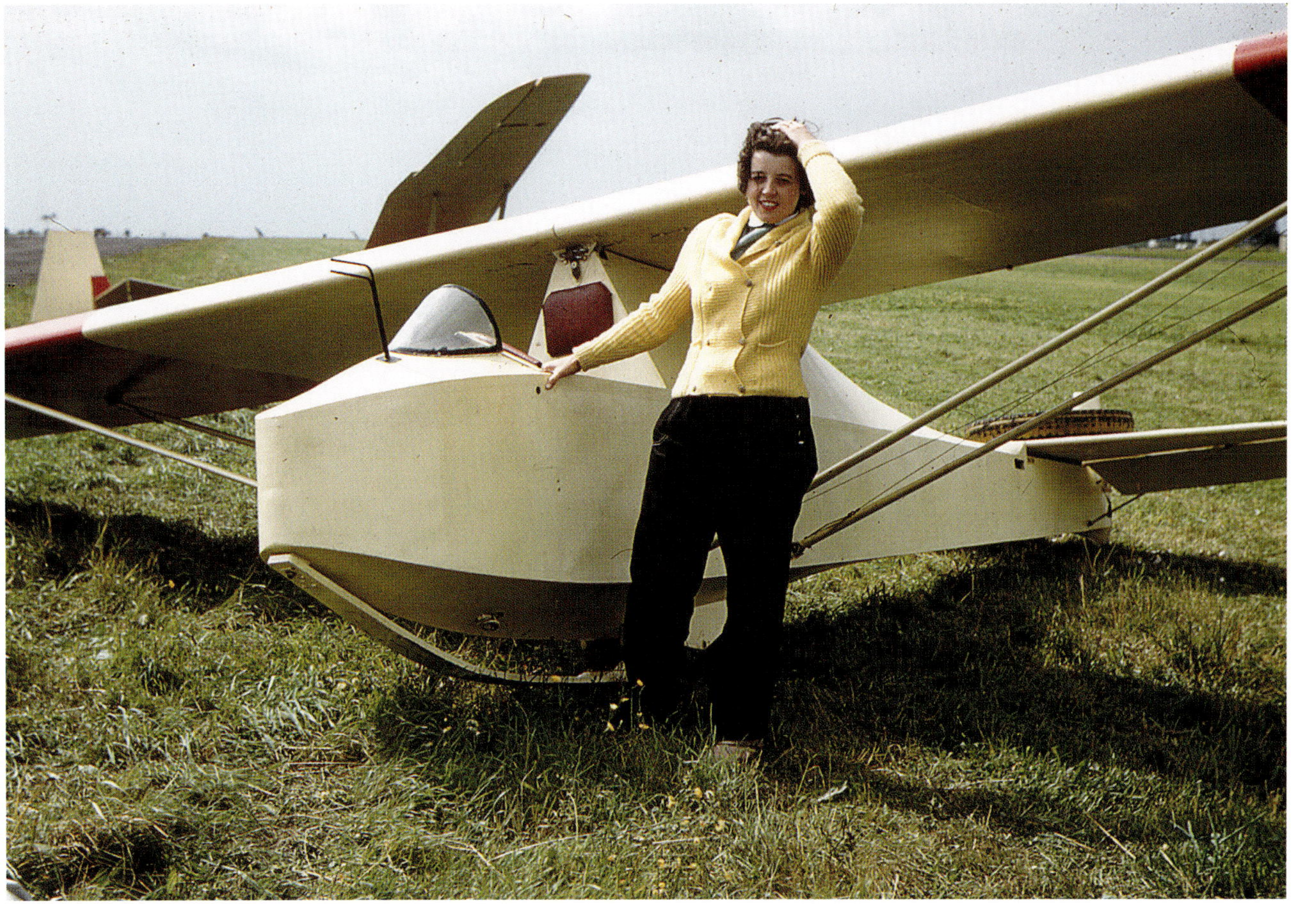

Silver C award

'The first cross country flight I made, in order to get my Silver C award, was from Baginton, Coventry, to Staverton airfield in Gloucestershire. The glider I used was little more than a wooden box with wings, called a Tutor made by the well known (at that time) firm of Slingsby. It had a glide angle of 1 in 12, which meant going forward 12 feet when you dropped a foot in height - marginally better than a brick! - especially when compared with my Schemp Hurth Cirrus Racer. This had a glide angle of I in 60, a significant difference! Well, the Tutor was no Cirrus and I had to work all the way to Staverton, especially over the Clee Hills, and then I could see the airfield. Being ultra-careful and aware that my progress would be under scrutiny from the ground, I joined the circuit waiting to land and, when my turn came, I peeled off and lined up the Tutor for my landing approach thinking that I'd give the hard surface runway a miss in favour of the grass. I'd just touched down and was aiming for the control tower when a Dakota hurtled past my wing tip as it touched down. I nearly wet myself! I'd no idea where he came from and was mortally embarrassed, especially as this had blown my chances of the Silver C. Walking to the control tower, I rehearsed my speech of apology to the staff, to the Dakota pilot, and to the passengers whose lives I'd inadvertently put in jeopardy. As the Dakota pilot signed in, I introduced myself, made my speech and was amazed when he put his arm around my shoulder and said, 'Don't worry about it, I saw you and as you looked like you knew what you were doing, I proceeded to land. No problem, congratulations!' He said this in front of the assembled officials who had been monitoring me. They just grinned and endorsed what he'd said! And that's how I gained my precious Silver C! I joined the elite that day.'

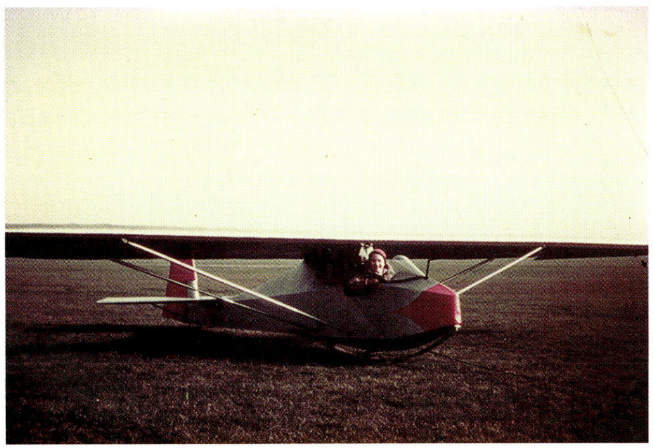

Eileen hangs onto her hairdo alongside a Tutor glider. She is an expert glider rigger and collector of gliders which are forced down miles from home.

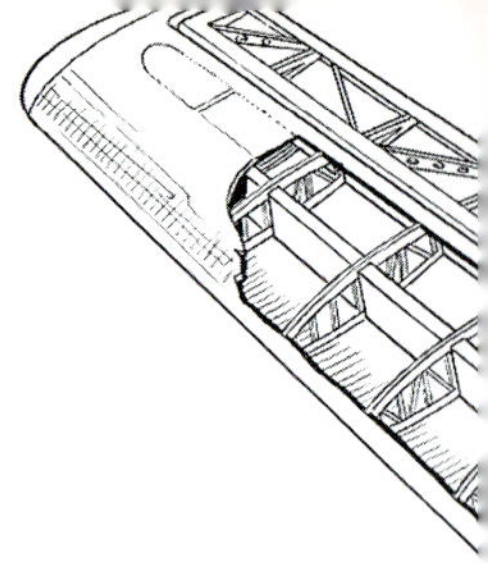

Balancing an A380 on a finger tip - at 630mph

In flight, everything balances out, becoming weightless, otherwise it would fall out of the sky. Which is why a 500/600 seater aircraft, such as the A380, achieves perfect balance at a comparatively early stage of its flight when it reaches its cruising altitude of around 30,000ft. 'At this stage, it is lighter than a feather,' says Lawrie, 'and, in theory, could be balanced on a finger tip - if you could stand the 630mph slipstream! The only reason for it to be going at this speed is to get from A to B quickly, it could just as easily fly at half that speed, or even less. The speed at which it becomes weightless is just above take off speed.'

Slingsby Swallow

In a new job the first big assignment has a special place in the memory and so it was that Lawrie remembered clearly the first time he was sent out to draw a complete aircraft for 'Flight'. 'I could not have been more pleased,' he recalled, 'because my special assignment was to draw a Slingsby Swallow glider. In those days, there were only two manufacturers of gliders in the country, and Slingsby was the frontrunner. So off I went to their factory in Kirkbymoorside, in North Yorkshire, to meet the man himself, Frederick Slingsby, later known as Fred. With my knowledge of gliders it was not surprising that from the start we were on the same wavelength and we ultimately became good friends. The prototype of the Swallow was unusual insofar as it was in kit form, designed to be completed by the buyer. It offered the potential of being bought by chaps who, by pooling resources and energies, could complete the assembly of the Swallow with the chance to fly it without the capital expenditure of buying their own aircraft. With a wing span of 12 metres, which is quite small for a glider, the completed craft was a tidy design. When it was built, I was given the job of test flying the prototype, and reporting to 'Flight' on how it handled and went together. I enjoyed that and my story also complemented the drawing of the finished article, and how it was assembled. Little did I realise at the time but I was to be asked later by 'Flight' to test fly every other glider which I drew. This, of course, was a very welcome bonus for me as an enthusiastic glider pilot. Over the following years, I flew all of Slingsby's models which, as they built only one model a year, each one with increased performance over the previous year's, must have included 20 different types in 20 (or more) years.'

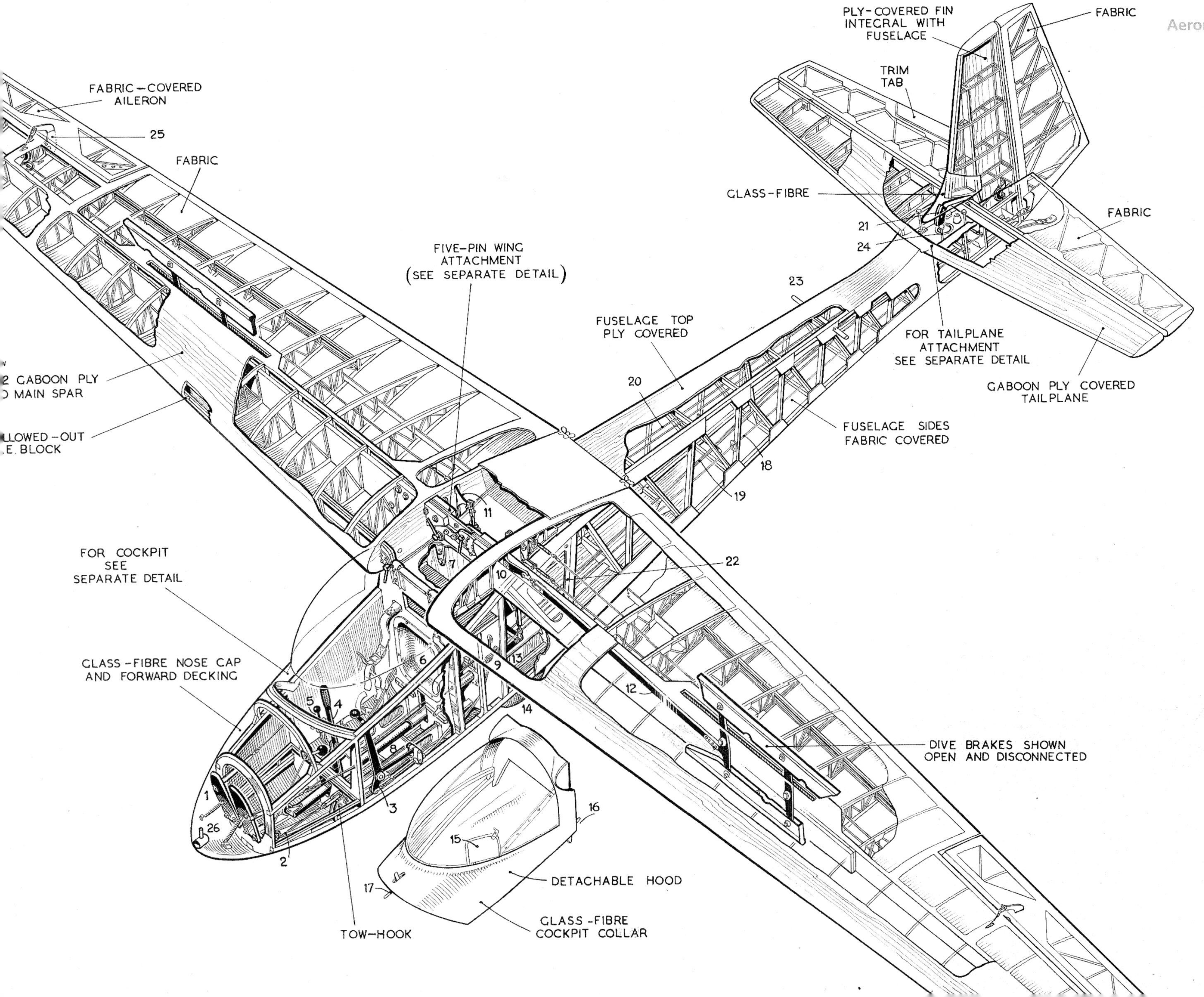

PLY-COVERED FIN INTEGRAL WITH FUSELAGE
FABRIC
TRIM TAB
FABRIC-COVERED AILERON
25
FABRIC
GLASS-FIBRE
FABRIC
21
24
FIVE-PIN WING ATTACHMENT (SEE SEPARATE DETAIL)
23
FUSELAGE TOP PLY COVERED
FOR TAILPLANE ATTACHMENT SEE SEPARATE DETAIL
2 GABOON PLY
) MAIN SPAR
20
GABOON PLY COVERED TAILPLANE
FUSELAGE SIDES FABRIC COVERED
LLOWED-OUT
E. BLOCK
18
19
11
FOR COCKPIT SEE SEPARATE DETAIL
7
22
10
GLASS-FIBRE NOSE CAP AND FORWARD DECKING
6
9
13
5
4
14
12
8
DIVE BRAKES SHOWN OPEN AND DISCONNECTED
1
3
16
26
15
2
DETACHABLE HOOD
17
TOW-HOOK
GLASS-FIBRE COCKPIT COLLAR

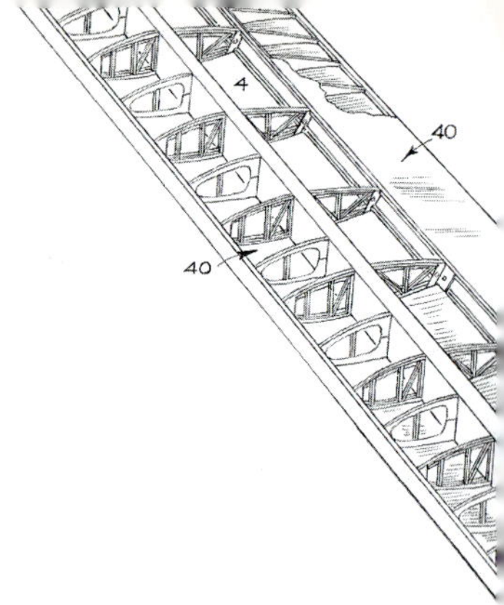

Slingsby Dart

In contrast with the Swallow, the model which followed, the 16 metre Dart, was a high performance sailplane mainly, but not exclusively, intended for competitions. As such, it was the opposite to the Swallow in having more state of the art construction, with high aspect ratio wings which create less drag and cleaner air flow. This all helped to make it an outstanding engineering achievement and a very desirable aircraft, judging by world demand for it. Made of wood, the fuselage used three ply 1.5mm thick, glued to beautiful wooden ribs, in the manner of a true racing glider. The fuselage was about six feet longer than the Swallow's. Another innovation lay in the nose; for the first time, it was made in glass fibre, thus facilitating the production of double curvatures.

'This raised some technical problems for me,' recalled Lawrie, of those far off days. 'With long wing spans, it soon became pointless to draw both wings simply because they could not both be accommodated on the same page. To attempt to do so would mean having ridiculously small details on the aircraft itself. So I got around this by drawing only one wing, usually at an upward slant, so that the layout man could accommodate everything I'd drawn without overlapping the page, or two pages should he chose to do that. Previously, both wings had to be shown, otherwise the aircraft was not complete, but you could see very little detail in those old drawings. The important thing to show in a drawing of a glider is the wing construction and assembly, and secondly, the controls. My other factor is to include as much information as a potential buyer would want to know so, if there's a choice what to include and what to leave out, I'd always include that which a future customer would find of interest. That's why the aircraft is drawn at an angle across the page, with the wing tilted towards the top right hand corner. This leaves me room to draw details in the spaces above and below the wing, details which potential buyers would want to see. In all my glider drawings, you'll find that I include details of the control system, that's what these buyers want to know because they will have to work on their own aeroplanes, so this knowledge is necessary in helping them make up their minds about buying. Each time a glider pilot flies, he has to rig and de-rig his aircraft whereas a pilot of a prop driven one merely collects his aircraft at the airfield. If a powered aircraft needs to be dismantled and re-assembled, it has to have an approved engineer there to issue a certificate of air worthiness. Not so with a sailplane. You form your own crew, those who will help you to rig your glider and these helpers are required to learn where everything fits, and sign for it, each time the craft is assembled for flight. So, in some ways, there is more control for safety reasons over a glider than a powered aircraft. Which makes sense as the glider harnesses nature to fly, while a powered aircraft drags you through the air and, if necessary, has reserves of power to get you home.' ▶

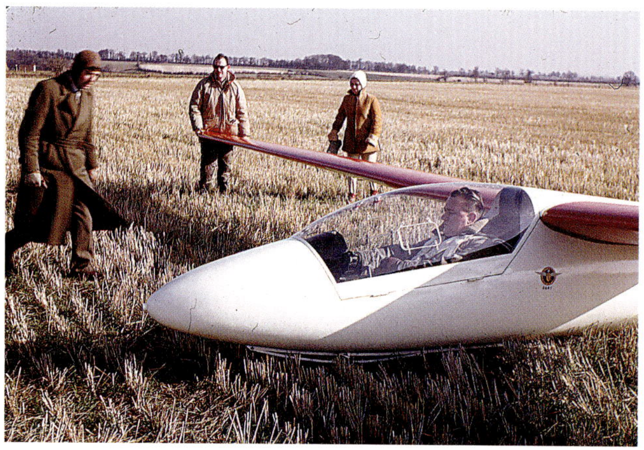

Lawrie about to undergo interrogation by the airfield security at Edge Hill before testing the Slingsby Dart.

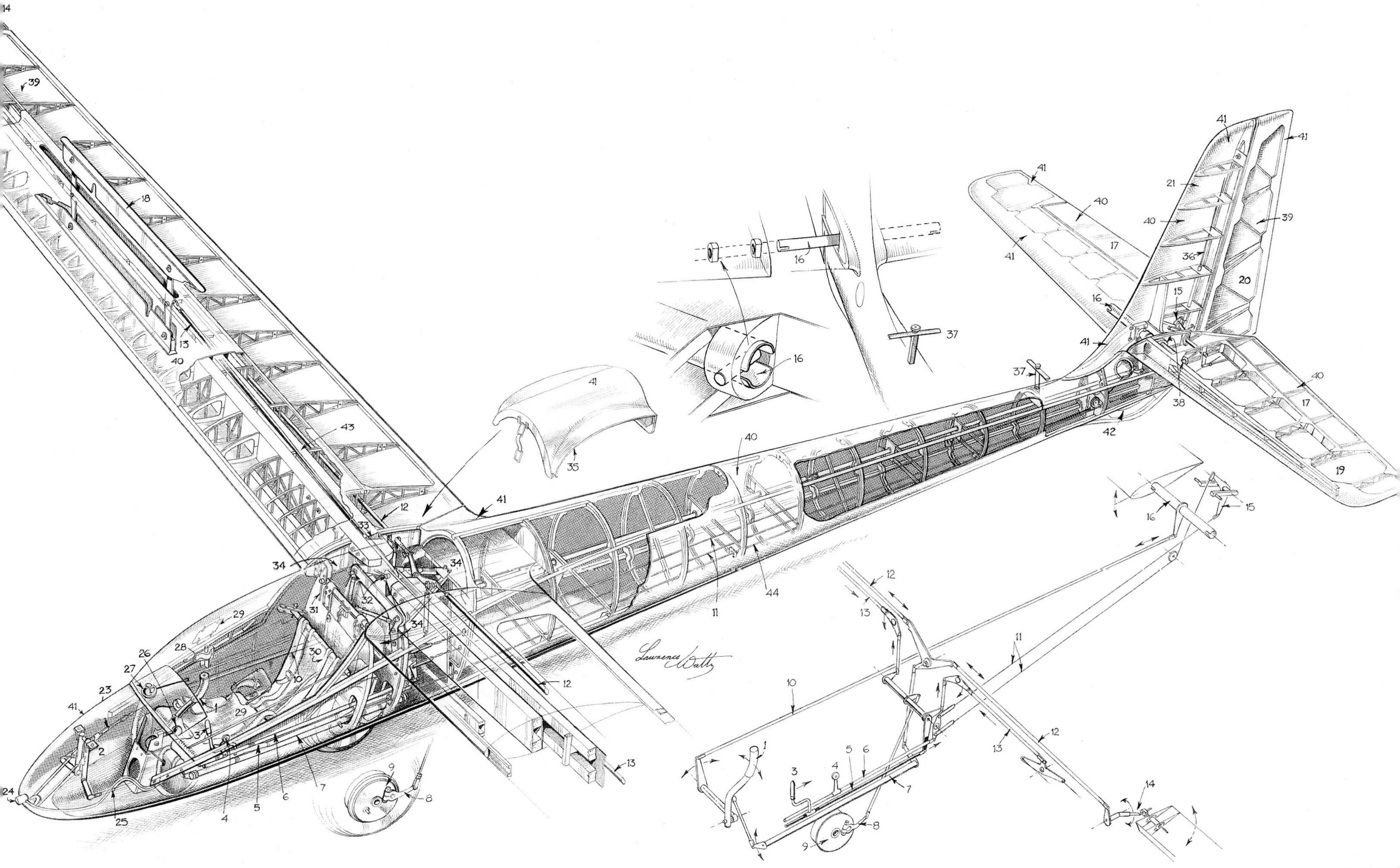
Lawrence Watts

'Slingsby was the greatest exponent of producing wooden gliders in the UK. Yet, they felt the need to progress and chose to make their first all metal glider, the HP14, for Sir Peter Scott. In addition to all his other interests, painting, wild life (especially birds), exploration, etc, Sir Peter also made time for gliding. Indeed, he was a very capable competitor and Slingsbys took on the challenge of building an all-metal glider designed by an American called, I think, Dick Schrader. One of the advantages of using lightweight metal was to secure a slimmer aerofoil section, flaps, air brakes, elevator trimmer, more sophisticated than the traditional wooden designs. But all this was at a price, the craft was very heavy to rig and it required a specialist crew. As it had not yet been constructed, I had to do my drawing from detailed construction drawings. Eventually, when it was completed, I had the opportunity of trying it. There's a vast difference between flying a high aspect ratio wooden sailplane to flying a much stiffer metal aeroplane like the ▶

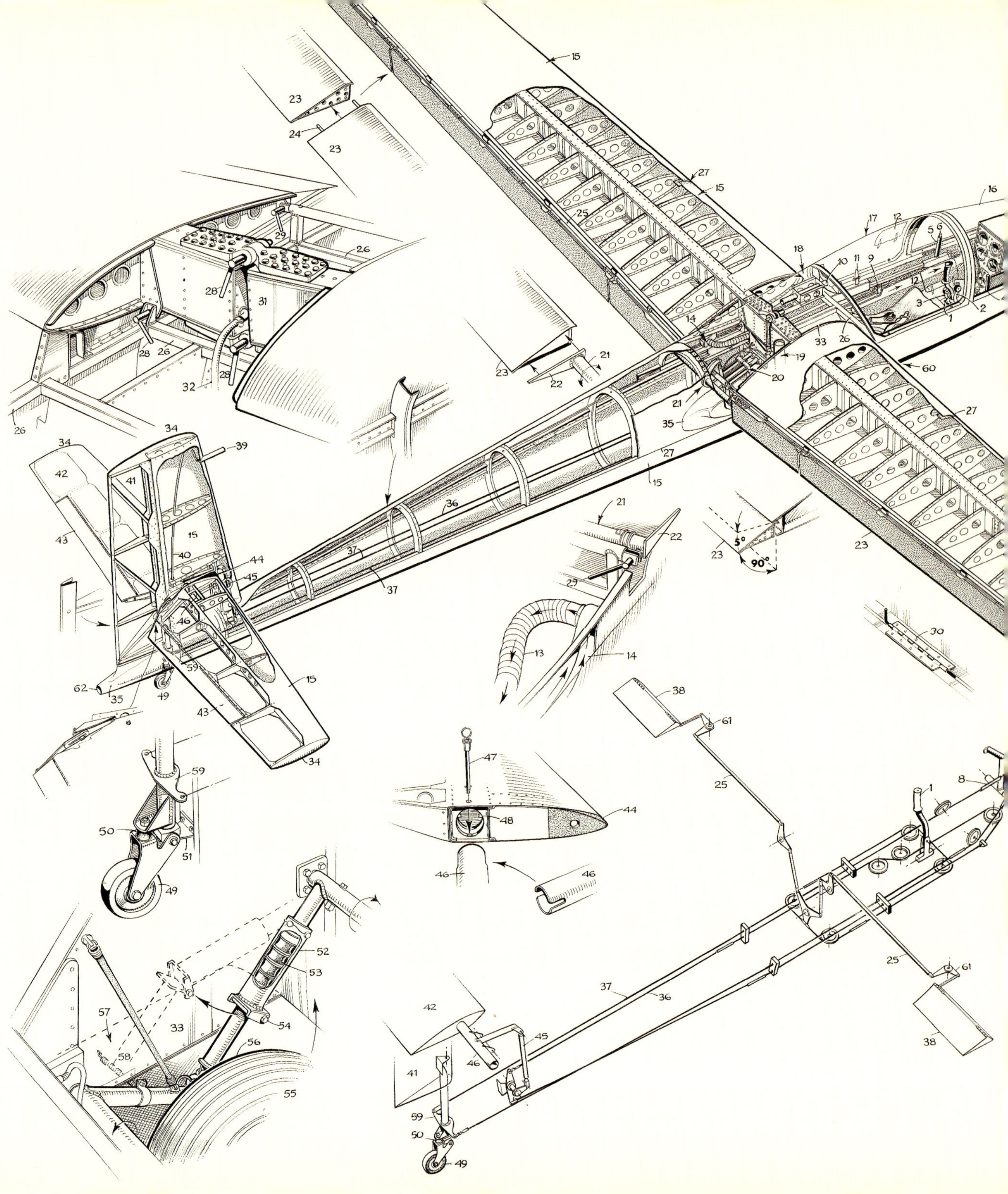

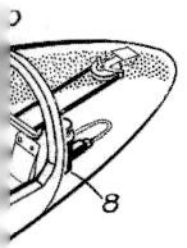

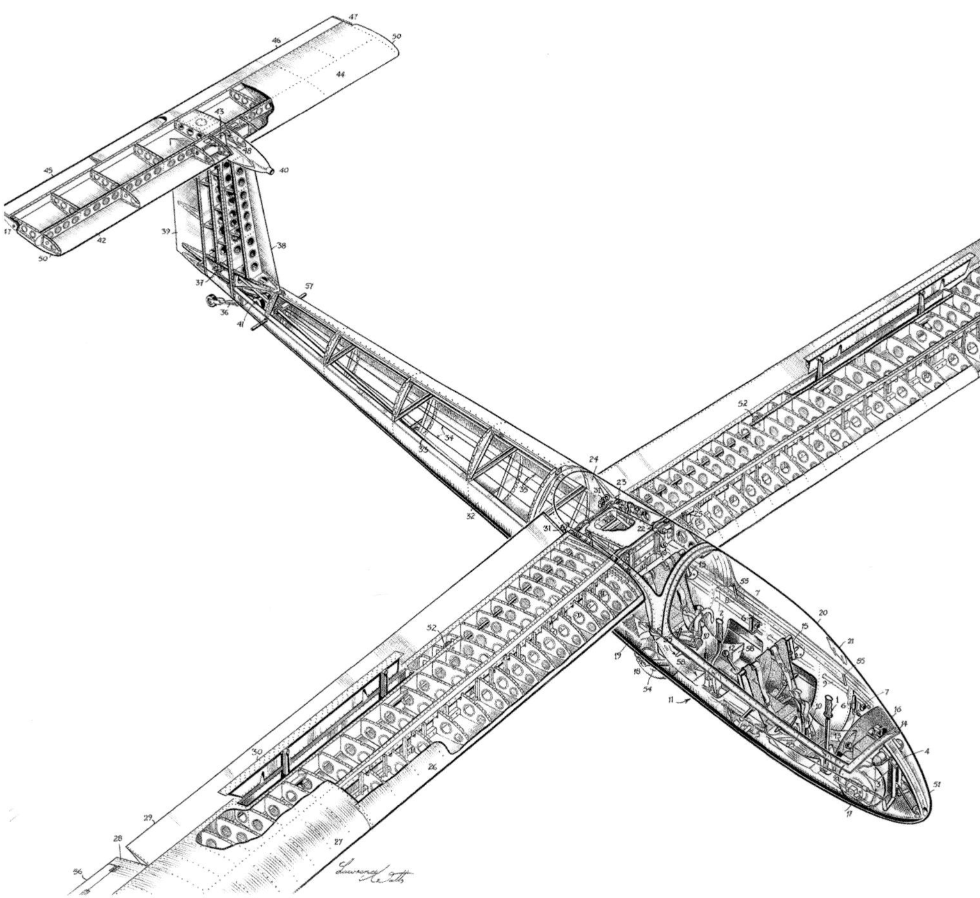

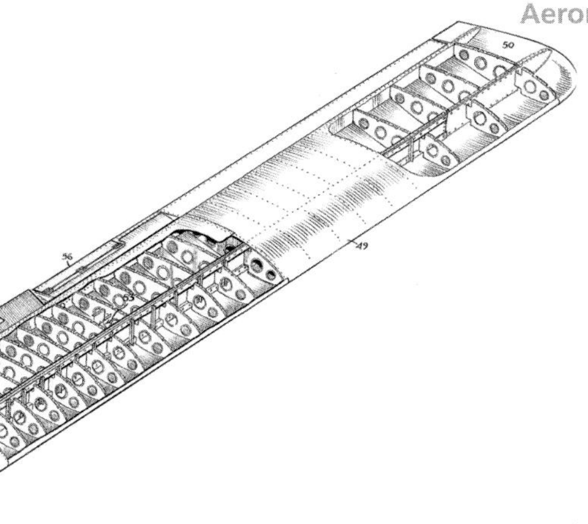

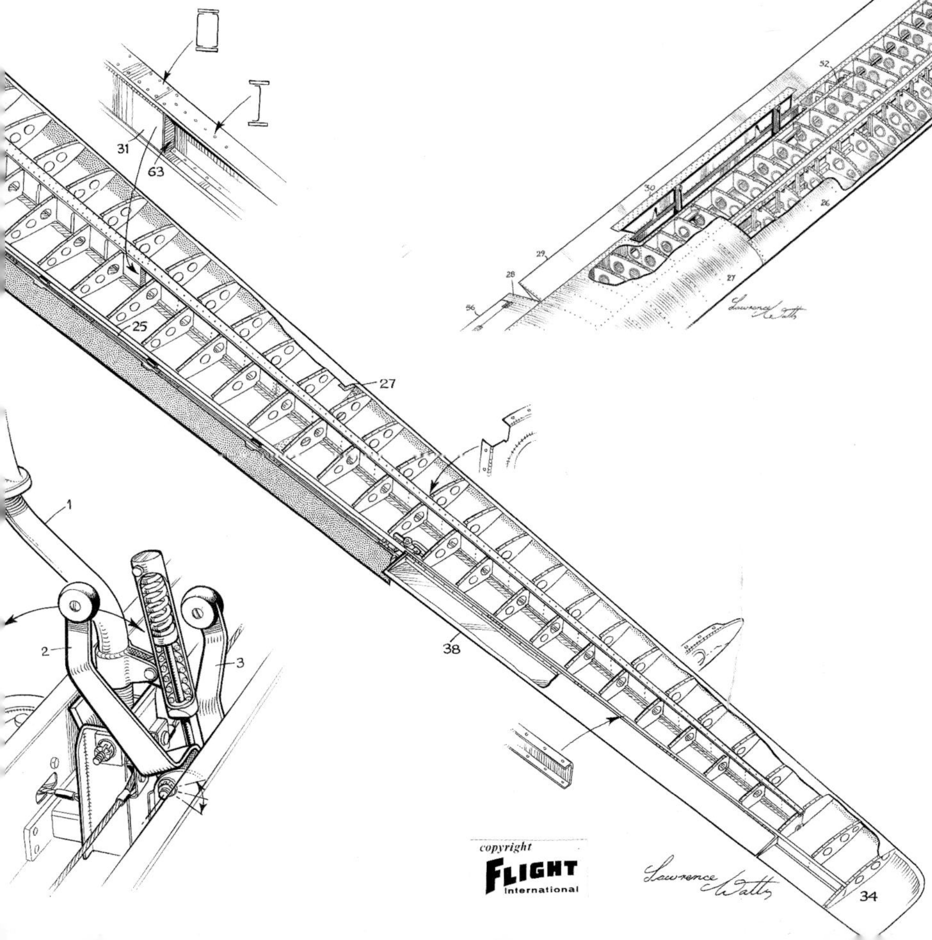

HP14. It's immediately noticeable when you first step inside, there's a complete lack of the smell of wood, resin and glue which you get in a typical wooden aircraft. There's an inert feeling about the way the metal craft responds, with little feedback through your fingers. Flying is a sensory thing, especially this kind of flying, and the lack of feel in a metal aircraft such as this isolates the pilot from the close involvement which he feels in a wooden aircraft. Unsurprisingly, I was not terribly impressed after my first flight and Sir Peter Scott seemed lukewarm as well. So it died the death, I don't think another was made.

'Undeterred, Slingsby followed up the non-success of the HP14 with the Sovereign 18 metre, their first all metal tandem two seater glider which was immediately distinguishable by its swept forward wings with flaps. This was a way of getting around wash out - to prolong stalling of the wing tip in slow flight, sweeping the wings forward has the same effect and reduces drag. In fact, it looked more like an aeroplane without an engine than a pure glider! The ATC loved it and bought some, the RAF also bought a few, ▶

Removable wing tip

'The 16m Dart was followed by the 18m Dart, of which I drew only the changed bits. One of these was a clever innovation. The wing tip, which was made of glass fibre, was removable. This allowed another section of wing to be interposed between the tip and the original wing section, thus increasing to 20m the total wing span. The purpose of this? It took the glider into another competition class, which gave the pilot an additional option in which to compete. Clever, eh?'

but it was unsuccessful in competitions and suffered the fate of the HP14, for the very same reasons, the rigging and de-rigging. In this case, having the wings swept forward created huge problems when rigging; imagine standing on one side trying to manoeuvre your half of the wing into position from an angle, while the rigger 60 feet away on the opposite side is trying the same manoeuvre, at the same angle, and timing it so that he slots home his half wing simultaneously to join with yours. It had to be done this way because the two halves of the wing, joined together, supported the fuselage, as shown in my drawing. It was a very clever design but frequently marred by involving nearly everybody on the field being called to lend a hand, all shouting, jostling, heaving, lifting, and keeping in touch with the opposite side. Quite a nightmare! And then, when you got into the air, the pilot's vision was seriously obscured by the forward sweep of the wings. Why this should be so I still can't understand except to say that Slingsby was exonerated because the design was an all-American one. And to think that they, the inventors and leading developers of flight, could have boobed like this. It's beyond comprehension. This aircraft, understandably, also failed to light a fire under any buyers and it proved one disaster too far for Slingsby and they went into liquidation. Which was sad, because they were an excellent firm and made fine wooden gliders which provided us with many happy, and safe, hours of flying.'

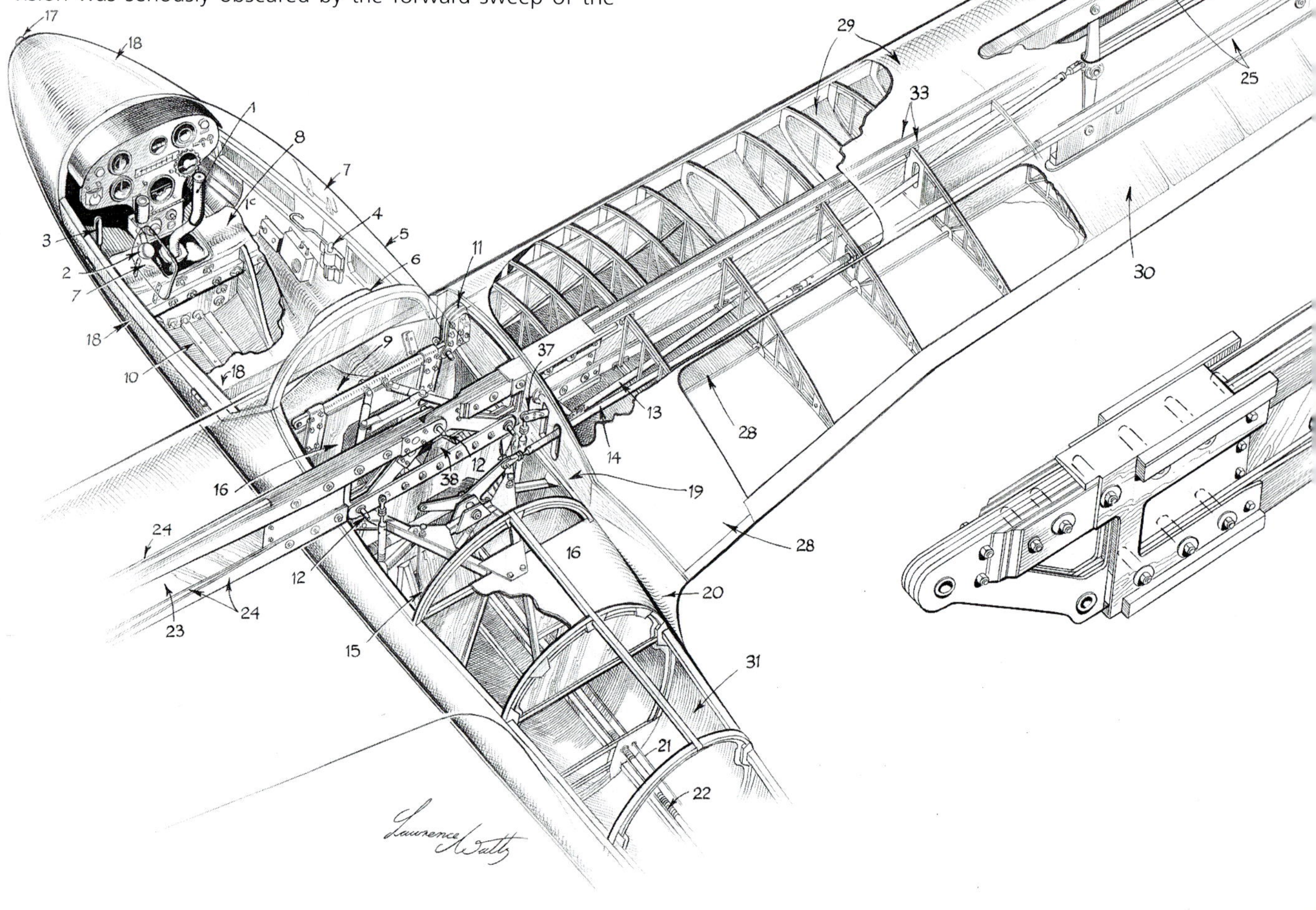

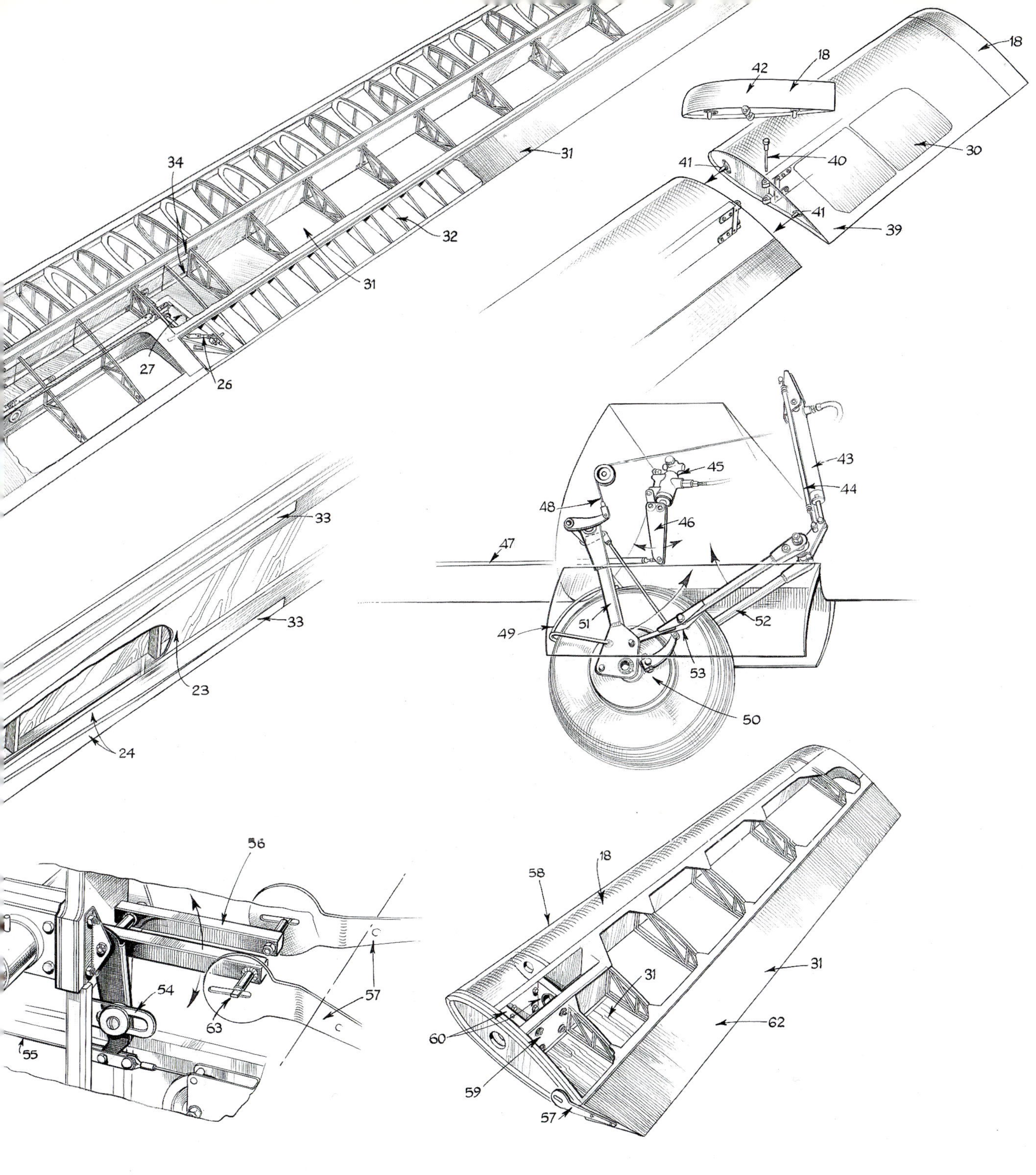

How not to brew tea

At the national gliding championships at Lasham one year, Lawrie had a near shave which owed little to gliding and more to providence. As he recalls, 'Fred Slingsby and Doc Slater, both good friends of mine, shared a caravan beside mine and we popped in and out of each other's all the time. But on this occasion it was a call for help, they had locked themselves out. 'Loll,' they said, 'You're about the right size to squeeze through a window (I must have been a few sizes smaller than I am now), so we want you to break in and open the door from the inside.' Easy peasy, I did as asked without too much drama, and returned to my caravan. I'd only just sat down when there was an almighty explosion which rocked our caravan. Jumping up, I was just in time to see Fred and Doc flying base over apex across the grass. They'd been blown through the door of their caravan. Apparently, after I'd left them they decided that a cup of tea would be a nice idea. So Doc filled the kettle and Fred lit the match. Bad move. Unknown to them, there had been a gas leak and the match did the trick. Luckily, not enough gas had accumulated to do a complete demolition job on the caravan but what it did was impressive enough. Our heroes escaped with singed hair, a comprehensive shake up, and a desire for something stronger than a cuppa. Which set me thinking; suppose I'd been the one to put the kettle on after I broke in...'

The Flying Plank

'The French and the Germans were experimenting with flying wings, high aspect ratio wings with a straight sweep - pardon the contradiction in terms! - and no tail. I flew one of those and a French flying wing called the Favel which was remarkable in having no sweep back at all. Basically, it was a plank with a fuselage in the middle. So how do you control the wing angle with no rear fuselage and tail? The French solved the problem by building wash out into the wing and made the wing into ailerons into what the French called 'elevons', so not only did they induce the turn but they could go up and down together. This made the aeroplane incredibly twitchy and called for very sensitive control. With no fuselage to control it from the back, it was bound to be twitchy. So Andy Goff, who had arranged for me to fly the Favel, suggested that I carry some ballast to help stabilise weight distribution. I set off with the ballast securely in place at the end of a towing line; I cast off and got to grips with the machine. After trying all the conventional situations one does with a glider first time up, I was getting used to the feel of it and eventually caught a thermal and started to soar the damned thing. Lots of other aircraft about, so I got in among them and, to my amazement, found I could out-soar them with the wing! After about an hour's fun, and thoroughly impressed, I went down to land, in case anyone else wanted to sample the wing. I landed, it stopped so very quickly that it almost turned over forwards! Understandable, because there was no fuselage aft, therefore nothing to restrain the wing when it stopped. Anyway, I needn't have bothered because nobody wanted to fly it! So I asked Andy if I could go up again and he agreed to my suggestion that I discard some, but not all, of the ballast. I took off behind the tug again and, without all that ballast, immediately shot up in the air, desperately trying to control the wing. The difference discarding that ballast made was unbelievable, it altered the centre of gravity, and thereby the machine's flying characteristics, so much that it was a full time job to fly it. Equally, it did not soar anywhere near as well without the ballast as it did with it, illustrating neatly the importance of ballast relative to the centre of gravity. The Favel never became popular, probably because of prejudice. It challenged too many pre-conceived ideas it could never work. Or so they thought. In fact it worked very well and repaid handsomely anybody who took the time to understand how it worked and to fly the machine to the best of its capabilities.'

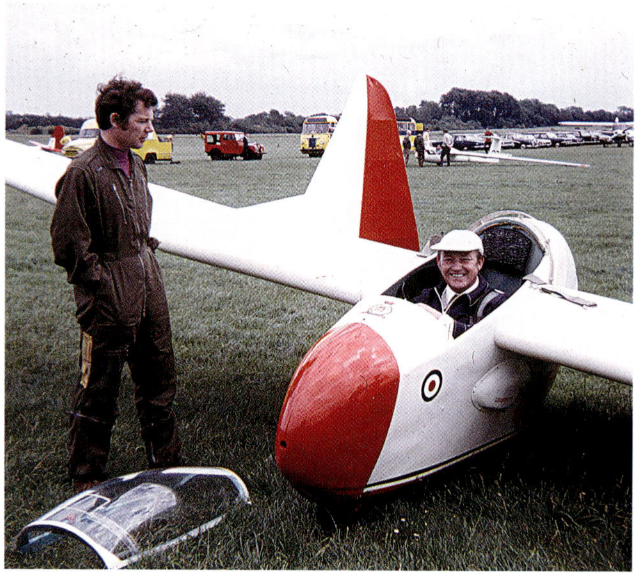

French designed Favel glider, The Flying Wing. Lawrie called it 'a plank with a fuselage in the middle', but its performance impressed him.

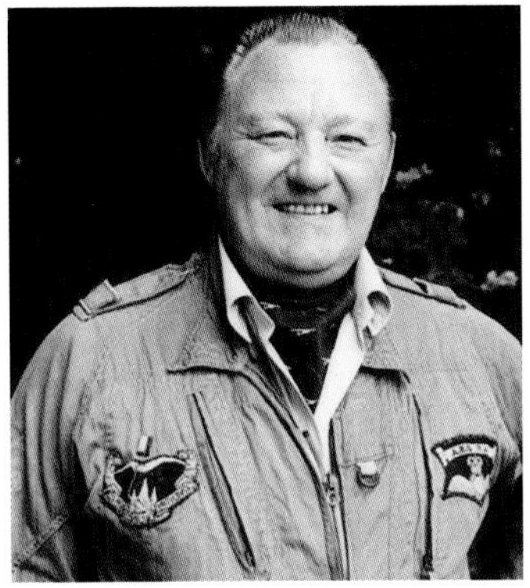

Practising to be an Air Commodore.

Under dodgy colours

Andy Goff and Lawrie were old flying buddies and it was no surprise when one day Andy called Lawrie to ask if he'd be interested in delivering a prototype Dart 17R (R=retractable undercarriage) to an RAF base. This involved a flight of about 40 miles from where the Dart was to the military base and, Andy added, 'You'll have to wear an RAF flying suit, it's a military base.' So they went into the office and came out with a smart looking flying suit for me. So I hopped into the tug aircraft I'd brought along, with its pilot, and we flew to collect the Dart - I needed the tug to get the Dart into the sky, and also to take me back home afterwards. We were collecting the Dart for RAF assessment, that was the whole point of the exercise. All the aircraft purchased for the RAF have to be assessed and, on this occasion, I was the assessor. As we landed and taxied up to the collection point, we were met by a very efficient looking sprog, 'Good morning, sah!' he said with a smart salute and clicking of heels, 'Your aircraft is ready for you, sah!'. I suppose I gave him a casual wave and said thanks very much, or something similar. 'Do you require briefing, sah?' he asked. To which I replied, 'Well, yes, please, I suppose I do.' So I went and sat in the cockpit, surrounded by a brand spanking new aircraft, and I was strapped in by my very helpful friend. 'Now this aircraft has a retractable undercarriage. Have you operated a retractable undercarriage before, sah?' To which I replied, 'Well, not in a glider'. 'Of course, of course, sah. Well, the undercarriage is down at this moment, of course, so what you do, sah, is you prime this levah here, nine times. That primes the hydraulic bottle which will take your wheel up into the fuselage and the doors will close, sah.' To which I replied, 'Of course, thank you so much.' And then he added, 'But remember, sah,' (you never, ever, said don't forget in the RAF), 'Remember, sah, once you're in the air you must prime your pump again, so it's ready for landing. Otherwise, it'll fly perfectly OK, sah.' I thanked him for his instructions, the tug pulled the cable tight, and off we went. Nice ride back, tugged most of the way, until we got to Oxford, which is where I pulled off. Not far from Oxford to Bicester, where I was due to land, so I primed the hydraulic bottle nine times, as instructed, ready for landing. As I flew around I knew I was going to have a pretty high level reception - air commodores, plenty of top brass - all standing around waiting to see this new aeroplane land. The expectancy had reached me, I began to worry about the type of landing I should make; spot on in front of where they were standing seemed like a good starting point. So I rehearsed in my mind how I'd do what we call a 'greaser'. As I did my circuit, I could see them all looking up at this beautiful aeroplane, it was truly beautiful, and then I leaned down and tugged the landing gear lever, which would lower the new nose wheel. Nothing. Not a sausage. It was obviously primed but was just not working, AWOL. The list of options on what I was going to do, could do, was now a short one. Then I had it. I'd do some self-induced longitudinal oscillations. Discreetly, so as not to attract too much attention. So I proceeded into the routine, which was designed to jolt the wheel bracket so that it would go over-centre and force the wheel out of its hiding place in the fuselage and down into a locked position. Nothing. No matter how hard I tried, no movement whatever from the wheel. By this stage I was so low that I'd run out of options and height and there remained only the formality of landing the pristine new prototype on its shiny belly, right in front of the top brass who were expecting to see an immaculate landing by a new glider with a retractable nose wheel. But not permanently retractable, as was now the case.... So I started my final approach, gradually getting lower and lower, handling the craft with kid gloves so as to minimise any potential damage when we finally belly flopped onto the runway. I leaned forward in my seat, and just wafted it down, like a feather, onto the runway, tensing myself for the excruciating tearing as the pristine belly scratched along the runway....automatically, I applied the brakes... why the brakes when there's no wheel down? but it started to slow... hey, that shouldn't have happened if the wheel was still retracted! No bumps, all smooth, probably the smoothest landing I'd ever made... Seemed that wretched wheel had come down, but no indicator to show that. The aircraft manufacturer had forgotten to fit a DI, a daily inspection facility! What a chump I'd been not to spot that when familiarising myself before taking off. And the guy showing me the controls had also missed it, in spite of his efficiency... More sprogs appeared, the cockpit cover was opened for me, I got a smart salute from the air commodore 'That was one of the best landings I've ever witnessed, it was superb!' Back slaps all round as I disappeared into the crowd and left it to Andy and the mob to discuss the new aircraft and its damned 'retractable' wheel. It was only then that one of the guys came across and said, 'You look very good in that suit, sir.' I asked why. 'That, sir, is an air commodore's flying suit you're wearing.' Which explained all that VIP treatment when I first picked up the aircraft. I even got more of it from the egg yokes themselves after I'd landed when they'd realised that I was not quite what I seemed. Really enjoyed that experience, the aftermath rather than the landing... Incidentally, the lack of a little green indicator light when the nose wheel came down led to the only major criticism in my assessment, a factor which was soon remedied by the manufacturer before the RAF got to grips with the glider.'

Grass cutting Bocian

Lawrie did much of his flying in a Polish two seater called a Bocian, including aerobatic displays all over the country. 'I worked up quite a repertoire of stunts which I did over the years. You could do anything in that aircraft, it was magnificent,' he recalled. 'On this particular day, there was an air show on. I was asked what I'd like to do, so I gave them the exact time I'd need the air space for my precisely-timed routine. I'd got a friend with me, Lorne Daniel. He'd never done aerobatics before, in anything, so he was in the back seat. My time came, I pulled off, climbed to 3,000 feet and did a clearance turn, which is a figure of eight, and which enables you to see that you've got enough space for your needs. Just then, I saw we had company, another glider, right in the middle of where I was going to do my aerobatics. I radioed the ground control to get him out of the way but as he had no radio, that was a waste of time. Eventually he moved on and I started my routine but, because of the delay, I was being forced to cram 2,000 feet worth of aerobatics into 1,500 feet. I tried hard to alter it but the last bit of the display was at zero altitude and maximum speed past the crowd, then pull up sharply to do a rate four turn, and come back and land in front of the crowd. I'd got to the point where I was about to go into the last manoeuvre, at about 30 feet and going quite quickly in front of the crowd, when I pulled the stick back to climb rapidly, do a tight turn, then round again. That's when Lorne Daniel screamed out loud. I thought 'what's the matter with him?' Then the aeroplane suddenly jerked. I stuffed the nose down, with a boot full of opposite rudder, flying sideways past the crowd who, by this stage were applauding and waving like mad... not realising that I was fighting for my life. I'd inadvertently hit the deck with a wing tip. By throwing the plane into opposite lock, as it were in car terms, and skidding sideways across the crowd, I'd just managed to prevent the glider from turning over...I learnt a lesson that day. When you've got an aircraft with a big wing span, such as the 60ft on the Bocian, and you're turning, it feels pretty high, but you can easily be deceived as I was that day. When I got out of the airplane, there, trapped between the aileron and the end plate, was a clump of grass torn out by the roots! I'd hit the ground around 70mph and the sheer speed, plus pulling the grass out, had caused me to partially turn sideways and face the crowd - plus my efforts to correct the slide. The crowd loved it, but they didn't know how hard, and for long, my knees were knocking...'

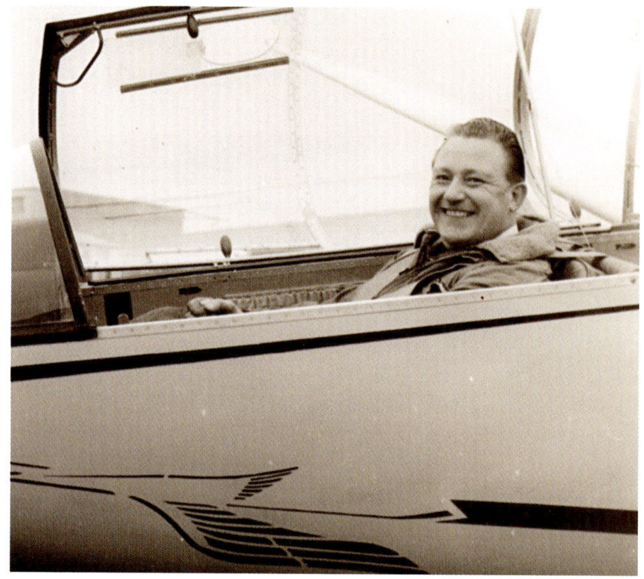

Lawrie proudly displays his first 2 seater Bocian glider.

My big crash

Lawrie led a charmed life in the air. By dint of good flying, excellent judgement and the luck which everyone needs, he avoided any major incidents. Those which could have been regarded as such had usually been brought on by someone else. And this is precisely what caused his one and only serious crash, a misjudgement by a trusted pilot on an innocuous cross country flight. Lawrie recalls the tale. 'A good friend of mine called John Haywood decided to base himself and his two-seater Czech Blanik glider at Talgarth, in the Welsh mountains. The glider had been sent up to Husband's Bosworth for a C of A (Certificate of Airworthiness) and then to return to Talgarth to continue as a training aircraft. The previous week, a two-seater Czech IS 32002 glider had been delivered to Talgarth for assessment, as the Czechs were keen to sell it in Britain. So John asked that I fly the Blanik to Talgarth after it had its C of A in order that I could try the IS. We had a tow from Husband's Bosworth to Worcester, where we pulled off and flew the rest of the way to Talgarth, a little airfield on the edge of a hill and delivered the Blanik, mission completed. It was 18th September 1983, the date is engraved on my memory, for a very good reason…
'Came the opportunity I'd been waiting for, trying the new IS 32002, not with my old ex-Luftwaffe friend, Hans Wolf, because he discovered that our combined weights were too much, but with John Balley, the chief flying instructor at Talgarth. Though I knew little about him, I thought well he must be OK to be the chief here, so I got into the aeroplane with him. With a massive 22 metre wing span, the IS was a massive sailplane, but beautiful with it. As we took off, the CFI at the controls, I couldn't help but notice how short the airfield was - we actually flew, behind a tug, over the roofs of houses in Talgarth at the end of the runway. The tug took us up the mountain and half way up we pulled off and got into the thermals and commenced our assessment flight. Chatting away to the CFI, I asked him to point out anything which he thought I ought to know. Shortly after he said 'You have it,' and I responded with the obligatory 'I have it' and took over the controls and then started to soar in the main lift above the mountains, back and forth, for about an hour and twenty minutes. I was still enjoying myself when the CFI said, 'Lawrie, I think it would be nice if we went over to the Brecon Beacons.' I was perfectly happy where we were and, besides, it was a long way back to Coventry so I felt uneasy at the prospect of a long flight to the Beacons and back again. 'Well it's OK with me if it's going to be a short one.' As he took the controls he replied, 'Yes, it'll be a short one.' as he set off down the line of the mountains towards the Brecon Beacons.
'So I sat back and began to enjoy the scenery, especially the reservoir over which we were then flying. Suddenly, I spotted something very disturbing; the sail boats on the reservoir were tacking into the wind, but we were not. We were sailing down wind. We were also very close to some of the peaks around us. Unless somebody leaves a car parked on the mountain, to be used for comparison purposes, it's difficult to gauge how high you are. The altimeter had not been changed, it still had the setting for the altitude at which we'd taken off. So there was lots of guessing going on! This really did unsettle me. So I said to John 'We are flying down wind and we are heading for the edge of the mountain, Pen-y-Fan.' Still no response. We flew very low over two big spurs and were approaching an even bigger one, Pen-y-Fan itself, at 886 metres the second highest peak in Wales. 'We're still down wind, John.' I shouted to him. Still no response. So I thought maybe he knows more about it than I do, so I'll sit and watch what happens. We drifted up towards the top of Pen-y-Fan until we were about 200/300 feet above the peak. Suddenly, he lost it, right on the edge of the mountain. The aeroplane rotated about its transverse axis and went into a vertical dive. I thought, he's trying to show me something really spectacular and we'll dive into the valley below. Heck, no! A glance showed that John was fighting like mad at the controls but nothing was happening, and we were going straight into the top of the mountain. And we did just that! We hit the top of the mountain with the bottom of the aeroplane, I heard the wheel tear out the back, the bottom ripped out, we were going down, John was pulling the pole back hard as he could, but without response. We must have been doing about 80/90mph bouncing down the side of the mountain towards the edge where it really dropped away… the machine came to a halt and slewed around until the tail was overhanging the precipice. Then it slewed around again, slipped over the edge, and slid tail first down the mountain. The fuselage crumpled up, breaking off in sections, and in so doing cushioned the shock for us in the cockpit. I remember thinking to myself 'I hope this isn't going to hurt.' while at the same time thinking about Eileen, the family, all the people I loved, because it seemed as if this was going to be curtains….then I saw the back of my own head, and in front of me was John, also bouncing around. I distinctly remember this, sitting outside my own body, watching myself as the plane banged its way down the mountainside. Then there was a colossal bloody BANG! What was left of the aircraft disintegrated about us as we hit something really solid and stopped.
'Everything went quiet. Silence. It was then that I realised we were trapped in what remained of the cockpit. I couldn't see

John at all.....so I shouted 'John, you all right?' A small voice answered, 'Hope so! How are you?' 'I'm trapped, can't move!' I replied. When the fuselage ripped open my leg had shot out. Then, when it banged against something else, the fuselage smashed the metal against my leg, trapping it outside the aircraft. My other leg was impaled on a metal stringer which had gone in one side and come out the back. So I was well and truly skewered. Apart from that I was OK, with sundry cuts and lacerations from being bounced about coming down the mountain. John, who was also in a bit of a mess, managed to get out and help disentangle me from the wreckage. Had he been killed, don't think I could have got out on my own. When John pulled the wreckage of the fuselage back to release my left leg, I was then able to pull my right leg off the stringer, which amazed me, because it looked a bad injury, but it never even hurt and didn't bleed much. Bandaged by tearing up our shirts, we found the radio, connected it to the battery and eventually got it working. But nobody heard us, probably because we were on the opposite side of the mountain from Talgarth.

'Nothing for it but to climb back up the mountain, taking the radio with us, and try again from the other side. Not easy in our state of disrepair. When we got there we succeeded in making radio contact, told them where we were and said we'd walk down the mountain and make contact at the base. When we got to the bottom, there was an embankment about 20 feet high over which we had to go. As we climbed up, bursts of gun fire greeted us! We'd blundered straight into a firing range used by the local rifle club! They stopped immediately, came over to tear yet more strips off us until they saw that they were dealing with two secondhand pilots who'd had a nasty experience. ▶

Year 1983 Total Gliding brought forward

Serial No. of Flight	Date	Glider Type	Place of Launch	Type of Launch	Crew Capacity	Time in Air H	M
9129	1983	PUCHACZ	H/BOS	A	P1	0	17
91.30	10-9-83	"	"	-	"	0	21
9131	"	"		"	"	0	18
9132	17-9-83	BLANIK	H/BOS	A	P1	1	40
9133	"	IS 32 MODEL 002	TALGARTH	A	P1	1	25
9134	2-10-83	BOCIANE	H BOS.	A	P1	0	23
9135		"	"	"	"	0	24

Glider pair escape crash on peak

TWO PILOTS from a Powys flying club escaped injury at the weekend when their glider crash-landed on one of the highest mountains in Wales.

The glider crashed 400 feet below the summit of the 2,900ft Pen-yr-Fan peak, the highest in the Brecon Beacons National Park.

One of the pilots was the chief flying instructor of the Black Mountains Gliding Club at Talgarth, 35-year-old Mr John Bally, of Glasbury, and the other experienced pilot was 58-year-old Mr Lawrence Watts, of Coventry.

The glider came down suddenly following a sharp change in air currents as it was crossing the north face of the peak.

Alerted

Officials at the flying club headquarters a few miles away were alerted to the incident because Mr Bally was in constant radio touch.

Rescue teams were on their way within minutes and met the two men walking down a mountain track.

Club officials have been examining the wreckage to decide whether it can be recovered and become airworthy again.

Brecon & Radnor Express
22 Sept. 83

Glider Crashes In The Beacons

The occupants of a two-seater glider escaped injury when it crash-landed on the slopes of Penyfan in the Brecon Beacons at the weekend.

The pilot, 35 year old Mr. John Bally, of Glasbury-on-Wye, and his companion, 58 year old Mr. Lawrence Watts, from Coventry, had earlier taken off from the Black Mountains Flying Club at Talgarth.

After circling over the peaks of the Beacons on Saturday afternoon, the glider lost height and dropped vertically some 400 feet below the summit onto the north face.

Although the glider was extensively damaged the two flying enthusiasts climbed out of the cockpit unhurt.

Never was a cup of tea so welcome as we told the sorry tale in the club house, by which time the rescue car had arrived from Talgarth and we had to relate the tale all over again. This beautiful £37,000 glider was now just a heap of battered metal spread half way down a mountain. When we got back to Talgarth, I thought that they would collapse at the sight of us. We'd set off in a brand spanking new state of the art glider on a short assessment flight and here we were dressed like a couple of blood-stained hobos out of some lurid Western, without horses or glider...Anyway, on the way back to Coventry by road, we stopped at every pub for a top up. It was great, I really enjoyed what I remember of that journey....Eileen knew we were late, because she'd been warned by one of the other pilot's wives that we had been 'delayed', but when I eventually got home, John went in ahead of me to tell Eileen there had been 'a bit of an accident', and then I walked in. Eileen had assumed that I'd been in a car accident, never dreamt it had been in a plane crash. She insisted on taking me to the hospital, where they patched up 38 lacerations. Truly grateful for no broken bones, I was nevertheless very, very, angry for an 'accident' which I felt then, and still do today, was totally unnecessary and which could have cost both of us our lives. In all the time I'd been flying, I'd never had an accident - yes, a few narrow shaves - but I wasn't flying that damned aeroplane. Later, I learnt that John had 700 hours' flying experience compared with my 7,000 hours. Had I known that before we flew, I'd have had no hesitation in taking the controls before we got into trouble. Another lesson learnt the hard way...'

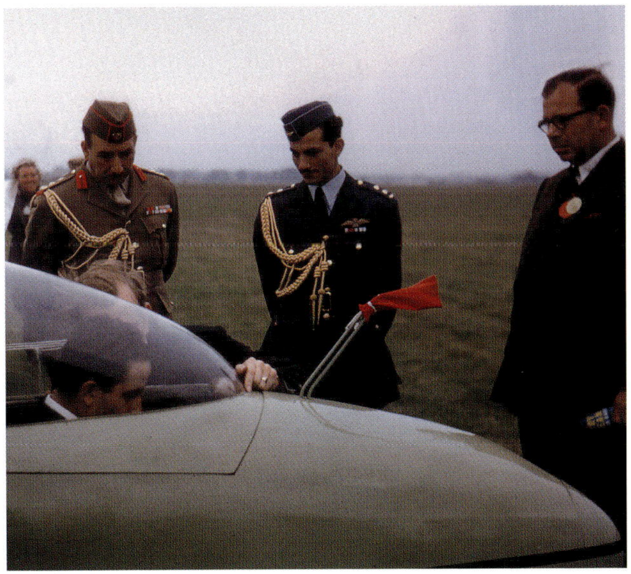

The Coventry Gliding Club was honoured to receive King Faisal during his visit to the Coventry Air Pageant in 1956. Seen here in Doc Gregg's Olympia, Michael Hunt the club secretary explains the cockpit details to his Majesty.

Baghdad to Baginton

Shortly after Lawrie and friends had founded the Coventry Gliding Club at Baginton, royalty dropped in for the day. No, not British royalty, but 21 year-old King Faisal of Iraqi. The occasion was the Coventry Air Pageant, the main event of which was Britain's classic of the air, the King's Cup Air Race at which King Faisal was the guest of honour. This high profile event attracted great interest from abroad in the form of high-ranking naval and air attaches, ambassadors, heads of multi-national industries, strongly supported by top brass from the host country. The MP for Stratford-on-Avon was also present with his wife. At that time his wife was the better known of the pair. Valerie Hobson was one of the biggest stars of the British film industry but the man she married, John Profumo, was to outshine her when, as War Minister in a Conservative government in the 1960s, he was implicated in what became known as the Profumo Affair, a huge scandal which ultimately led to the downfall of the Macmillan government.

Lawrie was there for King Faisal's visit. 'He was an exceedingly nice young man, interested in everything to do with aviation, and we were tremendously honoured to have him visit us. Then he asked if he could try one of our gliders for size. Luckily, a photograph of that magical moment survives which shows clearly the young king's pleasure as he sat in our glider, watched by Prince Abdul Ilah, Crown Prince of Iraq and Mr J.J. Parkes, chairman and managing director of Alvis Cars, which was one of the many motor car firms then based in Coventry.'

When only four years old, Faisal had succeeded his father, King Ghazi, when he died in 1939. Faisal attained his majority in 1953 and, as he matured, he displayed strong pro-western sympathies, which was evident on his visit to Baginton. But eventually this sympathy backfired and cost him his life when a revolution, led by Abdul Karim Kassem, overthrew the monarchy in 1958, two years after Faisal's visit to Baginton. Sadly, nobody guessed that day that they would be looking at the last king of Iraq...

Dolly Trouble 1

It's inevitable that in a long and eventful career Lawrie should have had his fair share of adventures '...more than enough,' he chirruped, 'but especially in the air, when indulging in my hobby of flying. For instance' (now settled comfortably in his favourite chair, cigar about to decant its ash on Eileen's immaculate carpet), 'there was the occasion when I was chasing Mike Costin and Ben Rude over the Cotswolds towards Daventry on the last leg of a race. Only one of us three could be the winner, everybody else was out of sight, way behind. Clouds had been building up steadily when I saw Ben disappear into the base of the cloud bank ahead, Mike followed him. I dived down through the same gap. Suddenly, all hell broke loose. Thought I was being machine gunned, deafened at the same time by the racket as giant hailstones battered the canopy and wings. All the lift had gone, I was coming down like a brick. Trying to go ahead was not an option. So I banked over the town beneath, Daventry, levelled off, and glanced around for somewhere to land. It had to be very close because by this stage I was low, very, very low. Ahead was an estate, the last thing I wanted to see, or hit. Then fate dealt me a tempting hand. Just over the rooftops, I could see a small playing field. It would be tricky, but I knew I could land the glider on that field if only I could reach it. And there, heading straight to the field, was a road. I could fly down that road if there were no telephone wires crossing it. So I gave it a crack, prayed harder than ever before, and flew down the road, over the gate at the end and plopped down onto the field. Soon as I braked, the hail gave way to the mother of all snowstorms. Within a minute or so I was in a white out, could see nothing, couldn't even open the canopy because of the snow build up.

'Nothing to do until the snow stopped, which it did after a time. Eventually, I eased the canopy open, dusted off the snow, and then it happened. Kids, hundreds of them! They'd seen me come down, took shelter where they could, then invaded the playing field to get a bit of me, my glider, anything as a souvenir, especially those who were trying to clamber onto the wings. I was surrounded, a prisoner in my own glider, shouting for HELP! It was at this moment that a lovely apparition appeared, a comely young lady who seemed to have some power over the children. In an authoritative voice, she commanded them to BEHAVE! Do not TOUCH that aircraft! Leave the pilot ALONE! It did the trick. I climbed out and radioed Eileen to let her know where I was, for collection by the rescue trailer. Then my saviour smiled sweetly at me. 'Poor man! Would you like a nice cup of coffee? And cake? These men here will look after your aeroplane.' Would I heck! As I was tidying up, she chatted away amiably. Then do you know what she said? 'I was lying on my bed looking out the window, wishing something exciting would happen, and then you flew past my window!' Eileen must have sensed that there was more than coffee in the air because at that precise moment she appeared on the playing field, on her own, with a 32ft trailer attached to the car, having made record time over snow-strewn roads from Husband's Bosworth airfield. Then the television people appeared, followed by radio, photographers, newspaper reporters. It was all happening, just because I'd been forced to put down in a playing field in a snowstorm. Fame at last! Not only that, but back at base when I was stepping out of my flying suit, I found a piece of paper in the breast pocket. Puzzled, I unfolded it and saw a phone number and a message 'Come and see me any time. You will be most welcome!' I showed it to Eileen and we had a good laugh. However, the following Saturday, the telephone rang and it was only her, asking Eileen if I could go and see her! What a nerve!! I put the note, and a newspaper photo of the Daventry happening, on the Club notice board. Somebody identified her. She had a husband, an American, who was based at Brize Norton. I'm quite sure that he was much bigger than me. Which is why I gave Daventry a wide berth in the future, on ground and in the air...'

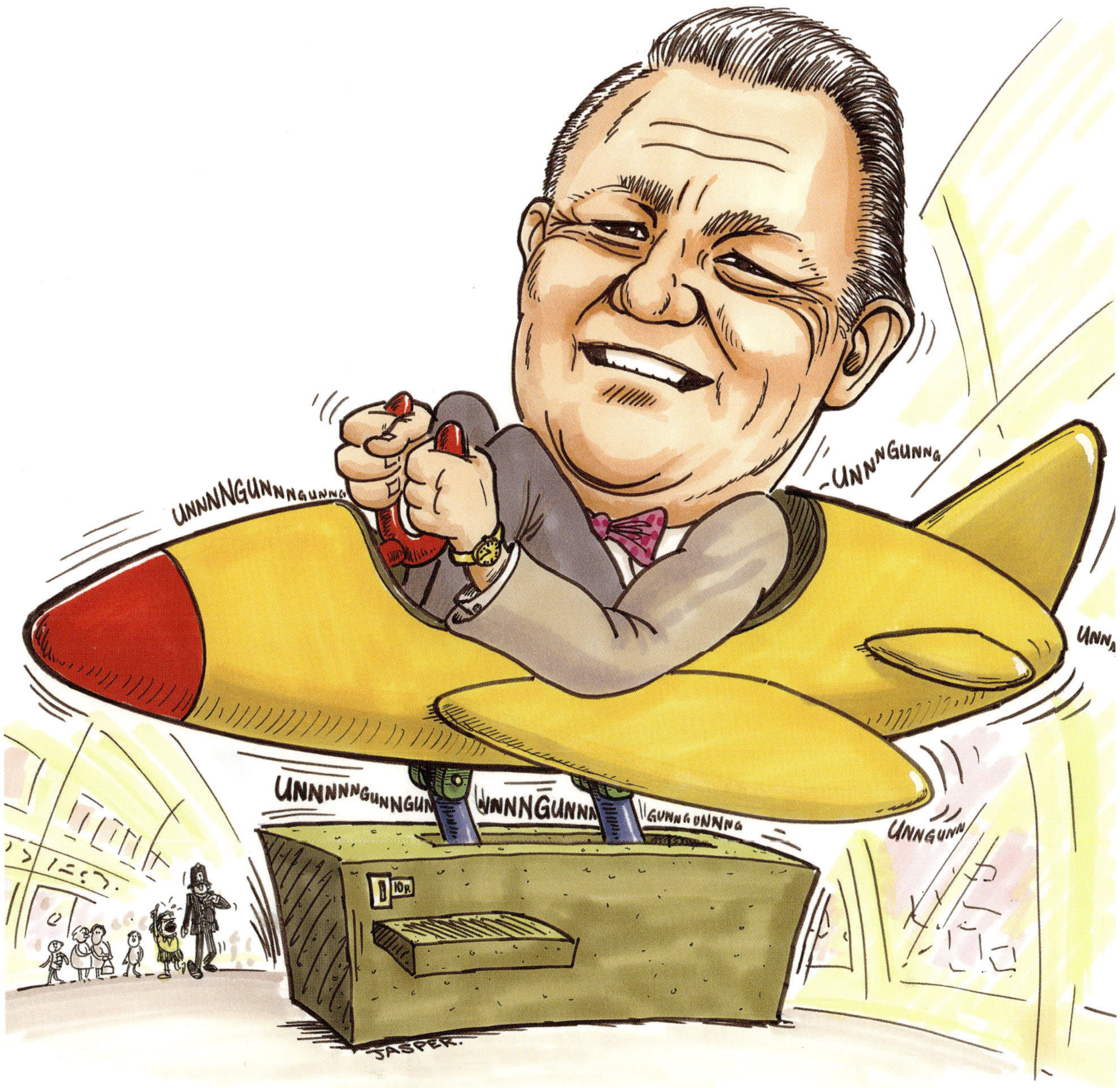

Nasty little nipper

Slingsby production was not all gliders, as Lawrie reminded me. 'Slingsby also designed and built some light aircraft, one of which was the Tipsy Nipper, which had a tricycle undercart, a single seater powered by a VW engine. Very small, a fully aerobatic aeroplane, you didn't know what you were letting yourself in for until you took off....then you couldn't see anything! Mainly due to the seating position between the wings, you looked either side and all you could see was wing, and more wing. As the nose came up during take off, you could see lots of nose, lots of wings but sod all else! Then, as you levelled off, the world (or some of it) revealed itself. Just as interesting when you were coming in to land. You could tell the angle of attack by looking at the horizen but you'd no way of sensing where the tricycle undercarriage was, just hold the nose up and pray that the two main wheels touched down first as you throttled off above the ground. If you misjudged it you just sat and waited, then......bang! You were on the ground (usually). For that reason, Tipsy Nippers became quite unpopular, and I don't think Slingsby made many. Two of my mates, Andy Goff, who was in charge at Bicester where I flew a very great deal, and Derek Pigott, both flew WW1 replica planes in the film 'The Blue Max'. Derek had experienced the 'Tipsy Nipper Feeling' - lack of vision when taking off and landing, in his replica Fokker triplane. His seat in that was completely surrounded by wings both above and below, a fatal design fault in those early days which was to cost many airmen their lives in WW1 combat because they were unable to spot 'the enemy' before being shot down. Which seemed to indicate that the message had not been absorbed by the Tipsy Nipper's designer, even if it was not intended for combat. Which is all the more remarkable when I tell you that six half size replicas of the RFC's (Royal Flying Corps, precursor to the RAF) original Sopwith SE5 aircraft were built at Slingsby's! They were big enough to fly, but because of camera angles etc, the difference in size did not show on the screen.'

Time for re-training on a flight simulator! This cartoon by Jasper was presented to Lawrie on his retirement by the editors and staff with whom he worked.

Dolly Trouble 2

Another adventure of Lawrie's also had unforeseen consequences.....'I was flying a Slingsby Sky, a large 18 metre sailplane built for the world gliding championships - which, incidentally, it won - and I happened to be at Haverfordwest. However, it was not going to be a good soaring day. On these occasions, it was the club's habit to go to the naval air base at Templeton, near Tenby. I was asked to take the Sky. This necessitated taking the dolly, which comprised two small wheels on a little carriage which hooked under the glider and was used for manoeuvring gliders on the ground. The plan was to tow me with the tug to the other side of the airfield, then drop me, whereupon I was to go into the circuit, fly down the runway, drop the undercarriage (the dolly), fly up and come back down. I went into a routine, did the circuit coming into my finals down the runway at speed and, feeling like a bomber pilot, I pulled the lever to drop the wheels. Then I saw the wheels in front of me. Hitting the deck had dramatically increased their momentum, so past me they went while still climbing! Instinctively, I ducked as I flew under the wheels. They missed the cockpit by inches. I thought that they would at least knock off the tail plane, but they must have missed that too, by millimetres. Breathing again, I continued the circuit as planned, then landed at roughly the point where I'd last seen the flying wheels. I climbed out of the glider and started looking for them. Nothing. The grass had not been cut for ages so it was now tall enough to hide a pair of wheels, and much else. Floundering around in the grass I was suddenly confronted by the biggest, most ugly and frightening man I'd ever seen in my life. He stood up in front of me, complete with woolly hat but not a lot else on. I thought he was going to flatten me. As I backed away, to my amazement a great big toothless grin spread from ear to ear and he pointed down into the grass. Gingerly, I walked over and there in the grass, was my dolly. But so was his dolly! Then I realised what had happened; the wheels had bounced off the runway into the grass and ended almost of top of my grinning friend and his lady. A classic example of dolly interruptus!'

The errant Dolly is held by the helper (right), as the man on whom it was (almost) dropped found his clothes and came to investigate the man who did the dastardly act. Somehow, he seems to have vanished.

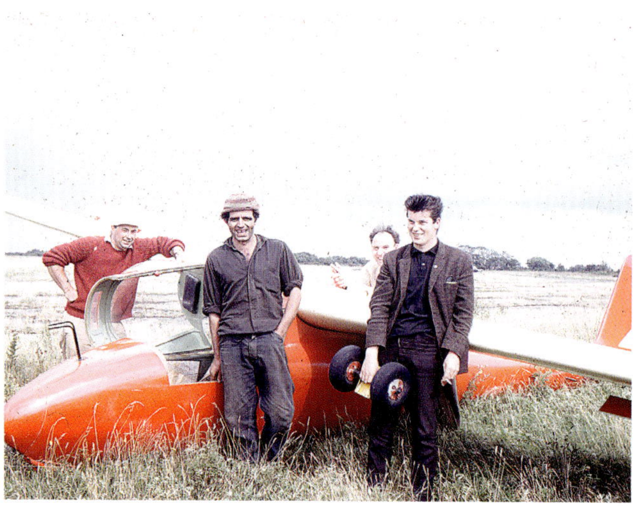

Test flying the Skylark glider.

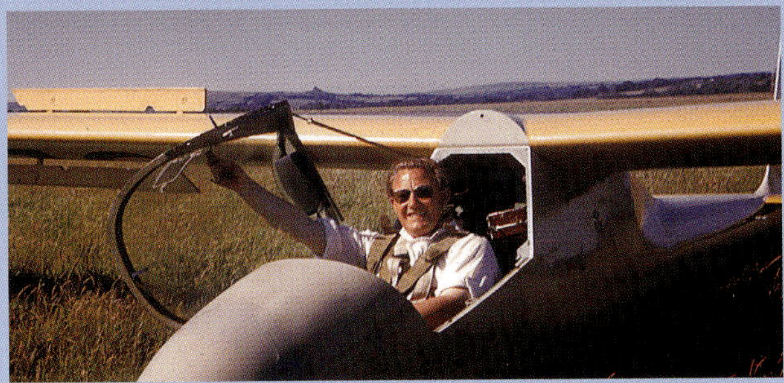

Inspector Watts

'When drawing the Slingsby Swallow, I had to work very closely with the chief engineer, Bailey Woods. One day I asked him if I could take an examination to become an airframe inspector. Which is how I spent a week in the factory doing training and so on. When that finished, the inspectors came in, examined me, and I got my ticket as an airframe inspector without too much drama. Many years later, having examined very many aeroplanes and gliders, I've still got my ticket, renewable annually. My employers, Iliffe Press and 'Flight' magazine, backed me all the way, as they always did. It means that I can do airframe inspections and issue certificates of airworthiness.'

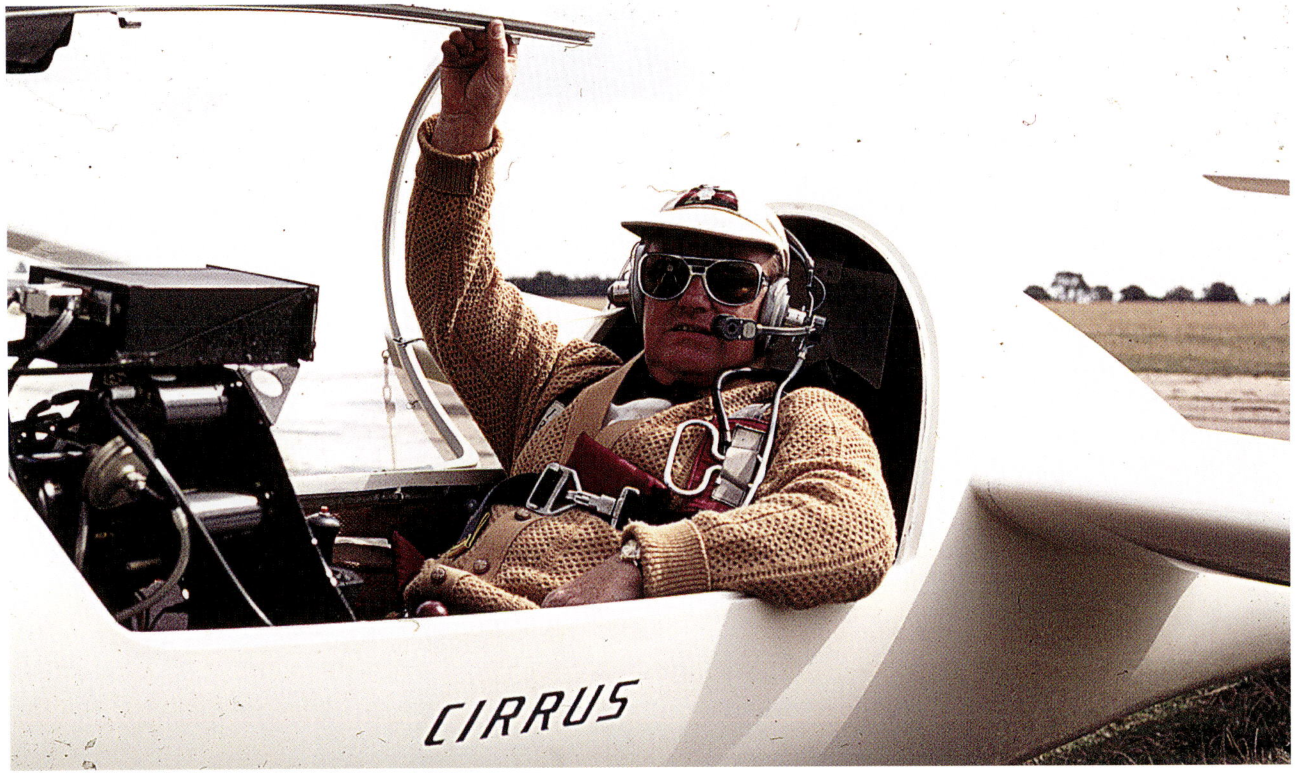

Above: Lawrie in his Cirrus 550 .
Below: Pretty lines of the Cirrus shown in this static shot.
Left: Taking a home-built aeroplane on its maiden flight near Haverfordwest, Wales.

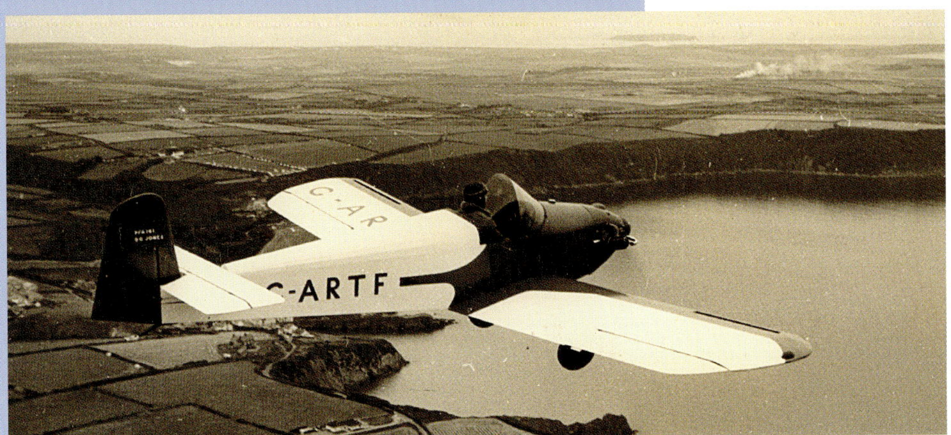

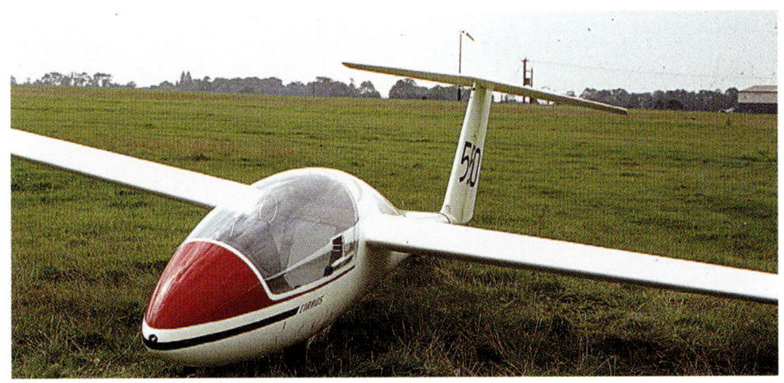

Watts' private air force

Lawrie's qualification to inspect came in handy one day when Andy Goff, who was in charge of the RAF's Gliding and Soaring Association at Bicester, asked Lawrie if he'd like to repair one of their gliders. He said that as the RAF was not allowed to pay for the work, he'd give an unserviceable glider in exchange for the work. Lawrie agreed, as he could see a means of getting his own gliders, once they were repaired, for nothing more than blood, sweat and no tears. As he explained, 'Seemed a fair deal to me, so I agreed. At that time the RAF were to-ing and fro-ing between Germany and the UK and anything which looked like being destroyed in Germany, they transferred to the UK. As I repaired each glider for the RAF I was given a broken one for myself to repair at leisure. So, after quite a long period of hard work, that's how I acquired five gliders, one and two seaters, of my own. They called it Watts' Private Air Force!'

That's how Lawrie started his own gliding club at Edge Hill, named after the battle which took place there many years before. His closest oppo at Edge Hill was Clive Bennett, the same Clive whose day job was development engineer at Ariel motorcycles. Lawrie explained, 'If we were to be successful, we needed a winch to launch our gliders. So Clive said he'd search around and, sure enough, rang shortly afterwards to say he'd found one. It was a balloon barrage winch, but without the truck or wheels, just the winding mechanism and 2,000 feet of 1,000 lbs breaking strain cable. Exactly what we wanted! Only snag was that it was miles away, in Birdlip, Gloucestershire. Somehow, we had to get it to Edge Hill and, when we got it there, it needed to be mobile. 'No probs, leave it to me,' grinned Clive, in his invariably cheerful manner. Few days later, 'I've bought a truck,' Clive announced. Great, I thought. But how do we get a ton of winch onto a flat back truck, especially as the winch, I'd now discovered, was on a concrete platform? 'No probs,' responded Clive, 'I'll crack it.' A week later he phoned, 'Job done, winch is on truck.' Flabbergasted, I asked how he'd ▶

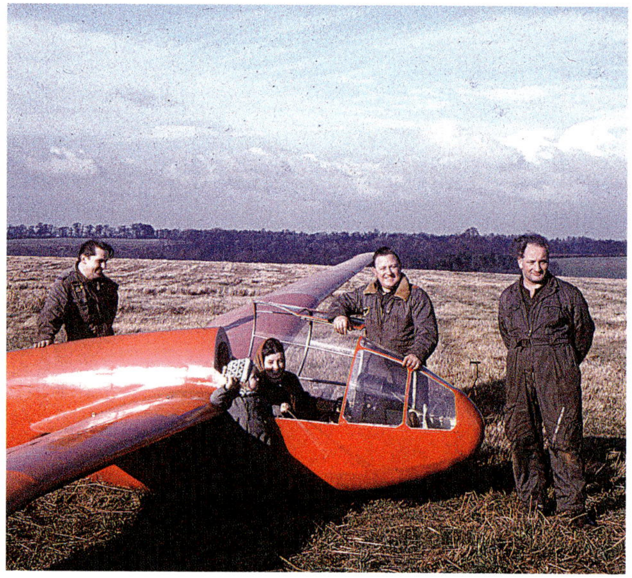

Edge Hill days; Lawrie (centre) with his Rhineland glider and great friend Clive Bennett on the right. Clive was a tower of strength, nothing was too much trouble for him. Lawrie taught him to fly and they enjoyed many happy hours in the air and on land.

done it all on his own, as I'd been expecting to help him. 'Simple,' he replied. 'I dug the ground out in front of the concrete base on which the winch was mounted, backed the truck into the hollow, attached the winch to the back of the truck and wound the winch forward onto the truck. No sweat!' An engineer's solution to a seemingly intractable problem. But that was Clive, nothing ever got him down, headaches were there to be solved. Nothing for it but to form our own club, the Edge Hill Gliding Club, and in no time we had attracted a lot more members to keep Clive and me company. Subsequently, a lot of flying was done from there, some of it by motorcyclists. I've always found that good motorcyclists make good flyers, it comes naturally to them because turning is the same on a bike or with a glider, you bank over. But someone who's not ridden a cycle or bike, they invariably lean the opposite way when they are banking in a glider, whereas motorcyclists automatically bank over with the sailplane. Clive is a good example of what I'm saying. He was an extremely good rider and he turned into an equally good pilot, learning quickly and with excellent co-ordination and reflexes. And all the while, he went about flying as he did on his bikes, smiling and getting a real thrill out of it. A marvellous bloke, a great friend, and I still miss him terribly, even though he's been dead more than 20 years...'

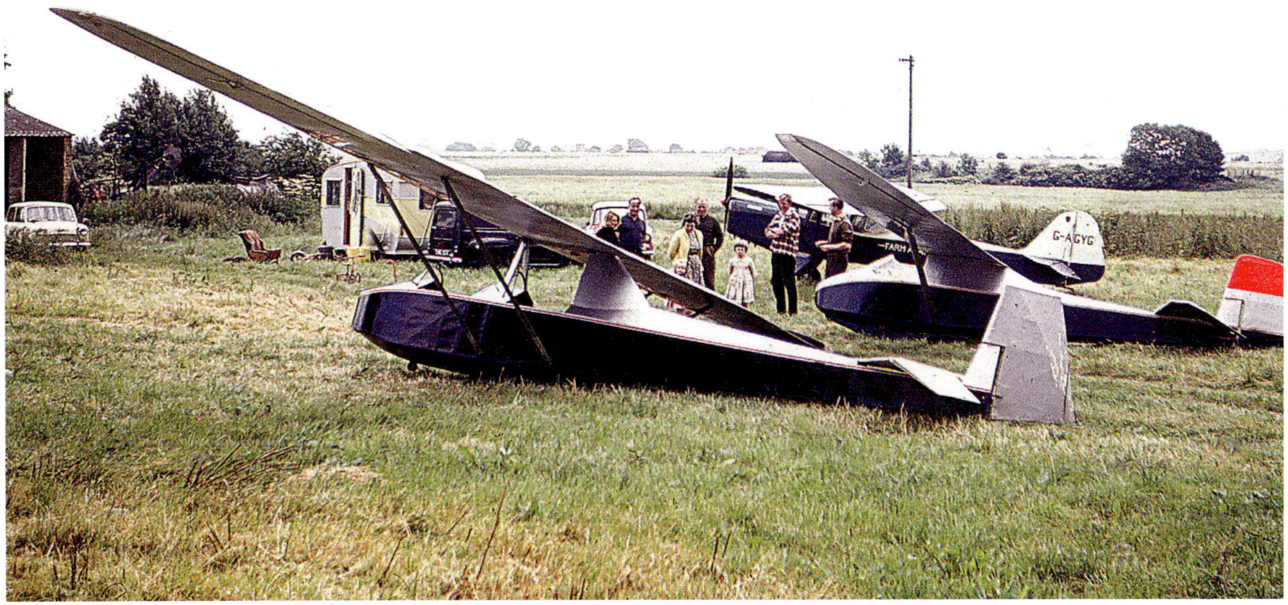

'Watts private airforce' at Edge Hill. Most of the aircraft were repaired RAF units, donated to Lawrie in lieu of payment for work done. The white Mini is one of the first production batch, Lawrie still has it.

Close shave in the Flying Barge

'One morning we decided that we were going to fly the Slingsby T21, a two-seater side by side open glider, unkindly nicknamed the 'Flying Barge'. We would have a tow out over St Bride's Bay, Haverfordwest, maybe 5 miles from the coast to the airfield. We were towed out another 5 miles, it's quite exciting flying a glider over the sea, more so when the tow cable snaps. You're five miles out to sea, the towing aircraft is rapidly disappearing towards land, his mission seemingly accomplished and ours just about to start. Because we were no longer behind him, the tug pilot assumed that we had released ourselves, little knowing that the release had more to do with a snapped cable than anything we'd thought of doing. My oppo, who was also a flying instructor, happened to be flying the T21. It's infra dig to take the controls off the person who's flying without mutual consent. In fact, that's the law of the air. So I let him get on with it. Soon, it dawned on him that we had to get back home, so he did a turn which, under the circumstances I did not judge to be particularly clever because we desperately needed to preserve every inch of height. It soon became patently obvious, from our rate of descent, that we were not going to return to where my oppo was aiming for. So, to wake him up, I said, 'Will you check what speed this aircraft stalls at?' He replied, 'Not bloody likely!' So I took control, pulled the pole (stick) back, still not stalling. Did it again, this time I noticed that the sea was getting farther away. Good, we were going up. Then I felt slight buffeting, the first signs of stalling. This was nearly 15/20mph slower than my oppo had been flying! In other words, he was diving and causing us to lose height hand over fist in his efforts to get back to land. What we actually, indeed urgently, needed was to keep as much altitude as possible. The instruments were obviously faulty, especially the air speed indicator, and we were getting duff readings. That's why I'd asked him to check our stalling speed.

'Closer and closer, we edged nearer the cliffs, shortly they would be within our range. Hopefully, we'd reach them. There we would find some lift which would enable us to climb up and over the cliffs. Once over them, we would have ground beneath us where, if the worst happened, we could land. By this time it was quite dark, the only way we could make out Haverfordwest was by its lights. Between Haverfordwest and Littlehaven, where we wanted to go, was a long straight road. With luck, we could follow the lights of cars straight down to the petrol station. There we turned left, and flew down that road until the next petrol station, then turned right. We were bang on line for the end of our runway. As the plan was unfolding, I was reminded yet again of a strange phenomenon; when you're flying in the dusk, you can see quite well, but when you descend lower, it is much darker and harder to see and very much riskier.

'The tug pilot had obviously warned my mates that we were in a spot of bother because, as we lined up for landing, the runway suddenly lit up like daylight! My mates had lined up ▶

their cars either side of the runway. Soon as they saw our glider silhouetted against the night sky, they turned their headlights on. This inspired me to make a cracking landing in thanks to their generosity of spirit. Once on the ground, I couldn't get my flying suit off, glued to my back with sweat! That was the nearest I'd been, or ever wished to be, in the drink. Only afterwards did I feel the full impact because at the time I was far too busy trying to nurse the aircraft back to land to worry about the consequences. When the chief instructor saw me the next day, he commiserated about the near squeak we'd had adding, as an afterthought, 'Oh, yes, that reminds me, I forgot to warn you that the air speed indicator was reading 15mph out....' In real terms, that's a 25% margin of error, sufficient for a charge of illegally driving on the road. In the air, it could be called criminal negligence. Anyway, being an inspector myself, I read him the riot act. Not trusting him to do anything about the matter, I removed the instrument panel myself, replaced the faulty instruments with new ones, and re-fitted the panel before anyone else was put at risk.'

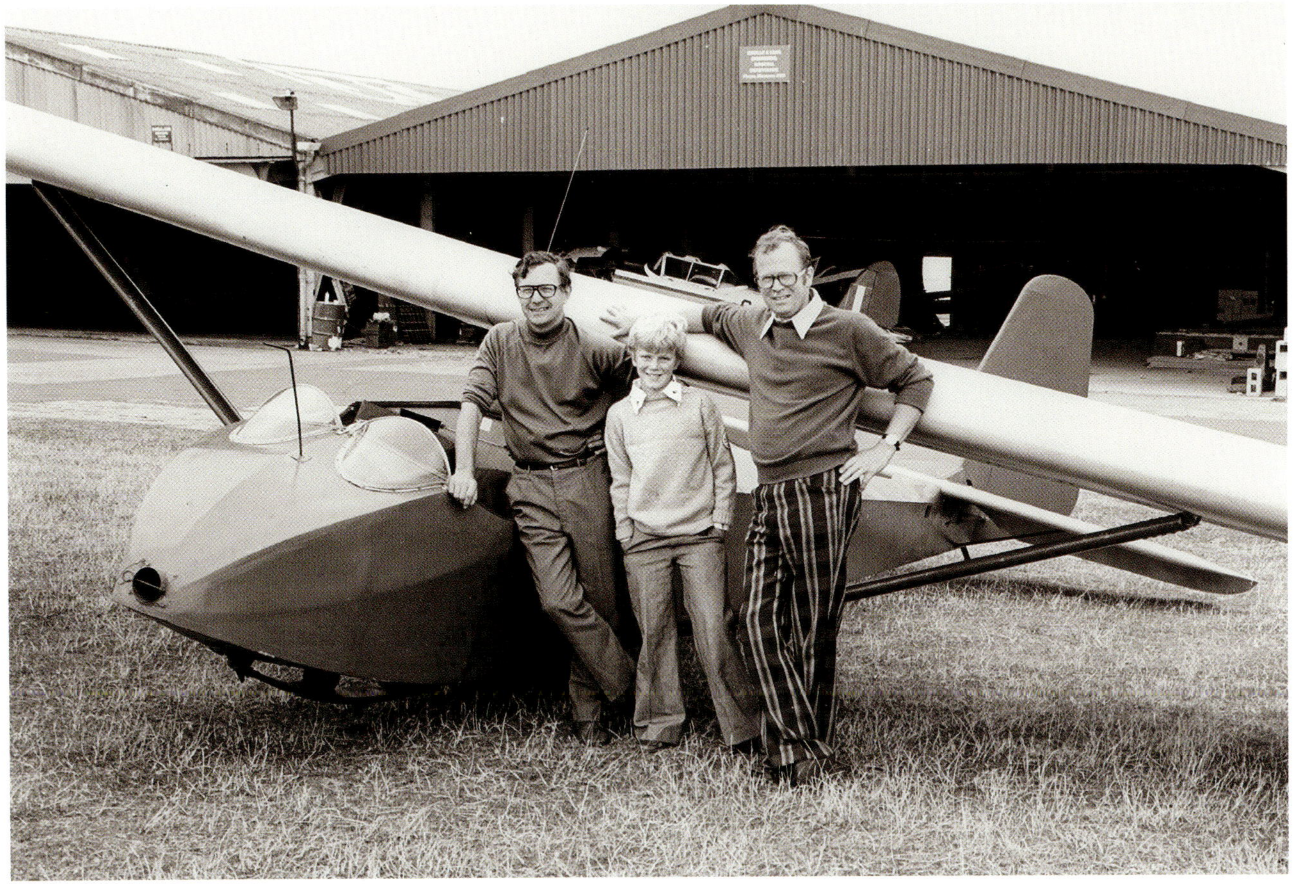

The Slingsby T21 'Flying Barge' two seater in which Lawrie had a harrowing experience at Haverfordwest. On the left is Guy Gothard his nephew and brother.

French frolics

Some of Lawrie's adventures happened while he was just sitting there minding his own business, an innocent abroad (if you can believe). One such incident was when he accepted an invitation in the mid '80s from Tony Turner for a trip to France. Tony wanted to sign a contract, and fly back next day. 'It was early morning when we met at the airfield to fly in Tony's four-seater Beechcraft Bonanza to Nancy, in France,' recalled Lawrie. 'We took off from Alcester in grotty weather, straight into cloud at 500 feet, on course for Southend where we would clear Customs before heading over the Channel. We remained in radio contact until we were out of Birmingham traffic control, flying south-east on instruments until we started our descent to Southend. We'd not seen any ground since take-off. I was watching the instruments as we came down, and down, and down. Nothing to be seen but cloud outside. By this time I could see from the instruments that we were only about 200 feet from the ground. So I looked around to see if I could spot anything

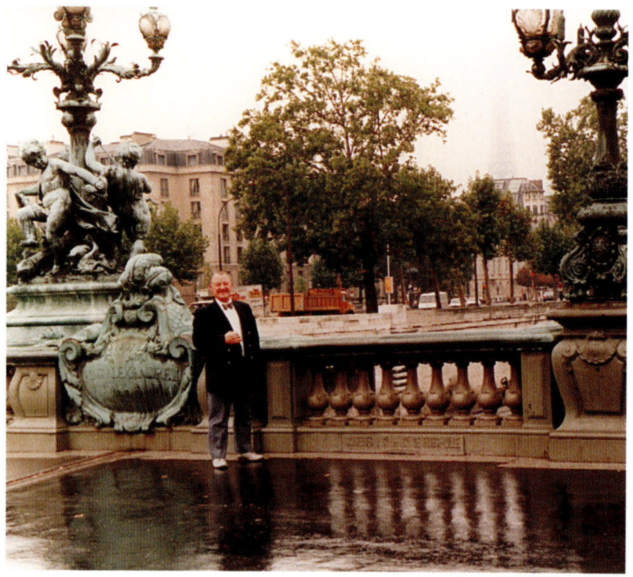

'I love Paris in the Springtime' - Lawrie settles for a drizzly day to survey the Seine.

and there, rushing past my wing tip were masts, lots of masts, of sailing boats at anchor! We went over the threshold lights, which guided us onto the runway, without realising that we were on the deck! That was how smooth a pilot Tony was. He knew exactly what he was doing while I, like a nervous car passenger, was fretting away.

'Anyway, we cleared Southend Customs, then took off for France and straight away into a thunderstorm which continued all across the Channel. The artificial horizon was doing its own aerobatics. We were going in a reasonably straight line to where we wanted to go but the attitude of the aircraft was anybody's guess. Eventually, we joined French aero space, got permission to fly a slightly different course away from all this grot, and in due course landed at Nancy. Tony did his biz, we stayed the night and left early next day. The weather had not improved much and getting clearance from the French to take off in such weather was not as easy as it is, or was then, in the UK. Eventually, when talking to the air traffic control lady, she offered an alternative plan. 'Gentlemen, I can give you clearance to fly to Calais if you are prepared to take 500 feet'. Being two experienced flyers, we did not wish to look too pleased at her offer, but in a state of the art aircraft, such as the Bonanza, it would be a fabulous experience. So we shrugged and said 'If that's the best offer, we'd better take it,' and signed out. Tony asked me to get the aircraft re-fuelled while he attended to other matters. So I took the Bonanza to the other side of the field and while re-fuelling I saw two red kites take off. I'd never seen one before and, in my excitement, drew the attention of the guy re-fuelling the aircraft. 'Ah, oui, we always stop flying while the kites are in the air, sir. They have precedence here.' So we had to hang about while the kites did their stuff. I was not too unhappy about this because I'd always wanted to see these magnificent flyers, a chance denied in England because of their rarity (they have since recovered and are seen frequently in Wales and have also spread east to parts of the home counties). Eventually, they landed and we took off, climbing straight to our 500 feet maximum. Because of hills, we chose to fly down valleys, with houses sometimes above us, all the way back to Calais. Before departing, we'd been warned about a non-flying area somewhere mid-way on our route. After a time I said to Tony, who was flying, that I thought we might be in the no-fly zone. He was too busy guiding us through the forests, valleys and hills to pay much attention. As we rounded a vineyard and went over a forest, I saw an area which had been cleared in the middle of the forest. There, in the clearing, were inter-continental missiles, all at the ready, raised to an angle of about 40 degrees, spread across the clearing. As I looked down upon the soldiers, they looked up, as startled to see us as we were them. Obviously, we were flying below their radar so they had no warning of our approach. We were also flying too fast, too low, for them to do anything, not even to get our identity number. As we flew on, I said to Tony, 'Did you notice which way those rockets were pointing?' He replied thoughtfully, 'Ohhh... yesss...' I said, 'They were pointing in exactly the same direction we are going, on course for Calais - London.' That they were pointing at the British Isles is without question, and London too. Nor were they on gimbals, they came straight up out of the ground and were on ramps, ready to go.....no question about it.' 'Let's get this straight,' I butted in, 'The French were, and still are, our allies so why in God's name were they pointing ICMs at us?' 'Don't know,' grunted an unhappy Lawrie, 'but they were, no question.'

After a long pause, Lawrie resumed, 'I still thought that we could be in big trouble when we reached Calais. We'd flown along the Belgian border, and came in down the coast, which was quite a good way to approach Calais. The runway there is east-west, and as we flew in over the coast to make our landing we passed over sand dunes. I couldn't believe it, there were craters everywhere, as if the war was still on. You can't see much of these craters on the ground but from the air they are clearly visible. Anyway, we landed, taxied up, and the control officer appeared with his papers, and asked 'Cup of coffee, gentlemen? You're not in a hurry, are you?' This was not the reception we were expecting, so we hastily accepted his friendly invitation, 'Yes, please!' We had our coffee with him, signed our papers, nobody bothered to check our aeroplane, so we climbed aboard and took off for England. From that day to this, not a dickey bird about us over flying their secret site with its ICMs pointing at England...'

Grasshopper dream

Poking my nose into Lawrie's garage one day I saw a contraption consisting of a backbone, what looked like a rudimentary tail fin, together with some wires and controls resting on the roof of his old Mini, one of the first production batch. Knowing Lawrie, I suspected that a story lurked around the corner. So it proved. 'That there heap of rubbish, as you termed it, is the source of the most pure flying experience you could ever imagine,' he explained. 'It's part of a Grasshopper, a primitive glider which once belonged to the RAF. Some years before I acquired this, I'd flown an operational one which provided me with the most exquisitely pure flying experience. Almost crude in its simplicity, the glider consisted of a single backbone with a rudimentary tail, a seat up front and detachable wings and tail plane controlled by wires. That was it. Because the pilot sat out in the air, with nothing between him and the ground except the spartan seat and footrests, it provided an experience like no other. Flying in the raw, like it was in the earliest days. To ensure that nothing was over stressed, the RAF imposed a speed limit of 14mph, yes, fourteen miles per hour! To ensure you knew when you'd reached that almighty speed, they had fitted a little flap within the pilot's vision. Never mind the fact that the tug which towed the glider into the air must have been doing 60mph to get airborne (!), I was totally smitten by the experience - not dissimilar to first experiences in a go-kart - and when the tug was finally released at 3,000 feet, I knew I was going to enjoy the experience. First problem was to get the speed down to the regulation 14mph, so I pulled the stick up and we shot up another 500 feet until I got the speed sorted. Then cruising along, I could not only study the world around me but also 3,500 feet below my feet. The Grasshopper responded well to the minimal controls and once I'd become used to the snail's pace, I really enjoyed myself. It was so different from any 'normal' sort of sailplane, and as it took me back many years to when I first flew, that I determined to own one. Easier said than done, 'very rare', 'much sought after', 'never come to market', etc. Years later, a letter from the RAF informed me that it 'might' be possible to have a Grasshopper, if I presented myself at a certain place on an appointed day. There was not one but two Grasshoppers there, one badly broken, the other repairable. My bid for that one was successful, which is how it joined the Mini, waiting for some TLC, and the day I can fly it again....'

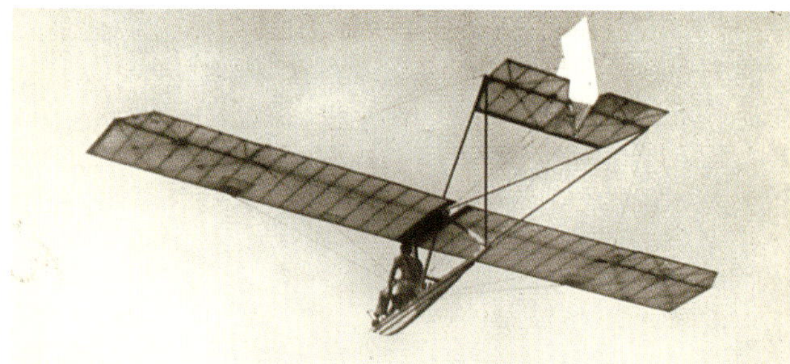

The 'Grasshopper' training glider referred to in the story, from which modern microlights have evolved.

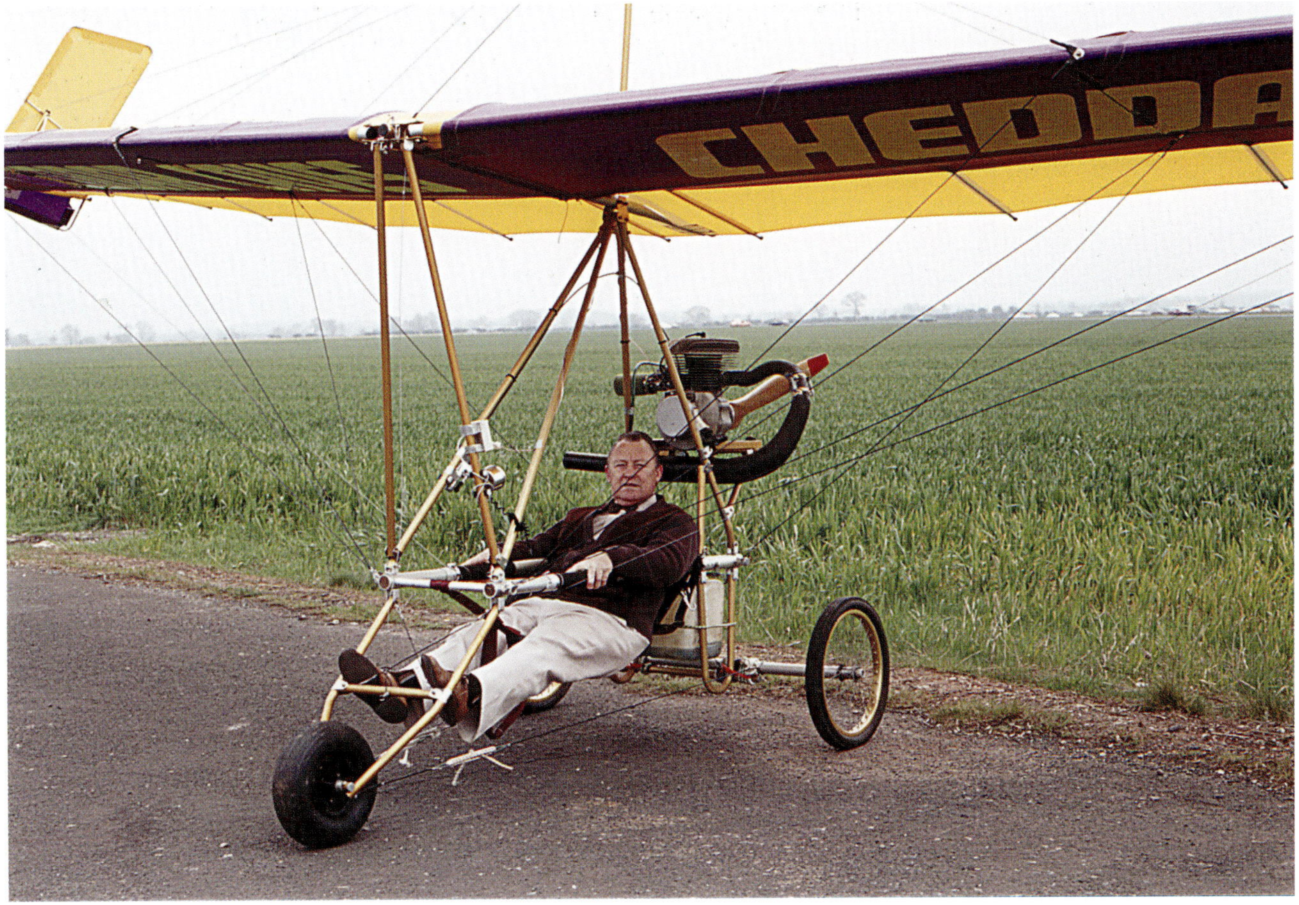

Successor to the Grasshopper is the microlight, seen here being sampled by Lawrie. It provides 'the most exquisitely pure flying experience', at modest cost, like in the early days of flying.

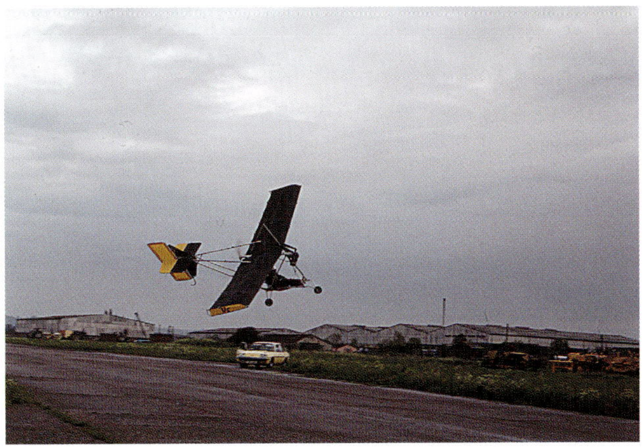

Right: A Triumph publicity photo taken at nearby Baginton Airport with an Olympia glider, Ron Glover (club engineer at controls), and an Israeli model talking to flying-suited Lawrie (left).

Devon nightmare

While instructing in North Hill, Devon, Lawrie was awoken one night by a terrible noise, as if somebody or something was banging inside his head. 'The pain was unbelievable, I woke up shouting for help. Then I lost my vision, couldn't see. I'd made such a racket that everybody in neighbouring caravans came to see what was up. The nearest neighbour, a police sergeant on a gliding course, had the presence of mind to pour some water in my affected ear, the left one. The banging stopped and the pain seemed to go, but it was still damned uncomfortable. Meantime, young daughter Rosemary had gone down to the main road to await the ambulance, which had already been summoned. They couldn't see what the trouble was either, so they took me into Honition hospital where a doctor had been called in to treat me. In syringing my ear, he discovered a pine moth which had climbed into my inner ear. He showed it to me and I could see that the pine moth has a long proboscis; this was what he'd been sticking down my ear and which caused so much noise and pain. The policeman's water had drowned the moth but it was so far inside that the policeman couldn't see it. The next day, I paid for it, this time with headlines in the local paper 'Moth-eared Pilot In Emergency', followed by television and radio interviews. It sounded funny, especially as the media went to town on it, but it was one of the most painful experiences I've ever had. It was only later that I realised North Hill airfield is bordered by pine trees…'

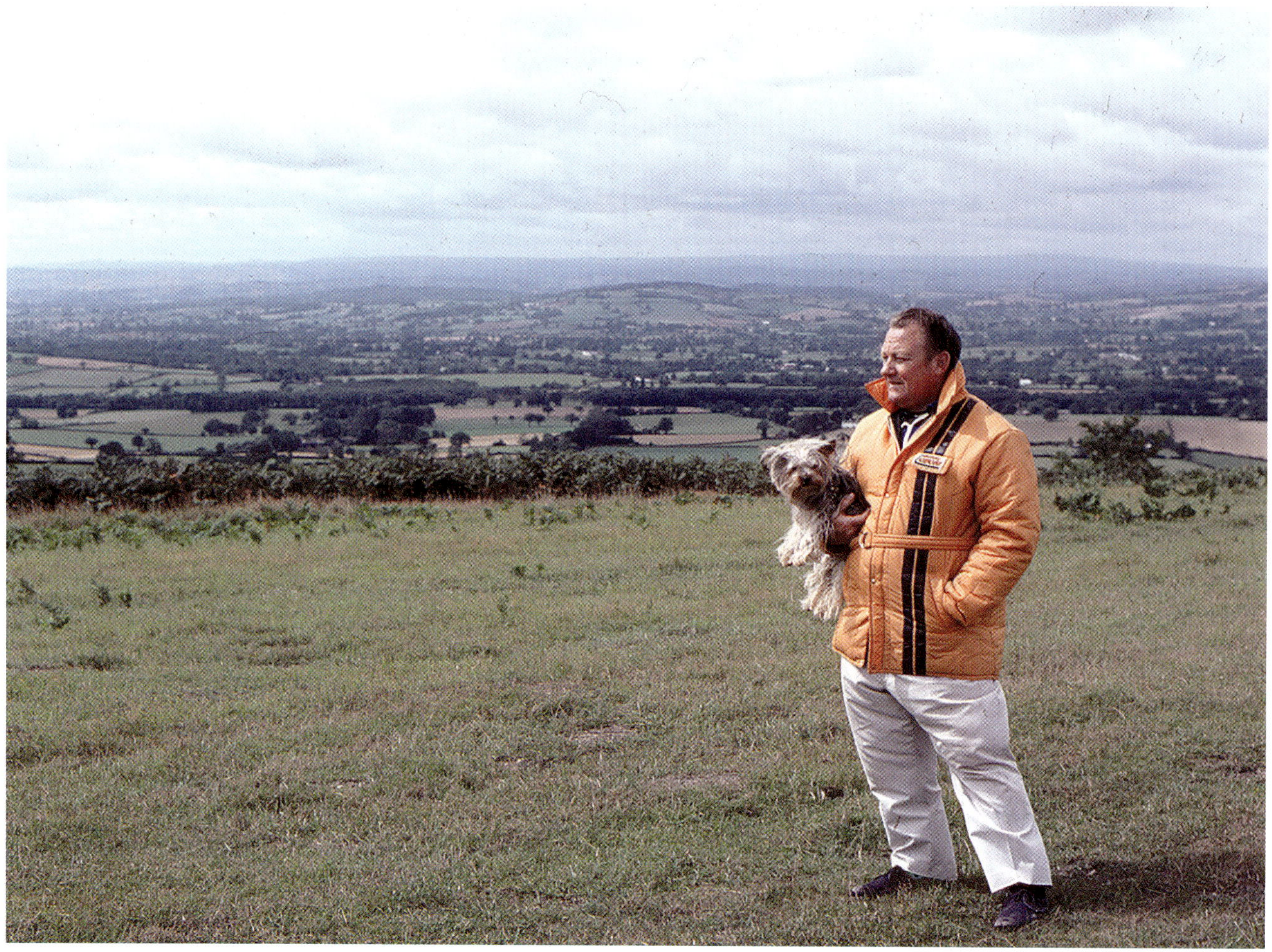

Before flying could start at North Hill airfield, 200 sheep had to be rounded up. That was done by Lawrie's dog, Pep, seen here with him overlooking the Devon countryside.

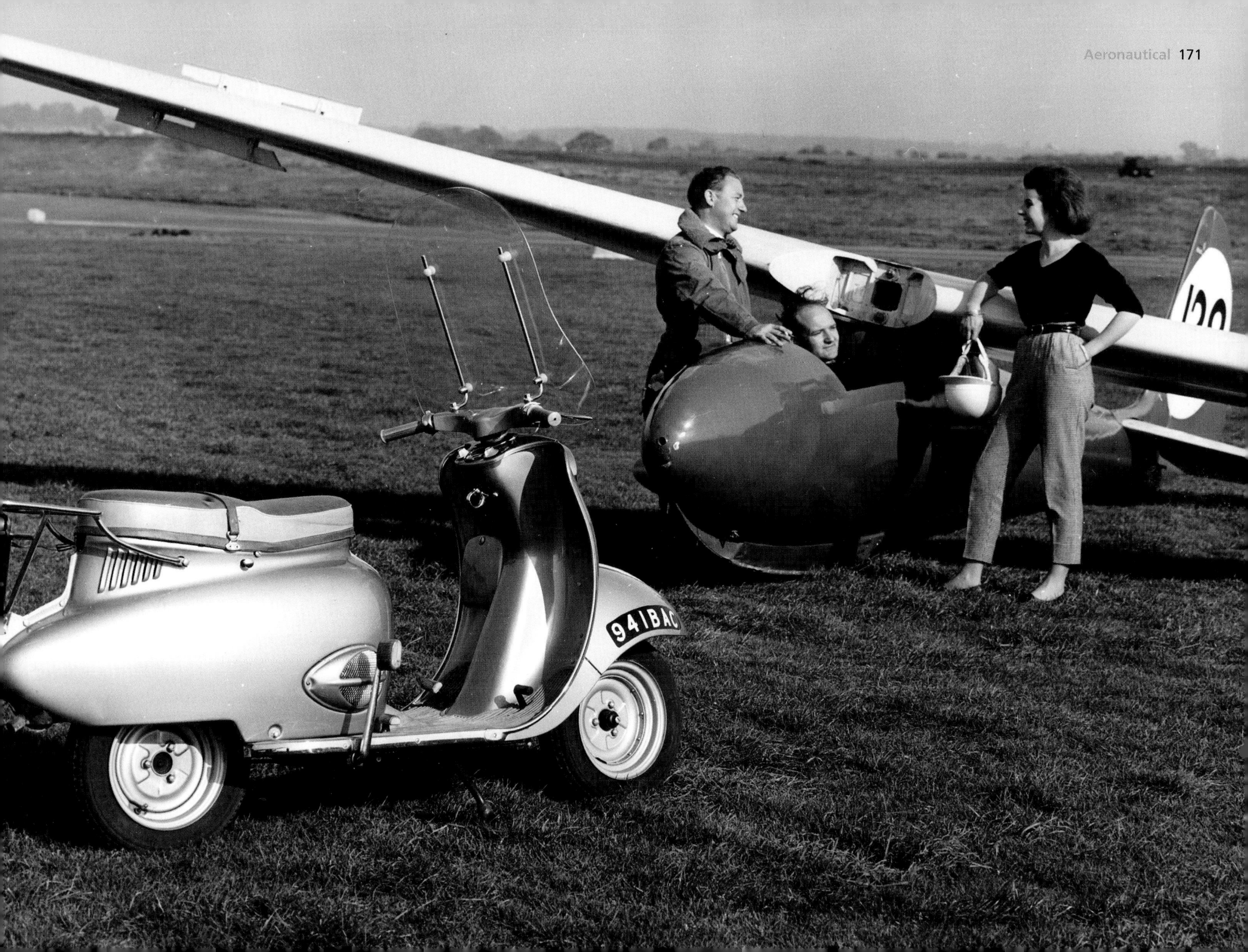
941BAC

Branson's epic balloons

'My association with Richard Branson really started with John Ackroyd. He was the designer of Richard Noble's Thrust 2, and John seemed to take his extraordinary talents between various places. I'd not spoken to him for some time and it was through Ken Sprayson that I'd heard what was going on with Thrust 2, so I got to know John Ackroyd very well indeed. The next thing I learnt was that he was working on Richard Branson's Atlantic hot air balloon venture. John was there primarily to design the capsule for the balloon because Per Lindstrand, the owner of the balloon company based near Oswestry, knew all about making balloons but nothing about making a pressurised capsule for the great altitudes when crossing the Atlantic. So I contacted Per and he invited me to his factory to see what I thought of the project. John had arranged to be there as well. So it was on that occasion that I met Richard Branson and we got on like a house on fire. We went out to lunch, chatted about the project and exchanged views and so on, and the association grew during that period while the capsule was being built and I was drawing it at each stage. I did a cutaway of the capsule which was the only sensible way to show the main features. I wasn't commissioned to do the drawing, which I did in my spare time. Subsequently it was shown in 'Flight' and various other flying magazines. I did the job because the Atlantic crossing and high-altitude flying interested me, it was a new challenge, and the technology was something I'd not experienced before. So the project held lots of interest.

'I didn't have another opportunity to work with John Ackroyd and Per Lindstrand until the trans-Pacific trip was announced. Altogether more ambitious, this one was to start from Japan and go east over the Pacific. Based on information gained from the trans-Atlantic crossing, the balloon itself was much larger with its own oxygen-making equipment and with much larger-capacity cylinders and, of course, a bigger envelope. By now, 'Flight' magazine had become interested in what others had already dismissed as a 'hair-brained scheme'. But, with a successful marathon trip already to his credit Richard Branson, relatively unknown at that time, was determined to face another tremendous challenge.

'You can't get any lower than the floor. Well, that's where the ginormous new balloon was laid out while being made! However, this factory floor was as spotless as any hospital operating theatre. As in the hospital theatre, the female operators knew that two lives depended on their expertise as they laid out each computer-designed segment of the fabric before hand stitching it into the main fabric. Of course, the fabric might have been stuck together with adhesive but in view of the envelope's huge size, the pressures it would have to withstand at great heights, and the distance it was intended to travel, it was felt that in the interests of safety each segment would be double stitched. Again, this was new territory because a balloon of this size had not been made before, and as a result there was no means of testing either for strength or leakage, nor would there be a test ▶

flight. The first time the balloon would be inflated would be the start of the trip, so the stitchers had one chance; get it right first time! That was a mind-boggling responsibility. I had the greatest respect for the women involved in this herculean task and, as a flyer, I did not envy either Richard Branson or Per Lindstrand flying this unproven flying cathedral across the Pacific. As a professional balloon pilot, Per knew the risk but Richard, in spite of his Atlantic crossing, was really taking a huge step into the unknown. That took real guts and he deserved a lot more credit than he got at the time, if my memory serves me right.

'I had to return to Per's factory a number of times to get everything absolutely right, also because I was drawing the light-alloy capsule as it was made, which was the norm for one-offs. Obviously, the sheer enormity of the canopy - it could have swallowed the entire factory a couple of times over! - precluded it from being drawn. No point anyway, sufficient to show how the retaining wires were attached to the capsule and assume the envelope was attached at the other end. I put a lot of effort into ensuring that every detail was included because firstly, this was completely new to me and, secondly it was an historic vehicle about to set forth on an historic trip. My job was to do justice to the mechanical concept of the capsule. For instance, the six huge butane gas cylinders secured to the sides of the capsule were double skinned to provide insulation; so I did a sectioned drawing showing this feature. Made in light alloy to save weight, the cylinders were pressurised and secured by explosive bolts equi-distant around the capsule in order that the cylinders would counter balance one another. When the gas was expended each cylinder was jettisoned by firing the explosive bolts which retained that cylinder - a tricky operation because it required these bolts to detach simultaneously, otherwise it could upset the balance of the craft. Of course, once a cylinder was detached, its weight did cause a momentary loss of balance but the equalising process resumed by using gas from the cylinder opposite the space created by the jettisoned. When the gas was exhausted in that cylinder, it too was dumped as were the others in sequence, thus preserving the capsule's balance.

'You had to be very goods friends to endure day after day cooped up in that capsule! Entrance was through one small door in the side, room was at a premium, so the pilots sat almost facing one another in offset chairs with a working table in front. The pilots had to remain in their chairs throughout the trip, work, eat, sleep, and er, answering nature's call from the same chair. Getting its ergonomics right was essential, which is why it ended up looking like a dentist's chair. I can vouch for its comfort.

'My only contact with the team after they set off was through radio and media reports, same as anyone else - a frustrating time for me, trying to keep track of what was happening. The team set up base on an island south of Japan which they judged was the best place from which to take off. A lot of space was needed to lay the huge canopy on the ground so that it would rise up cleanly during the filling process. The ground also had to be smooth to avoid tearing the fabric; made of a single layer, that fabric was so thin the slightest puncture would be enough to cause massive problems in space. It took an army of people to lay it out on the ground but during the afternoon and evening when it was being laid out, it froze to the ground. Being so thin and fragile, it could not be unstuck from the frosty ground without tearing the material. So the horrible truth dawned on the crew that they would not be taking off that day, or any other day, until the ground frost thawed and released their priceless balloon. That evening, while sitting on the beach surveying the scene before him, and surrounded by his disconsolate crew, Richard had a bright idea. 'Ring Lawrie!' Where was I? Tucked up in bed at home in Coventry. After all, it was 3am... So the phone rang. It was John Ackroyd. 'We're sitting here wondering what to do. Richard thought that you should be the first to know that we're stuck to the ground and going nowhere.' So I replied, politely as I could under the circumstances, 'I'm much obliged for the information and I hope that you're unstuck and airborne by the time I wake up in the morning, Richard.' It was nice to be involved, even at 3am...

'The principle of flying a hot air balloon, such as Richard Branson's and Per Lindstrand's, was established in the 1700s by the French, who popularised balloon flight. Nowadays flight is made easier by sophisticated gas burners which heat up the air inside the balloon, but the principle remains the same. We've established that hot air rises. To start with, a balloon is lying flat on the ground, in Richard's case frozen to a Japanese beach, and to open the envelope it's necessary to blow in some cold air. Once the material starts to fill, time to pump in some hot air which will cause the inflated balloon to rise up. Gas from butane cylinders is released into a series of pipes arranged around a central burner situated above the basket. The pipes pressurise the gas before releasing it into the burners where it is ignited by an electronic switch controlled by the aeronaut, the 'captain' of the craft. He's licensed to fly and is responsible for the safety not only of the craft but also of its passengers. Anchored to the ground by ropes while being loaded, the balloon is released by the ▶

helpers holding the ropes on a command from the aeronaut, who simultaneously releases more hot air into the envelope to assist lift off. He will have determined wind speeds and directions and, if favourable, will follow his course by adjusting the amount of hot air inside the balloon, more hot air to rise or burners off for a gradual descent until the desired air stream in the opposite direction is reached. Still air at ground level may give way to an easterly breeze at 50 ft which in turn may change to a westerly wind at 100 ft. How to ascertain the wind direction? Look up and you'll see which way the balloon is leaning! Having determined the various factors, the aeronaut flies the balloon into the desired air stream by activating the burners and in this manner he navigates a route across country. It's different from glider flight as you don't have the same measure of control but it's similar insofar as you make your own thermal by means of the burner.'

I asked Lawrie about the jetstream, if this would have played a part in Richard and Per's progress across the Pacific. 'The jetstream is an unpredictable 120/130mph phenomenon in the northern hemisphere. There are periods in the year when you know it's going to happen sometime, but it can change the weather in a matter of hours. Certainly, it can be a big help if it's blowing in the direction you want to go, when you want to go, you've got no option but to go with it! But there are other winds blowing around the earth which are more predictable and which can be helpful but they are not necessarily at the altitude which you want to fly. So it's decidedly tricky trying to arrange all the ducks in a row. There was a very close association between Richard Branson and meteorologists who were advising him and Per on the best routes to choose. At the end of my year studying meteorology at Birmingham University, the professor stood up and said to us 'I've enjoyed teaching you… but I have to inform you that you have been studying an inexact science.' As with one voice we replied, 'Thank you very much!' By that time, we thought we knew everything there was to know. ▸

Hanging on in space

'On top of the capsule was an oxygen-making machine. At some stage in the trip this stopped working and, to save weight, it was decided to jettison it. But how? No provision had been made as it was supposed to be foolproof. Anyway, Richard decided that he'd have a go. The only way out was via the astral dome, which was made of clear perspex and whose sole purpose - apart from providing an emergency exit - was to allow the pilots to monitor the burners on top of the capsule. So out through the dome he went, a tight squeeze for a big man, and with the right tools he was able to detach and jettison the oxygen-making machine. All this was achieved at very high altitude while clinging to whatever hand grip he could gain on the capsule's exterior fittings - - without the safety of having a parachute! One slip and he'd have followed the jettisoned machine. A remarkable feat by a very brave man.'

Basket case

'I had an insight into the world of ballooning when I was at Per's factory. There, in a corner, were the traditional open wickerwork baskets which you see attached to most balloons flying overhead. A throwback to the earliest flying balloons, these baskets are a revelation in construction, being hand woven in natural materials such as willow or hazel and, in comparison to their size, they weigh surprisingly little. They are also extremely strong and I had much admiration not only for the craftsmen who still make them but also for those intrepid balloonists of 200 years ago, the forerunners of modern flight.'

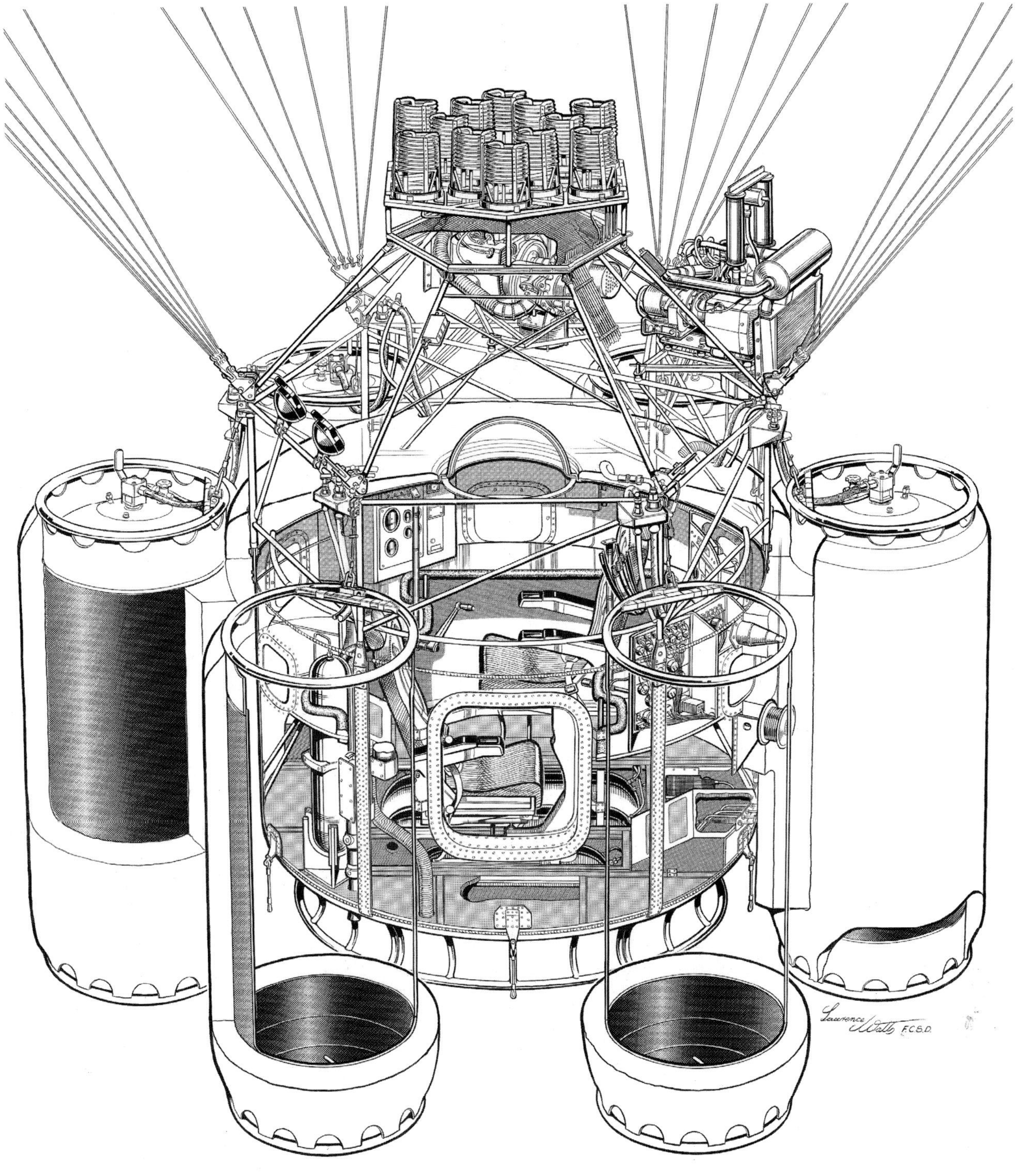

Well, the jetstream is part of that inexact science, it's part of the guessing game. But I wouldn't like people to think that when you're up there in a balloon that you're entirely at the mercy of the elements. You're not. You have in a balloon a means for controlling its flight, unless Sod's Law or a great thunderstorm you didn't expect, intervenes. I've been struck four times in the air by lightening and am still here! Branson and Lindstrand needed luck on their side to succeed; in a thunderstorm, lightning could have triggered any, or all, of the explosive bolts holding on their gas bottles. The same type of bolts also retained the capsule to the balloon. An unlucky strike which activated those bolts would have spelt catastrophe for Richard and Per....Equally, if all those bolts did not activate simultaneously to release the capsule at the moment of touch down at the end of the flight, then the capsule could have landed on the ground lying on its exit door....In spite of the most fastidious pre-flight calculations, and piloting of the highest order, a large slice of luck was needed for a successful flight.

'Crossing the Pacific Ocean must have been quite an experience, staying in the right part of the vast sky to catch the most favourable winds. How to know if that's the right part? Could there be more favourable winds at a different height? These, and many other variables, were faced by Richard and Per every day, every night, every waking moment. And they had to react quickly and decisively to maintain momentum and height, especially height. Losing height in a balloon over the Pacific Ocean does not sound a very clever trick to me, so it would have concentrated their minds wonderfully. Keeping cool under those sort of conditions would take an extraordinary type of person.

'Modern tracking aids let the ground force know the exact whereabouts of the balloon at all times. Which was just as well because after it had successfully crossed the Pacific, the balloon lost height and was eventually forced down in the snowy vastness of Northern Canada. Their control centre was able to pinpoint their landing area and a rescue party found them without too much delay. Luck was on their side again; without those modern electronic tracking aids they could have been lost forever in the Canadian vastness... Sad end to a magnificent adventure! Richard and Per deserve the highest praise for having challenged the unknown and for coming so close to achieving their aims.' ■

Percy Pilcher, aviation pioneer

Talk to Lawrie for any length of time and it won't be long before Percy Pilcher's name crops up. Percy who? You may well ask. Mention the Wright Brothers, you'll know them as the Americans who first flew an aeroplane in 1903. But for a tragic accident, Percy Pilcher would almost certainly have flown years before the Wrights. As this stopped him from becoming one of the greatest names in aviation history, Lawrie has dedicated much of his life to ensuring that Percy's name is not forgotten, if not as the first man to officially fly a powered aircraft, then as Britain's leading aviator of the late nineteenth century.

Born in Bath on 16 January, 1867, Percy Sinclair Pilcher lost his father when he was eight and joined the Royal Navy as a cadet five years later. During his seven years in the navy, Percy travelled the world becoming a lieutenant at the age of 20 before resigning in favour of an apprenticeship with Glasgow ship builders Randolph Elder & Co. It was here that his abilities as a designer and inventor took root and he became interested in the possibilities of flight, of which very little was then known. He struck up a friendship, and exchanged views on flight, with the German Otto Lilienthal who, in 1881 had made his first gliding flight near Berlin. Lilianthal had made a detailed study of bird flight and, based on this, successfully designed his own gliders and flew them years before the Wright brothers. Intrigued, Pilcher had joined Lilienthal in Germany to study the shape of birds' wings in flight. Unfortunately, this all came to an end when Lilienthal was killed in a gliding accident in 1896. Pilcher had designed, developed, and flown his own man-carrying glider, the 'Bat', in 1895. A series of gliders followed, each one an improvement on the last, until the Hawk, which proved the most successful. By 1897, Pilcher had turned his mind to powered flight because it would give him more time in the air and better control of the aircraft. Fitting an engine with a pusher propeller in the Hawk seemed the logical thing to do. However logical it seemed, it was unsuitable so a new design, a tri-plane, emerged. With three wings, one above the other, the tri-plane offered much more wing area, and therefore lift, than the Hawk. To power the tri-plane, he used a horizontally-opposed twin cylinder four stroke engine made by the Wilson Pilcher Car Company. To build the new craft, he gratefully accepted the facilities offered at Stanford Hall by its owner, Adrian Verney Cave, later the 6th Lord Braye. However, developing the new craft was expensive and to gain some investment in it, Percy announced a demonstration - now termed an 'open day' - for the 30th September 1899. Prospective investors in his project headed the guest list. Word got around and local interest groups and schools were organised to see this spectacular event. The day dawned wet and windy but in the afternoon conditions improved and Percy set about demonstrating his aged Hawk glider before moving on to the powered tri plane. The Hawk was to be launched by tow line but due to degradation of the tow rope, through having been left on the wet ground overnight, the rope snapped a number of times. On the third attempt, Percy had climbed to about 60 feet before the glider stalled and crashed to the ground. He sustained injuries from which he died two days later. A tragic blow to British aviation....

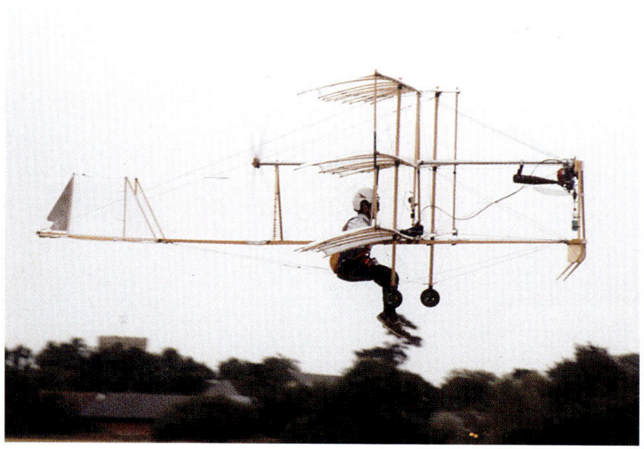

A replica of Percy Pilcher's tri-plane being flown above the runway at Cranfield during filming of the BBC documentary in 2004.

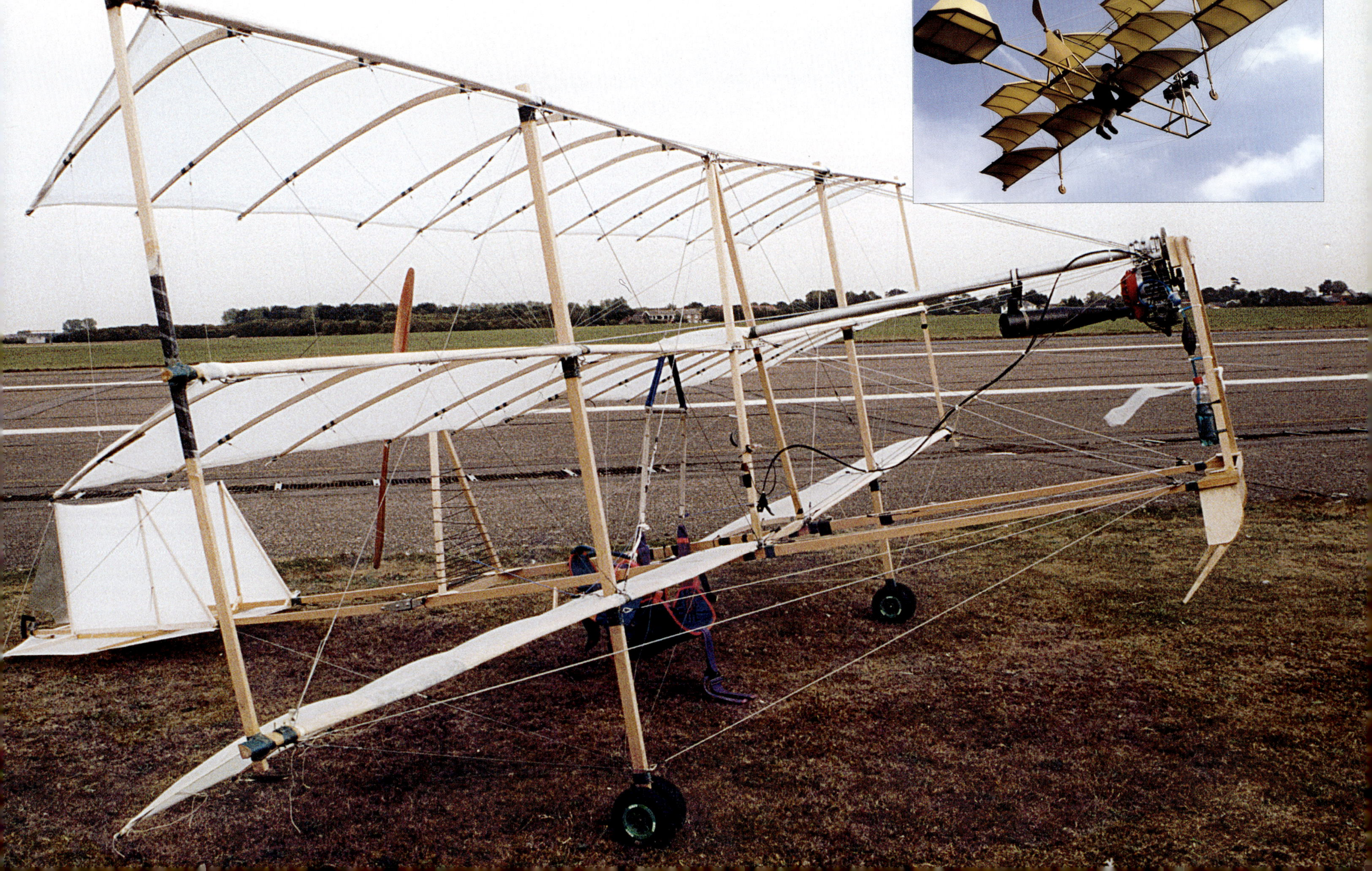

Cranfield apprentices reconstructed the 50 year-old replica of Pilcher's powered aircraft for the BBC documentary and used a modern Anzani two stroke engine in place of the more powerful four stroke Wilson engine which Pilcher used to power the original aircraft .
Inset; the tri-plane takes to the air. Pilot reported it was 'fantastic' to fly.

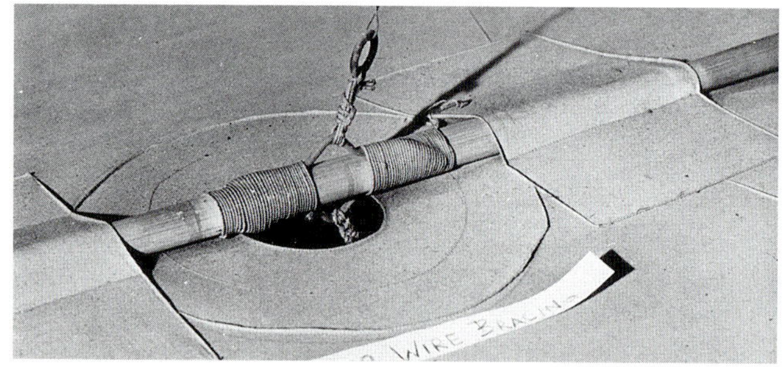

Historic restoration

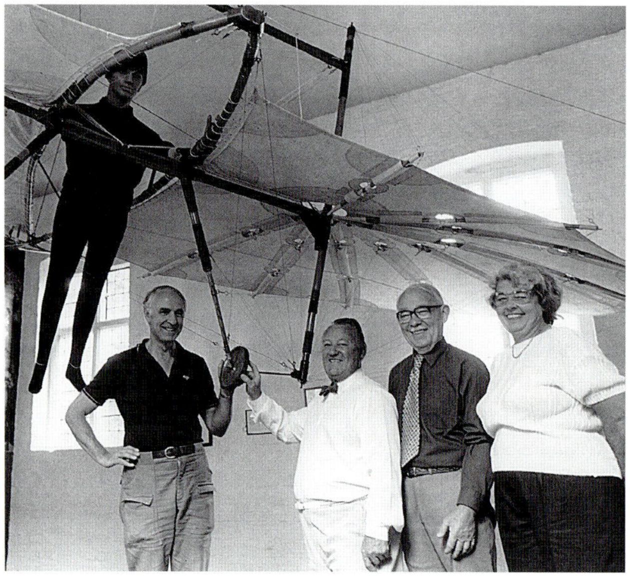

Lieutenant-Colonel Aubrey Fletcher (Lady Braye's husband), Lawrie, Dick Eales and Eileen Watts, at the formal handing over of the restored Hawk and model of Percy Pilcher 'flying' it, at Stanford Hall museum.

How Lawrie became involved with Percy Pilcher goes back to his days when he was a designer at Armstrong-Whitworth. 'The apprentices were detailed to make a replica of Percy's Hawk,' explained Lawrie. 'When it was finished, the replica was presented to Lady Braye to be hung in the stables at Stanford Hall near where Pilcher had had his workshop and where he died of his crash injuries. As a member of the Royal Aeronautical Society, and an employee of Armstrong-Whitworth, I became involved with the restoration of the Armstrong-Whitworth apprentices' Hawk which, by this time, had deteriorated into a dreadful state. The fabric was pea green with algae, the woodwork had warped, it had been totally neglected. So people had stopped going to look at it in the Stanford Hall museum. The Society's Coventry branch asked if I was prepared to take on the Hawk's restoration. Examination showed that a total strip down and re-build was necessary if it was to be restored to a condition of which Percy Pilcher would have approved. Stripping down, cleaning, realigning, glueing the bits together, and varnishing the woodwork was time-consuming but relatively straightforward. The fabric, of which there was a great deal, spread over 36ft wide by 7ft deep wings, was altogether a tougher nut to crack. A monumental lot of fabric! I had samples under a microscope and the degradation was such that within a couple of months it would have been full of holes. That's how bad it was. Luckily, the fabric was not glued to the air frame, it was secured by sleeves through which the ribs pass. But how to clean it? That's when Eileen saved the day; 'I'll wash the fabric and see how it turns out.' I don't think she realised what she'd let herself in for! The tail plane and the fin were made of the same material so we started with them. If the washing damaged that material then we knew it wasn't going to be worthwhile washing the wings. ▶

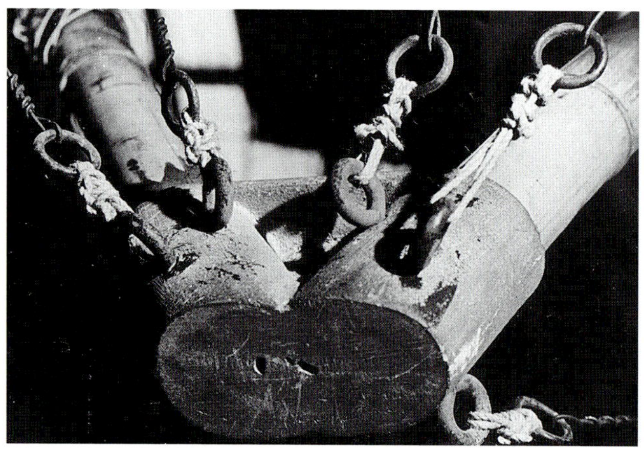

Above and left: the 50 year-old rigging of the Hawk before being restored to 'as new' condition...

Rescued by Daz and Fairy Liquid

'So we made up our own cleaning compound, Daz washing up powder and Fairy Liquid, a mixture which would clean anything on the face of this earth! We soaked the tail plane fabric and the fin for a couple of days in our bath at home. Within that time, the bath turned green, but the fabric came clean! We rinsed the material a number of times to really test it, but it seemed to have retained its original strength. So, after 50+ years of hanging around Stanford Hall without ever being cleaned, the material seemed good as new. Which is more than could be said of the cleaners; while the bath was occupied we had to go without baths or showers, so it was quite a luxury to have one again (and our friends came back to visit us...). Buoyed by the revelation that the material was capable of being saved, I dismantled the airframe into a heap of sticks, making suitable drawings for re-assembly purposes. The only major items which required replacement were the original tensioning wires, of which a great deal had been used in the aircraft's construction. So I bought a roll of wire approved for aircraft purposes and a roll of locking wire which was non-corrosible. We brought the wing fabric back home; one wing's worth of fabric filled the bath, we made another batch of cleaner and left the material to steep for a week. Even after that time, it required a couple of washes before it came as clean as the original. Eventually we had a snow white set of fabrics for the aeroplane - and our daughter Julie had recovered the use of her bathroom from our daily shower visits....

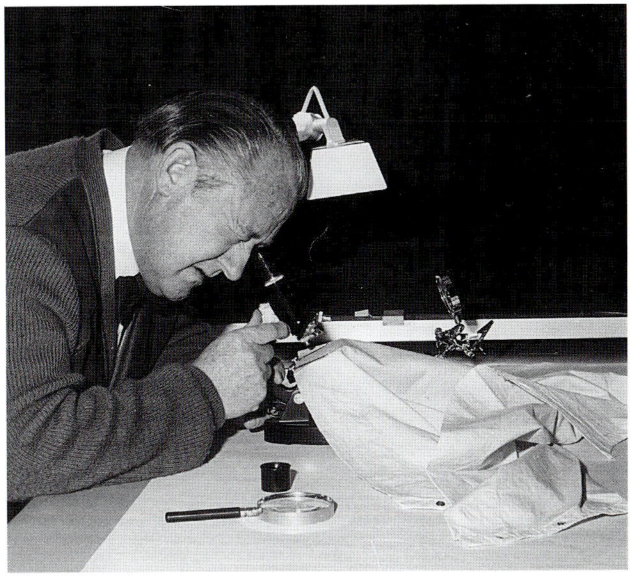

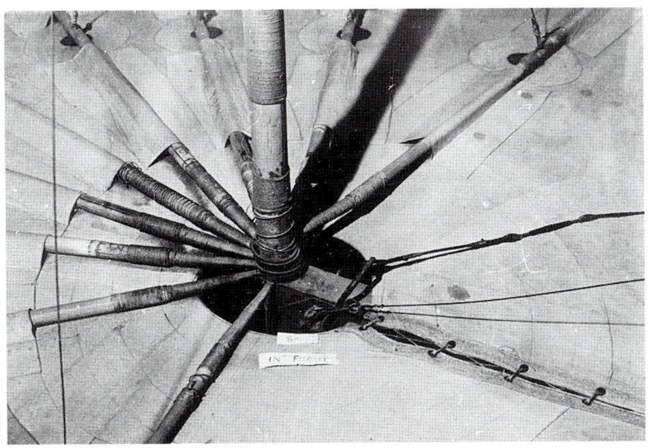

Lawrie examining the wing fabric with a microscope for signs of deterioration. Left: a section showing the method Percy Pilcher used for attaching the canvas to the main structure.

'In the meantime several gliding club friends had got stuck in and helped me clean up the various bits and pieces. Included in the party was Dick Eales, who had been in charge of the Armstrong-Whitworth apprentices when they built the replica of the Hawk presented to Lady Braye. Dick, also a member of the Royal Aeronautical Society, worked very well with me and put great effort into the tedious job of rubbing down and varnishing the wood work until it was Percy-proof, meaning it would have passed his inspection. Altogether, we put in more than 600 hours' work pulling it into shape. Then came the time that the binding had to be done; all the little brackets and wire loops had to be bound into place using a special cord. We'd found the manufacturers of the right cord and they very kindly sent us spools free of charge. Already I had the sketches which showed how many loops and bindings were required on each joint. This was meticulous work because when the aircraft was finished, it had to be exactly as Percy had made it.' ▶

Shadowy help

'Our gliding base hanger at Husband's Bosworth provided space for the restoration. Working there alone one night, and concentrating on the bindings, I heard the door open and thought that it was Eileen, who was staying in our caravan nearby, bringing me a cup of tea. But it wasn't. I then surmised that it was Ernst Spek, an ex-Luftwaffe pilot who used to give advice I didn't want, and who usually came in and sat in the shadows on the other side of the workbench from where I worked. His great interest, and why he was intrigued by what I was doing, was Otto Lilienthal, Pilcher's friend and mentor in the early days of Pilcher's aeronautical career. Ernst was unaware that Percy, a fluent German speaker, and Lilienthal had been close friends so he was babbling on about them as I worked head down, concentrating on what I was doing. After about 20 minutes, I stood up, straightened my back and started to say to Ernst that I'd had enough for one night. For the first time, I looked at him... no one there! I'd been talking to myself! Yet I'd definitely heard someone come in. I'd heard that person walk across the hanger floor. I'd also heard a seat being drawn up and somebody sitting down and talking. Yes, somebody was there, or had been there, and I'd been talking to this person, a man. It was not a frightening experience, it was unusual, but not all that unusual because previously I'd had an occasional feeling that somebody was close by, as if they were helping me. But not talking to me. That was the difference. Occasionally, I'd been aware that I was muttering to myself as I worked, such as 'Wonder how Percy would

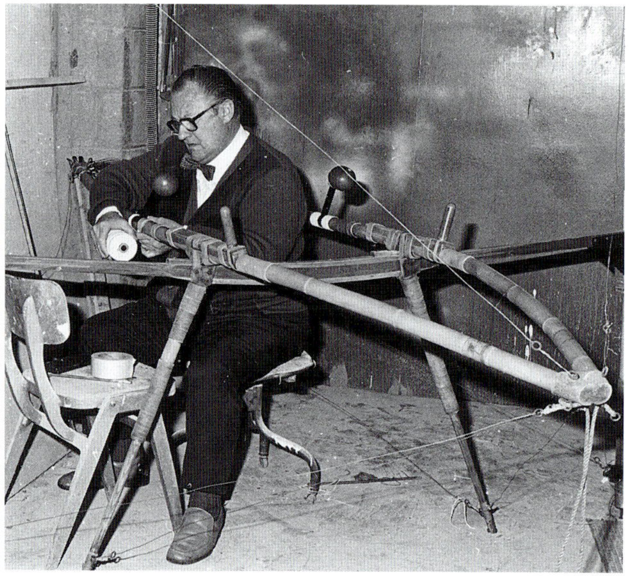

Lawrie binding the structure 'helped' by a shadowy figure of 'someone' in the background...

have done this, or that,....' Or perhaps I'd be saying 'I wonder if Percy would approve of what we're doing....' That sort of thing. It was a positive influence. It helped me to understand the man, to get under his skin if you like, and it definitely helped me to restore that aircraft to what I like to think was its original standard, as Percy would have done. Eventually, the work was completed in the hanger at Bosworth, I took photographs for record purposes and, with the help of some friends, I moved the craft out of the hanger. A light breeze was coming down the hill into the hanger as I stood in Percy's place inside the space between the wings. I could feel the craft lifting in the breeze '.....this thing wants to fly' I told myself. So we wheeled it up the hill and turned into the wind... and I was just able to grab onto it, ape like, as it caught me unawares and tried to lift off. The next time we tried, I was able to stay with it, just a matter of little hops for it, but blooming big ones for me. But fly it I did.

'The BBC got to hear of this and wanted to film the Hawk before it went back to Stanford Hall, where it would have to be semi-dismantled to get inside the building, then re-rigged once inside. They came, did their shooting, and it made a cracking short film on BBC Midland News. They made another short film before I dismantled the machine and returned it to Stanford Hall where it was put back on display in its original place, suspended from the ceiling. Eileen and I now ensure that it gets an annual spring clean and is maintained throughout the rest of the year in pristine condition.'

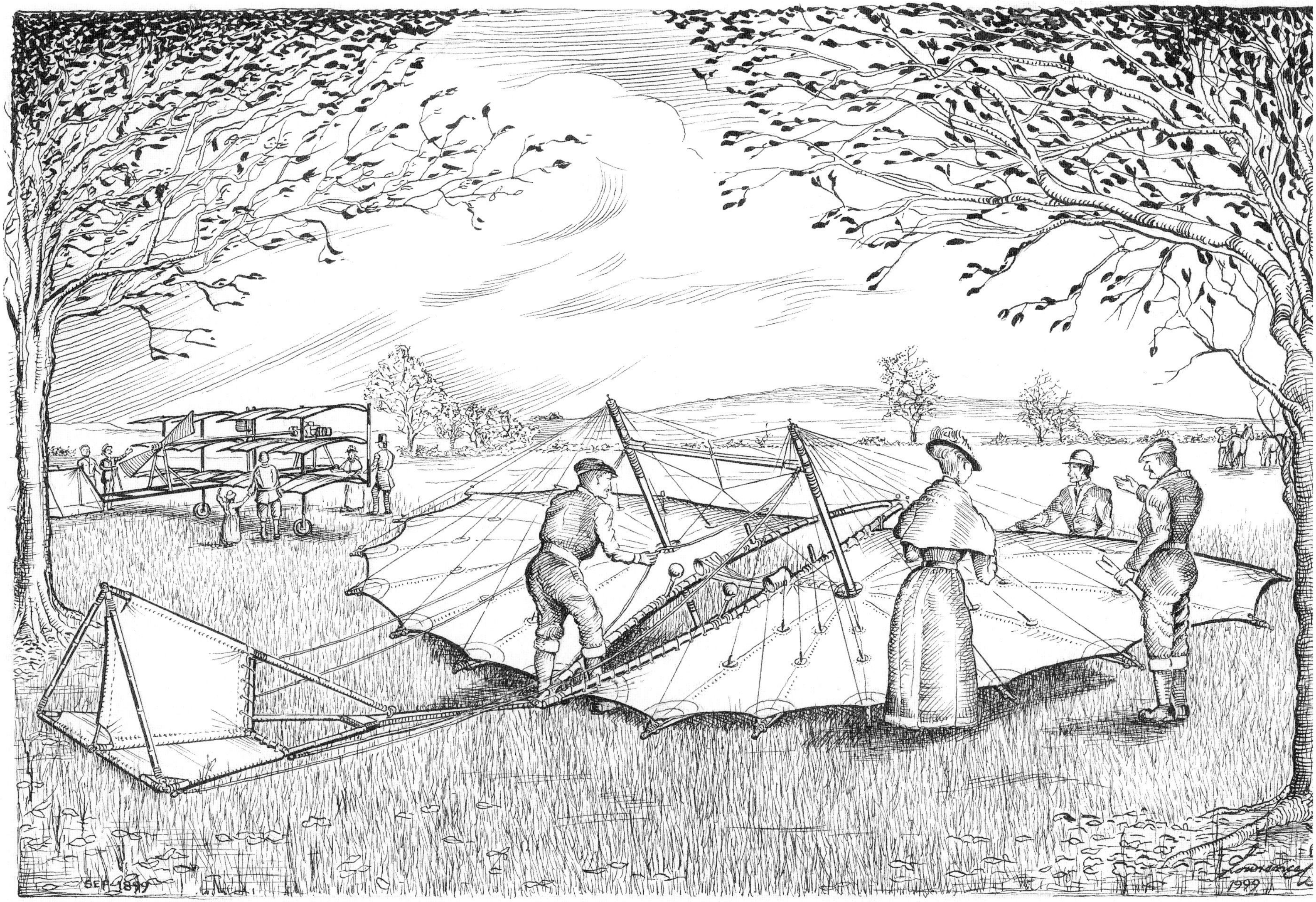

30th September 1899: a Watts interpretation of the fateful day as Percy Pilcher, centre, prepares the 'Hawk' assisted by his sister, Ella, Lord Braye (arm raised) and the farm manager (in bowler hat). In the distance two horses wait to be attached to the rope which would tow the Hawk to take-off speed. Parked on the left is the powered tri-plane which Percy would demonstrate later that day. Sadly, he died before he could do so. Available evidence shows that he had almost certainly test flown the aircraft, three years before the Wright brothers in America.

Ill-fated demo

In order to try and raise funding for the ongoing development of the powered triplane, the first such aircraft in this country, Percy and Lord Braye decided to have a demonstration in the park at Stanford Hall. Firstly, the Hawk glider would be demonstrated for the benefit of the assembled press, followed by the powered tri plane. Verney Cave, Lord Braye's son, organised a pair of horses from the village to attach to the rope which would pull Pilcher's Hawk into the air. Crucially, this innocent departure from the normal was to have disastrous consequences. Previously, Percy did all his launching downhill but, because the demo area in the park was flat, some assistance would be needed to launch the Hawk. Hence the horses, which were to be connected by rope to a block and tackle anchored to the ground. In Lawrie's opinion, 'this was an unnecessary complication because the rope would cancel out the drag factor without being pulled by two horses. You don't need speed, just a gentle pull to give lift only against the wind. The Hawk was so well designed and built that it would fly anyway'. It rained all day, the aircraft had been covered but not the hemp cable, which was left in the rain. Two attempts were made to get the glider airborne by horse power but each aborted when the rain-sodden rope broke. Third time was not to be lucky for Percy, though he must have felt his luck had changed when he climbed to around 60 feet. The rope broke again, the Hawk stalled. In other words, Percy could not control it any longer due to the low speed and the pull of the heavy rope. As it stalled, the tail plane folded forwards, hit him across the head and smacked the cross bar between the two vertical posts which were holding the wings on. As a result, the Hawk collapsed in the air and crashed to the ground. The unfortunate Percy broke both his legs, cracked his pelvis and suffered head injuries. He died two days later, aged 32, in Stanford Hall from loss of blood, which wouldn't have been allowed to happen today.

'So that's why the tri plane was not demonstrated that day,' explained Lawrie. 'I spoke to the gardener at Stanford Hall when I first went there - he was in his late 90s then - and his mother was there the day Percy was killed and saw what happened. It's my opinion that knowing Percy from all that I've seen, read and heard, that he was not the type of man to take unnecessary risks. The demonstration was not intended as a fun day but was a serious attempt to show the assembled potential investors just what his flying machines could do. Logic suggests that he would almost certainly have tested the powered tri plane to ensure that it performed as he wanted before displaying it publicly, especially to an invited audience of notables. I also know from the records that the engine in the tri plane had been tested and that it had broken its crankshaft. I know that this had been repaired. I'm almost certain that Percy, being the sort of perfectionist that he was, would not have risked failure by attempting to fly a repaired engine without first testing it. Without today's test facilities the only way that Percy could have been sure of things working OK was to fly the tri plane. And this, I'd wager my life on, is what he did, three full years before the Wright Brothers.'

BBC's 'Percy Pilcher' flies the 'Hawk', which is attached to a cable (invisible) suspended from a crane. The model dressed for the part and impressed everyone by quickly learning how to guide the 'Hawk' by moving his body weight, just as Percy Pilcher would have done.

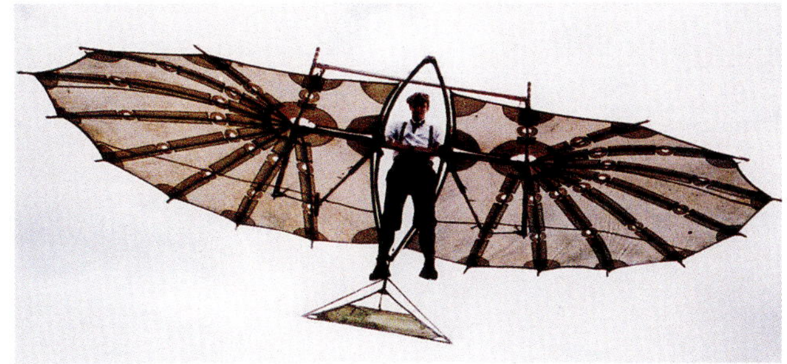

The book, and films

Involved in a BBC film about Percy, Lawrie explained how that came about. 'An extremely well written and researched book, 'Another Icarus', by aeronautical specialist, Phillip Jarrett, chronicled Percy's life with accuracy and upon this book, or the latter part of it, the BBC based their 2004 documentary about Percy,' said Lawrie. 'But they went further. They wanted to see if, had he lived, Percy could have been the first man to use powered flight. So they got Cranfield apprentices to reconstruct the tri plane as near as possible to the original, and install a hang glider engine of similar layout and power output as Pilcher's, and put it to the test. When completed, the 'new' tri plane was wheeled out of its hanger, a qualified hang glider pilot took the controls, the motor was fired up and, with a hop skip and a jump, the tri plane took off! Not only that, it stayed up for nearly 30s, which is three times longer than the Wright Brothers' first flight! The flight ended when one wing tip dropped in a side breeze and it tipped up, damaging the wing. Repaired, it took to the sky two days later, this time to circle Cranfield airfield without any problems. The pilot reported that it was 'fantastic to fly', the exact replica of Percy Pilcher's original, except that the replica used a lighter two stroke Anzani engine which produced slightly less power than Pilcher's original Wilson four stroke. No question, the original could have flown in 1899! Further evidence to this effect is revealed in a letter from Pilcher to Baden Powell, prior to the ill-fated demonstration, when Pilcher wrote that he 'needed better lateral control'. Could he have said that if he'd not already flown it?'

BBC documentary

That BBC documentary was screened in 2004 and was much more extensive than anything before. This time they enlisted the help of Eileen and Lawrie to advise them on the technicalities and help them to do the filming at Stanford Hall, where Lawrie was now Curator of the Museum. As Lawrie recounts, 'For eight days we were getting up at 6am to be at Stanford Hall, 30 miles away, for 8am, when work started on the production. Our job was to get the Hawk out of its stable and into the courtyard. Simple? Not on your life! To get it into its overnight secure home we had to de-rig it, which meant taking off the wings to get it through a narrow entrance. In the morning, we re-rigged it. Not the job of a moment but Eileen, who has helped to rig all my gliders, is now an expert rigger and she and I got the Hawk's procedure down to a fine art. The BBC were so security conscious that they employed a night watchman to stand guard over the Hawk. They overlooked the fact that the aircraft had hung in the museum for more than 50 years bothered by nothing more than gathering dust...

'Eileen and I worked hard not only looking after the glider but also giving advice, answering questions, all the things you do on a specialist programme. One of those questions concerned the rope which was to be used for the enactment of towing Percy's glider into the air. Instead of following my advice, the enthusiastic young lady in charge of the production got a rope which would not have looked out of place anchoring the Queen Mary in Southampton Dock! I'd already told her that the 'cable' used by gliders, and that of Percy, was not much thicker than a piece of string, but she over-ruled me! Which was a pity because she got most other things right, and played Percy's beloved sister, Ella, in the film. Another nice touch was the guy who played Percy Pilcher, he came onto the set in period costume and was so like Percy in appearance, that he gave an air of authenticity to the production. He was also up for it when he felt the Hawk trying to fly. We had a big crane, with a 180ft

Lawrie helps to rig the 'Hawk', with the actor in place. Helping is the crane driver who later had a go and successfully grasped the flying technique.

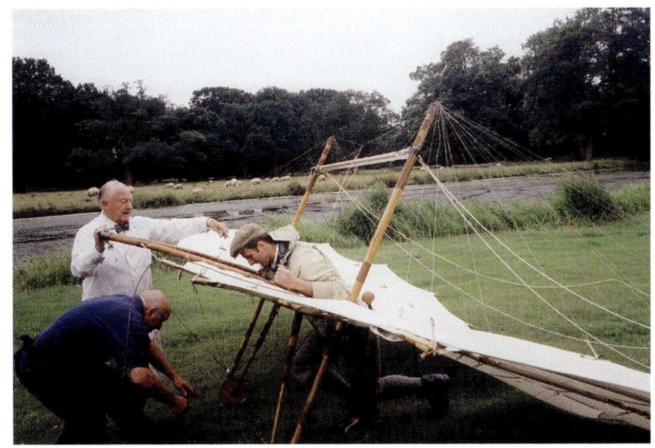

Flying crane driver

'Though not in the cast, the crane driver was a flying enthusiast and after filming was over, he had a go at flying with his assistant working the crane. He got in without the harness, without the rope from the crane being attached, and promptly took off! But the producer was having none of it; he had to wear the harness blah, blah, etc. So he put all the gear on, his mate lifted the Hawk with him aboard off the ground and quite high into the air. I couldn't believe my eyes, the rope went slack! The Hawk had started to fly! The crane driver was going through the right motions to control it, diving, climbing, levelling off - the only manoeuvre he couldn't do was side slip, which the Hawk had not yet been modified to do. It was an incredible sight! And it proved, yet again, the correctness of Percy Pilcher's calculations. He'd worked it out from what little knowledge was available and then applied his own common sense and intuitive reasoning to come up with the sort of stunningly correct answers only a genius could have provided.'

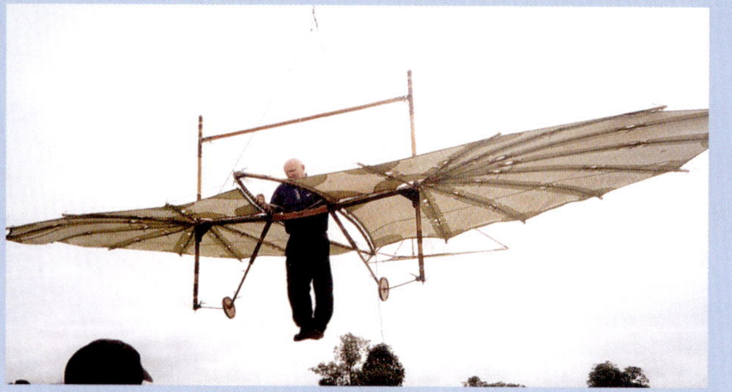

The crane driver also 'flew'!

Lawrie (centre) found that the 'Hawk' was willing to fly into the wind (but LW ran out of that before he could become airborne...)

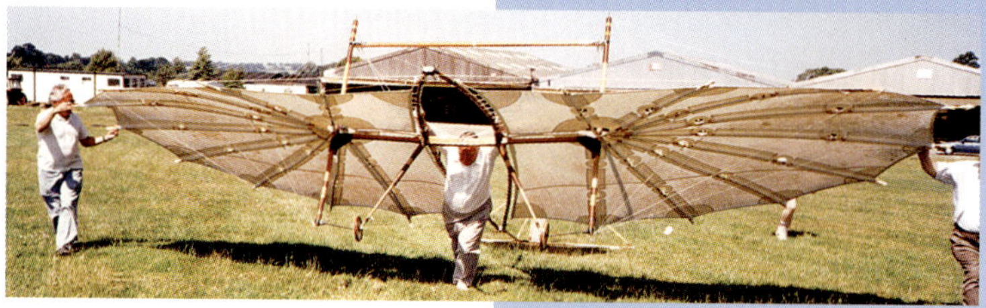

extended jib, and the Hawk and 'pilot' were suspended from this for the flying scenes. I taught him how to fly it and at the first opportunity, suspended by the crane, he was using the correct body movements to guide the Hawk. In fact, he not only looked the part of Percy, he flew like him as well when the rope went slack in the strong wind and he rode it out! When eventually we were able to get him down, he was grinning from ear to ear. 'That really gave me a big buzz...' An excellent piece of casting!

'I have to confess that the crane driver's performance inspired me to have a go. Imagine raising your weight off the floor by leaning on the backs of two chairs, and then moving your body to either side. That's how you flew the Hawk. Percy must have been a fit fellow! Supporting my body in this manner, with the side rails between the wings of the Hawk replacing the backs of my chairs, my mature years lacked sufficient arm and shoulder strength to maintain the flying position for more than a few seconds. But that was sufficient to imagine myself as a latter day Percy Pilcher reaching for the clouds... an experience that will remain with me forever....' So, full credit to the BBC for what they did and for putting right what I have long felt was a grave injustice, not recognising Percy Sinclair Pilcher as Britain's foremost aviator and the man most likely to have been the first to fly a powered aircraft but for misfortune overtaking him at the 11th hour.'

Grandfather and Percy

Lawrie's grandfather was a fine engineer, as mentioned earlier. What's not common knowledge is that he was also interested in flight and actually built three of his own flying machines. Lawrie recalled, 'During this stage he met Percy Pilcher and they communicated, as many of the pioneers did - and there were a lot of them around at that time - who were interested in flight and exchanged information. Unfortunately, my grandfather had to abandon his flight plans in 1902 because of poor health. But not before he'd flown his own aircraft at night! Seen by a local bobby, it went down on the report sheet as 'a very large bat, species unknown!'

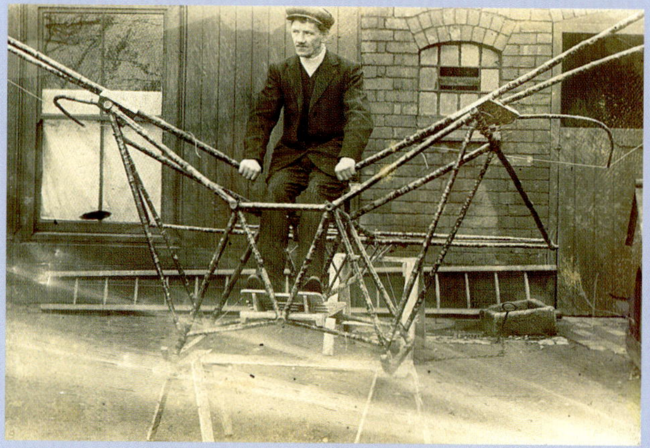

Wilson-Pilcher flat four-cylinder water-cooled car engine which also embodied a Wilson pre-selector gear box.

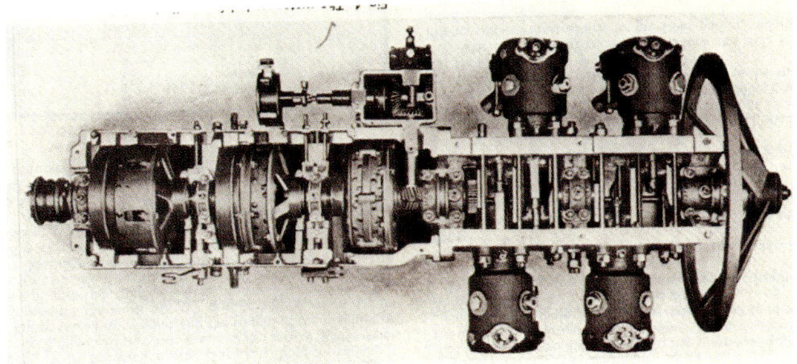

Wilson-Pilcher car

In addition to his flying activities, Percy Pilcher had been involved with the Wilson Car Company, the firm which supplied the flat twin engine for his aeroplane, on new designs for the burgeoning automobile market. One of these was a horizontally-opposed four cylinder with longitudinal crankshaft which drove the Wilson pre selector gear box. Lawrie explained while looking at a photograph of the old engine, 'An interesting feature was a flywheel, mounted at the front of the engine, which was of annular construction ie: a cast-iron hoop with spokes, efficient as a flywheel because it's most effective as a flywheel at its outer perimeter where the lever arm is longer. A clever idea which is not often seen. Then there are the four horizontally-opposed cylinders, with the central crankshaft support between the two sets of two cylinders. Attached to that is the self-changing Wilson gear box. Eventually this developed into the gear box which we all knew much later, and which set the standard for fully-automatic gear boxes 60 years later.

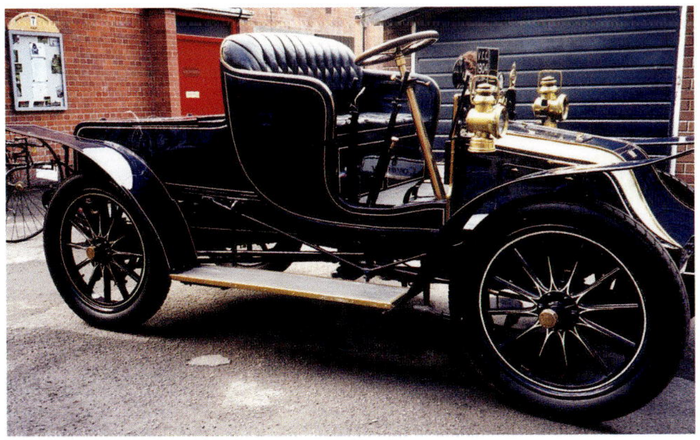

The stylish (for its day) Wilson car in which the engine was tried.

It's interesting to see that Wilson also used the annular concept for the sun wheel and planets in the gear box. I can only think that he had a bee in his bonnet about reducing weight where possible.

'Wilson and Pilcher were partners in these developments and also in the tri plane. But towards the end of Pilcher's young life, it looks as if there was a cooling between them because there's no record of Wilson being on the scene - he was not present at Percy's fatal demonstration, nor was Wilson at the funeral. Don't know what was happening, but perhaps a clue lies in the failure of the twin cylinder Wilson engine in Percy's aeroplane. Perhaps Percy had found a weakness in the crankshaft which Wilson did not want to get involved with? Maybe he felt it might reflect badly on his car engines? What is certain is that Mr Wilson vanished from the aero scene and did not reappear until years later and then it was with my grandfather. Wilson had provided some support for my grandfather's aircraft but after a while that support vanished with Mr Wilson leaving my grandfather to get on best as he could, and he eventually succeeded in flying his own aircraft.'

Eileen

'Eileen has been with me since I met and fell in love with her when she was 14 years old. From that day to this, regardless, she has been by my side unselfishly helping. I've always been very conscious of that. Now that we're both getting a bit longer in the tooth, Eileen has taken to flower arranging and is currently chairwoman of the Coventry Flower Club, which is rather nice for her and her circle of friends. It seems that it's my time to support her. So I'm going to sit back, just a bit, and help her as much as I can. Seems the right thing to do. I loved her when I married her, and I love her now, more than 62 years later!'

Homewood Bound L. Watts 1949

**Homewood Bound after a long day's toil
The air is sweet with the scent of fresh tilled soil**

**Softly blows the evening breeze
And birds return to their roosting trees**

**Spent are the clouds that fade away
But they will return another day**

**I peer into the purple haze
Where cattle rest after a long day's grase**

**The wise old river flows to the sea
With a calm that comforts me**

**Tender are the lambs at rest
As I return to my love nest**

**Ghostly shadows slowly creep
Like beckoning fingers invite the world to sleep**

**Though silent is the fading light
The evening changes into night.**

Ones that (almost) got away

I'd noticed for some time that Lawrie, when going through his archives, occasionally put aside a drawing, always in a special place. He never offered an explanation. One day, as we neared the end of our tape recordings, he harrumped a couple of times, shifted uneasily on his chair and gave a convincing impression of a man with a problem on his mind. Finally, he blurted out 'I know nothing about these' and handed me a bunch of drawings from the 'special' place where he'd been stashing them all those months. 'What do you mean?' I replied in astonishment. 'You obviously drew them for a reason so you must know why you did them', glancing at some of his finest work. 'Honest, mate, I wish you were right. For the life of me I just can't remember when, where or why I drew them. Of course there was a reason, probably to illustrate an article Bob (Currie) wrote, or maybe Vic (Willoughby). But I can't say. I do remember occasional details which impinged on my memory as I did the drawing. You're welcome to that, if it's a help. It might point you in the right direction for a couple of the drawings but as for details of the others, sorry mate, you're either going to have to find somebody who recognises the drawing for the full story, or else ditch them'. After examining each drawing, we both reached the same conclusion; they were too good artistically, too interesting technically, and as there was almost certainly a good story behind the drawing, the book would be the poorer without them.

Which is how, for months after I thought the book was complete, I searched for leads to people who had been connected to the subject of each drawing, in some cases up to 50 years previously. Many had died, some had left sons to take over the business, some had been employed on making the subject of the drawing. Regardless, it was entirely due to their generosity in providing me with invaluable information that I eventually pieced together stories behind what I regard as some of Lawrie's most significant motor cycle drawings. They provide an insight into some of the technical innovations of the old British motor cycle industry in its pomp, and as such, they are part of that industry's history. Men who provided technical innovations illustrated in these drawings deserve to be commemorated as they were seldom in their lifetimes. As clever innovators, they provided the expertise which, handled differently, might just have influenced the course of the UK industry...

Four cylinder Norton Racer

Sat on his little stool in Norton's Bracebridge Street race shop in the early '50s, Lawrie found himself faced with a conundrum. He had been sent there by 'The Blue 'Un' to do a sectioned drawing of Norton's top secret new 500cc four cylinder racer. Slight problem; it was so top secret that it had yet to be made! A pile of castings, some part-assemblies and designer drawings was all Lawrie had to work with… and his fertile imagination. The resultant drawing shows a complete engine, integral gear box and shaft final drive installed transversely in a duplex loop frame bolted to an unmistakably Norton Featherbed rear sub-frame and wheel. That much might have been interpreted from the material in front of him by a competent artist. But it would have needed the touch of a genius to show the contents of those castings before they had been machined and bolted together. That the young Watts possessed the genius to translate a designer's intentions onto drawing paper just four years after joining Associated Iliffe Press is illustrated brilliantly in this drawing of a unique machine which, had it been developed, could have met the challenge of Gilera and MV Agusta throughout the 1950s.

Unsurprisingly, after more than 50 years, Lawrie's memory of the exact circumstances of how the drawing was executed is hazy, except that 'it was a little difficult to piece the jig-saw together. I'd seen the MV fours and took my cues from their layout, as I'm sure that Nortons had. So that was my starting point. I could see from the castings where the designer was coming from and where he intended things to fit. As the bottom half of the engine had already been machined and assembled there followed a certain logic in matching up the other items. For instance, as the unit was to be water cooled, the radiator would have to be mounted in front of the

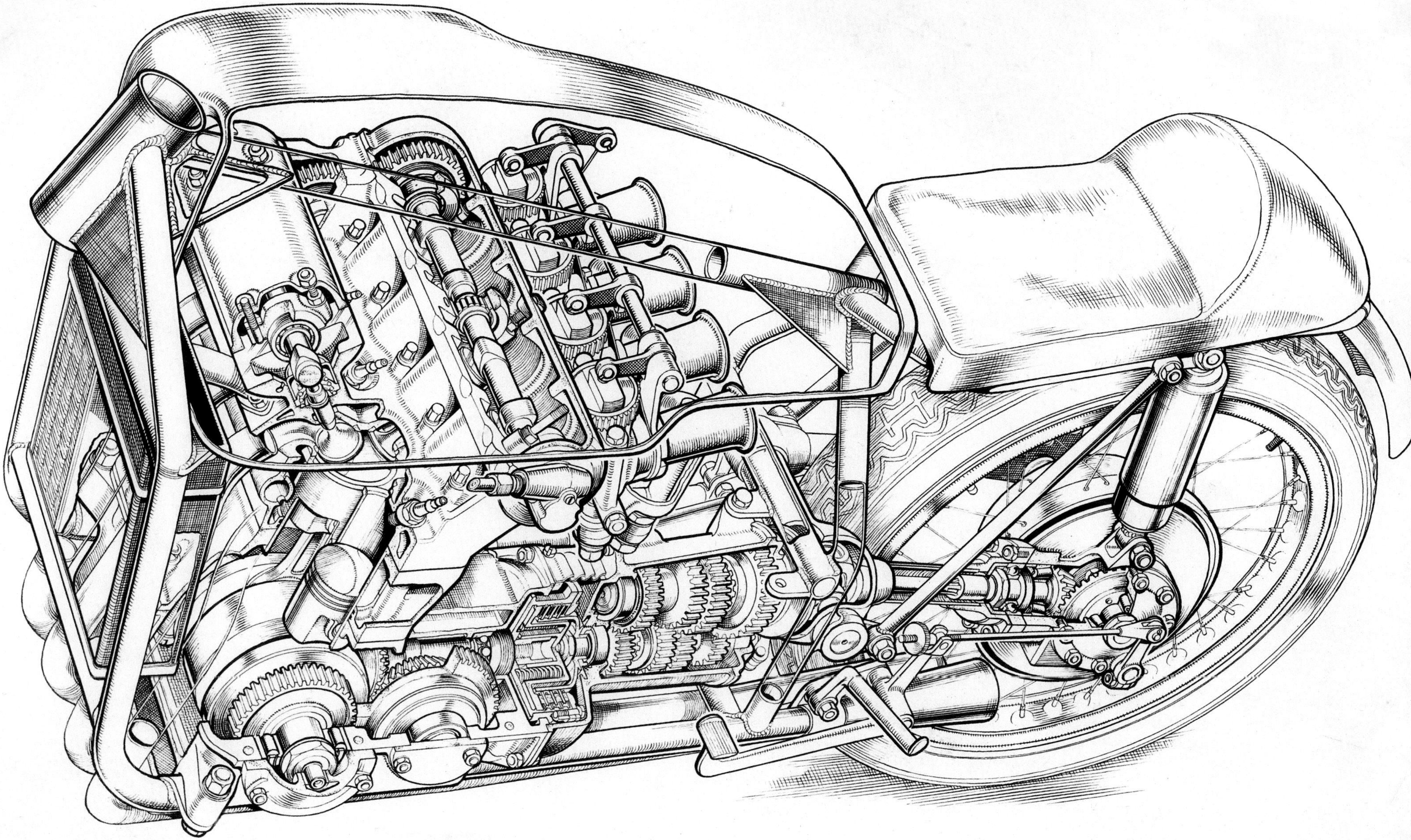

engine between the front down tubes. Because of the engine's width, the tubes would need to be splayed farther apart than on the existing Featherbed frame which was designed around Norton's single cylinder race engine. The frame would also need to be stretched because of the integral gear box on the four cylinder, and room would have to made for the take-off and installation of the shaft final drive. From the designer's drawings I calculated how the complete engine would look and then drew a frame around it utilising a single top tube instead of the Featherbed's twin tubes, because they could not be wrapped around the transverse engine. As you can see, I added gussets wherever possible to stiffen the frame and counteract lateral twist which the more powerful engine would impose on the Featherbed frame. Whether Norton would have adopted the same strategy for the complete machine we will never know because the project was aborted before completion.'

So many questions were raised by the drawing that I needed to find someone who could supply answers. Easier said than done! Over the intervening 50+ years most people with knowledge had died. Everybody had heard about the 500-4, there were even some brief articles published, but little of substance. Two people who would have known refused to return my calls (don't ask!), adding further to my frustration. The breakthrough came during a conversation with that arch enthusiast Keith Blair, 'Why don't you ask Fred Swift? He worked in the race shop and is still around'. Fred knew about the four but it was before his time. So he suggested Charlie Edwards, Norton's race mechanic and Joe Craig's right hand man. Charlie was not only still very much alive, aged 87 years, but bubbling over with knowledge and enthusiasm. He confirmed that the only part of the project which had been completed was a 125cc slave engine built during a spell when Joe Craig, Norton's legendary race chief and Charlie's boss, was at BRM the racing car manufacturer at Bourne in Lincolnshire. Charlie believes that the four cylinder was designed while Craig was at BRM, but not by Craig. However, he returned to Nortons with the 125cc unit but refused to allow it to be tested on the race dynamometer. 'So we ran the little engine on the dyno that we used for production engines and it produced 11-bhp at 13,000rpm', reported Charlie. 'The projected four cylinder was generally believed to be designed by Eric Richter after he left HRD. His design was a water cooled in-line four cylinder with twin overhead camshafts. Two valves per cylinder were set in the head at a wide angle, following the fashion in those days. Unusual features included wet valve guides, as used in some aircraft engines, which assisted cooling, and cylinder liners which were screwed into the cylinder head. Another departure from normal was the gearbox, which was built in unit with the engine. The drive was turned through 90 degrees to accommodate shaft final drive to the rear wheel.' In answer to my query if Norton had copied the first MV Agusta 500 four cylinder, he told me an unusual tale. 'After ▶

Right: Charlie Edwards prepares to put the stand under Ray Amm's kneeler Norton during a pit stop at Monthléry. Joe Craig stands behind the bike while co-rider Eric Oliver, left, prepares to take over the next stint.

Charlie Edwards on the kneeler Norton with his hero, Artie Bell, preparing the prototype bike for record breaking runs at Monthléry, France. The front part of the fairing shows signs of an involuntary encounter with the road...

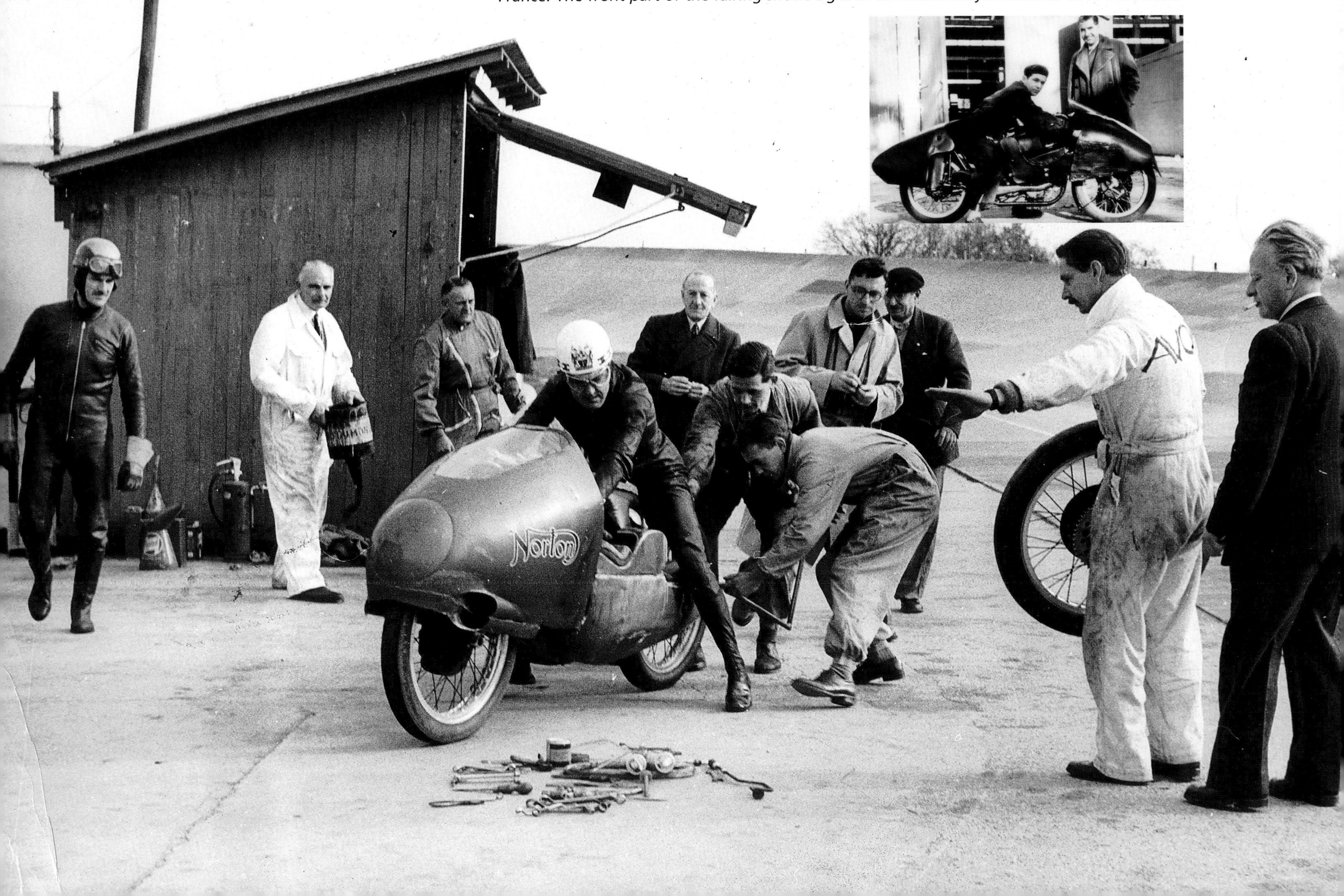

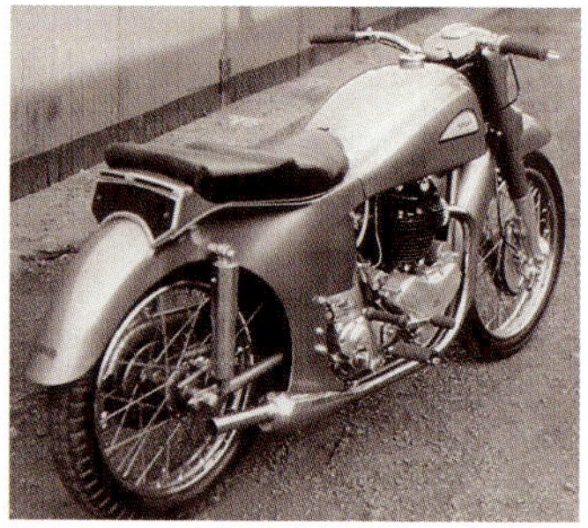

The prototype Dominator, with sheet steel bodywork, into which a tuned S8 Sunbeam engine was fitted to see whether a longitudinal layout would be preferable to a horizontal installation for the four. A catastrophic crash provided the answer, leaving Geoff Duke shaken but not stirred.

the last Grand Prix at Monza one year, the factory Norton and MV teams stayed on for testing. MV and Norton riders were good mates, as were the factory personnel. So, in the security of private practice, riders swopped bikes one day. Our Norton lads were screaming around Monza on MVs and their lads were banging around on our bikes! MV had the power but they badly needed better handling. As Norton had been for years the yardstick for road holding, what better way to learn than by sampling Norton handling on MV's home track? The Norton riders could find out at first hand the size of the mountain they had to climb when competing against the powerful MVs. It wasn't just the MV engine which was more powerful, so too were the brakes, as Norton veteran Johnny Lockett discovered when sliding down the track after locking the front brake on 'his' MV! Luckily, nothing more than pride was bruised'.

Charlie suspected that that day's private testing influenced Norton's use of shaft drive as shown in Lawrie's drawing. But other ideas were tried before signing off the design, as Charlie explained. 'We thought about mounting the four cylinder in line for a narrower frontal area and for the cornering clearance it would provide. It would also have suited the shaft drive. So, using a Featherbed chassis, we made a mule to test the experiment by extensively modifying a 500cc S8 Sunbeam, probably the last machine in the world to be associated with racing! But, installed in the Featherbed frame, the S8's in-line twin cylinder engine and shaft drive were ideal for our experiment. The single overhead-camshaft engine's docile performance was transformed by tuning tweaks to produce a comparable performance to the prototype 500cc Dominator in a Featherbed frame which we were developing at that time. So, on a quiet Sunday, Geoff Duke and I took turns comparing the two bikes against one another at the deserted MIRA test track near Nuneaton in Warwickshire. Initial results were encouraging, but then Geoff took one last run on the S8 to check out something. ▶

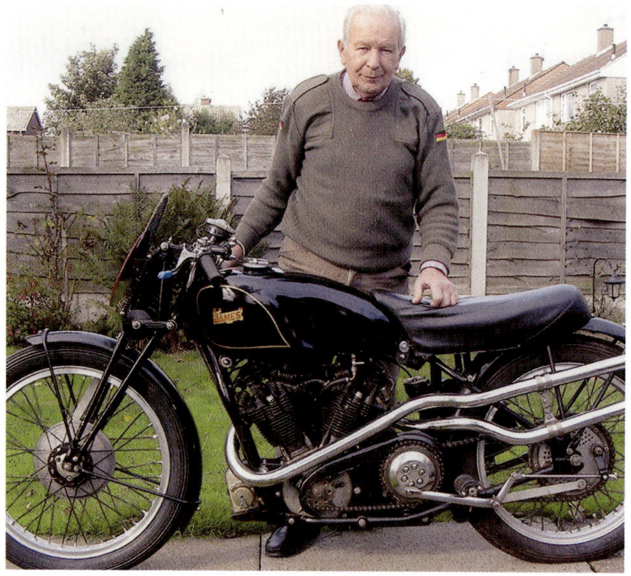

Charlie Edwards poses alongside a 500cc James he made, using an engine which first saw service on speedway but which he adapted for road racing. He says 'it goes like the clappers!'

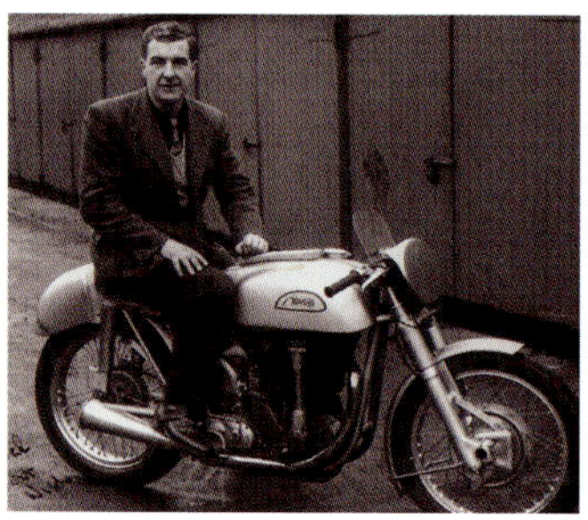

Artie Bell on the prototype Featherbed Norton which was first tried out along a twisty and narrow road alongside the River Lagan in Belfast. The front fork was borrowed from a racing sidecar, not used on the finished bike.

As he approached maximum speed the overworked Sunbeam transmission disintegrated, locked up everything, and pitched Geoff off, luckily without serious injury. The bike totalled itself. That's when we decided that an in-line shaft driven engine was not such a good idea. So we opted for the transverse layout. Shaft final drive was dropped in favour of a conventional chain, which also absorbed less power and provided a quicker means of altering overall gear ratios, which is essential for a racing design'.

Charlie soon felt that Joe Craig's heart was not in the project. 'Maybe it was because retirement was on the horizon. After all, he'd been at the sharp end of racing since long before the war, taking the Norton singles to the limit of their development and all the while battling with the board for more funding,' explained Charlie. 'That was always a problem at Norton, lack of funding. In the race shop, we knew the way forward was with the new four cylinder but its complexity would have been a whole new ball game. Maybe Joe was too set in his ways. Also, the four had not been his baby and he would have had the responsibility of developing it to be a world beater. Nothing less would have satisfied the directors. I'm not sure that he was capable of that. Joe had a reputation, which he cherished, of being something of a tuning wizard but this had been gained almost exclusively with the Norton single cylinder race engines. I'm not aware that he had the additional knowledge required to develop an unproven four cylinder capable of beating the successful Gilera and MV fours, which had many years' development behind them. Even if the board had raised the finance for such a programme, which is extremely doubtful given their past performance, it would have been a huge task for Joe in his twilight years. All the parts, except for the cylinder head which had still to be cast, were ready in the drawing office waiting for permission to complete the engine… now that's another curious thing. The drawing office worked independently of the race shop and, although the new four was a racing engine, and therefore our responsibility, all the castings and other bits were delivered to, and kept in, the drawing office. We were not allowed near them! Except that we'd already assembled the crankshaft, with the connecting rods on, and installed the assembly into the crankcase. But that's as far as it got. We had to wait for the cylinder head to be made before we'd be given permission to complete the engine. It never happened. I personally feel that this was down to Joe's disinterest. Without him driving the project forward the directors lost heart and the four cylinder project was abandoned. Our leading riders, Geoff Duke and Reg Armstrong, had departed to Gilera at the end of the 1952 season, to be replaced by Ray Amm for 1953 and the advent of the 'kneeler' Norton shortly after. Norton as a factory withdrew from racing at the end of the 1954 season to concentrate on the development of production Manx Nortons.

'What happened to the four cylinder? Well, the parts were cleared out and to make sure they would not become an embarrassment to the factory in the future, I believe that most of the bits ended up on the far side of the world. Would I attach blame to anyone? I'm not into that but it has occurred to me that somebody within Norton management, perhaps the entire board of directors, did not plan the 500-4 project properly. It was one thing to design a four cylinder, but another entirely to build, develop, take on, and beat the Italians at their own game. That would have required not only a massive leap of faith but also a huge financial commitment. Norton would have required a commercial partner to share this burden but obviously they never found one… even if they bothered to look. The factory staggered on at Bracebridge Street until 1962 when owners AMC took over and transferred everything to their HQ at Woolwich. Even today, I think of what might have been if things had worked out differently. For example, what might have happened if a race engineer with multi-cylinder experience had been given the 500-4 to develop?' ■

Sir William Petty - 17th century English philosopher

'When a new idea is first propounded, in the beginning every man objects, and the poor inventor runs the gauntlet of all petulant wits, every man finding his several flaws, no man approving it - now not one out of a hundred outlives this torture, and those that do are at great length so changed by the various contrivances of others that not any one man can pretend to the invention of the whole, nor will agree about their respective shares in the parts.'

BSA Wankel, the industry's last stand

Though Lawrie was not to know it at the time, this drawing of the BSA Wankel engine is one of the most significant of his drawings because, firstly, it shows the British motorcycle industry turning from piston power to rotary power. Secondly, it represents the last attempt of the industry to stem the Japanese take over of the traditional British big bike market. Throughout the 60's, Japanese lightweights got a stranglehold on the lower-capacity market, leaving the heavyweight sector to British and European manufacturers.. But that changed dramatically in 1968 with the advent of Honda's stunning new four cylinder CB750. The immediate success of the CB750 showed that British dominance of this market sector was about to end. All three mainstream manufacturers, BSA, Triumph and Norton, knew they had to fight to retain their traditional heavyweight market sector. All three had already marketed new twin, and triple, cylinder models but it was not enough to meet the challenge from the east. Market response was not encouraging and the three major manufacturers knew that they had to pull something spectacular out of the hat if they were to survive. They probably recalled what Lawrie's old friend and colleague, Vic Willoughby, had forecast some eight years earlier after a visit to the NSU factory in Germany, that the future lay not in the reciprocating engine but in the rotary Wankel, as pioneered by NSU.

A radical approach was required by the British manufacturers. So nothing could have been more radical than their adoption of the revolutionary Wankel concept. Invented by the German Dr Felix Wankel, and then patented and developed in partnership with NSU under the direction of Dr Walter Froede, the Wankel attracted the British manufacturers for technical and commercial reasons. The rotary engine was remarkably light and compact and eliminated vibration. With only two major moving parts, a three-flank rotor and the output shaft, the Wankel had significantly fewer components than the new generation of multi-cylinder engines from Japan. Hence, the Wankel was potentially cheaper to manufacture and, as Willoughby had already astutely observed, had the potential to be the new cutting edge of technology. That tantalising possibility would enable British manufacturers to leap-frog Japanese technology. The Wankel's favourable power to weight ratio and lack of vibration were also vital factors in a competitive package for meeting the Japanese challenge in the lucrative heavyweight sector. No wonder that the Wankel seemed so attractive to the beleaguered British industry!

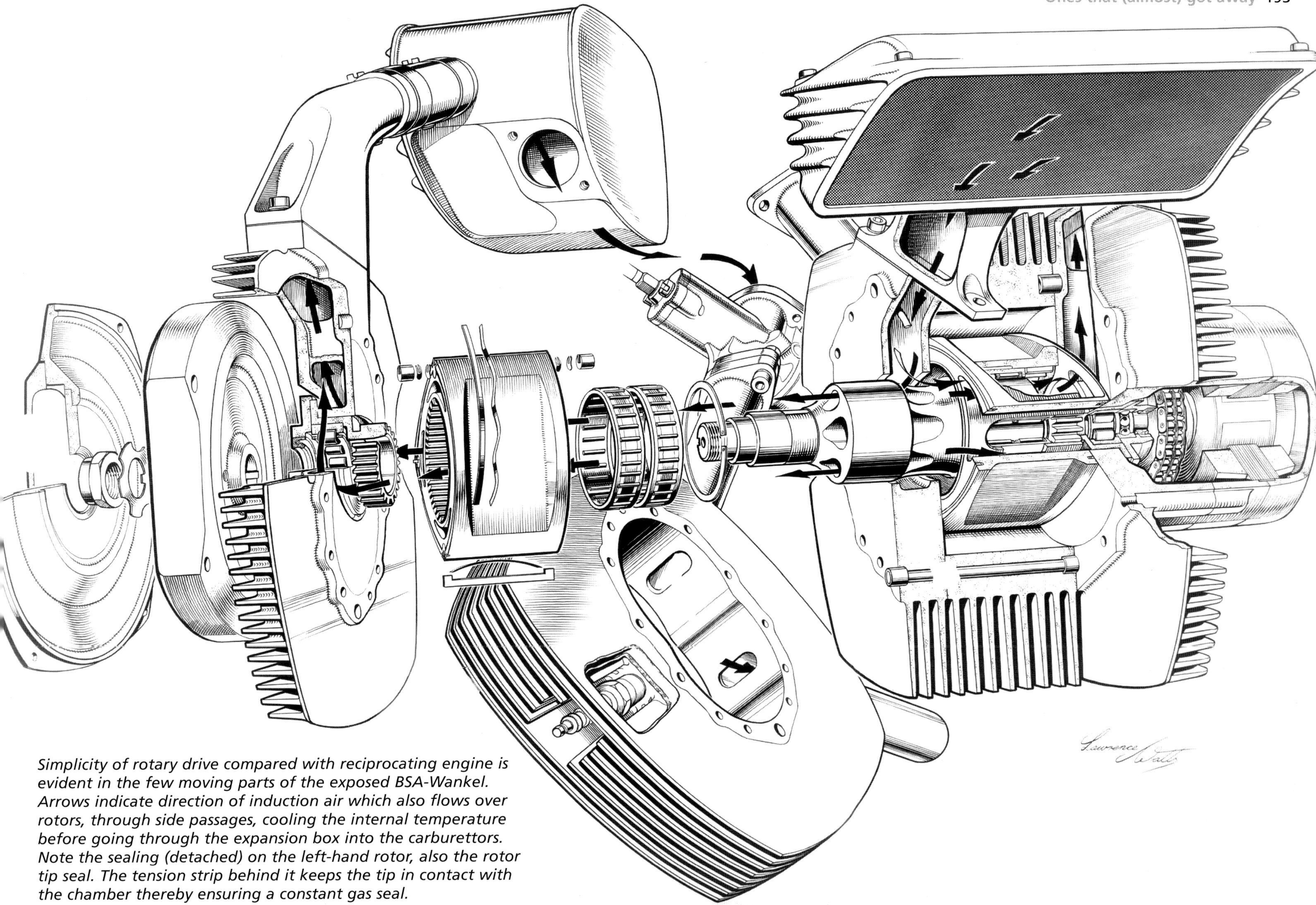

Simplicity of rotary drive compared with reciprocating engine is evident in the few moving parts of the exposed BSA-Wankel. Arrows indicate direction of induction air which also flows over rotors, through side passages, cooling the internal temperature before going through the expansion box into the carburettors. Note the sealing (detached) on the left-hand rotor, also the rotor tip seal. The tension strip behind it keeps the tip in contact with the chamber thereby ensuring a constant gas seal.

How it worked

No one described the operation of the Wankel principle better than Willoughby did in 'The Motor Cycle', in February 1960, when he wrote: 'With only two major moving parts - the output shaft and a three-flank rotor - the NSU Wankel engine is just about as simple in layout as an internal combustion engine can be...' He went on to explain, '...formed on the shaft is an eccentric boss which engages with the bore of the rotor, thus power impulses to the rotor turn the shaft. The shape of the chamber in which the rotor works is a two-lobe epitrochoid - somewhat like a lumpy figure of eight with a very shallow waist.

'The triangular rotor moves inside the chamber in such a way that all three apexes maintain contact with the wall, so forming three compartments of varying volume in all of which the four stroke cycle of induction, compression, expansion and exhaust takes place. This special motion is imparted to the rotor by internal gearing. In one end of the rotor bore (alongside the bearing track) is an annular gear which meshes with a stationary gear having two-thirds the number of teeth and rigidly attached to the inner face of the adjacent chamber side plate. So, for each revolution of the shaft the rotor turns forward one flank on its own eccentric axis. Gas sealing is effected by metal strips in grooves across the apexes and along the sides of the rotor.

'Three turns of the shaft brings the rotor back to its original position and during that time each flank in turn has received a power impulse. The inlet port feeds one lobe of the chamber and the exhaust port evacuates the other; the ports are fairly close together, one on each side of the waist, so that each compartment, at one of its two minimum-volume positions, spans both ports to give a small degree of overlap comparable to that in a four-stroke when both valves are partly open at top dead centre. The sparking plug is located roughly opposite the ports, where their compartments again reach minimum volume.'

Lawrie's drawing is of a type 'B' second generation air-cooled engine under development by BSA around 1973; it clearly shows the difference between a rotary and a traditional reciprocating engine. The eccentric drive shaft (the equivalent of the crankshaft in a reciprocating engine) on which the rotors are mounted, is shown with the left-hand rotor detached in order to illustrate the internal gear. This meshes with the stationary gear, which is secured to the left hand side plate. Also shown are the large-diameter needle roller bearings which support each rotor. For clarity, Lawrie has detached the metal sealing strips which fit into annular grooves in the sides of the rotor, and also the metal sealing tips. Behind the tip is a spring-steel strip which ensures that the tip maintains a gas-tight seal between the rotor and the wall of the chamber. Lawrie has also shown the drive shaft; it's the one with the eccentric boss which engages within the bore of the rotor. Power impulses delivered to this rotor cause the shaft to turn.

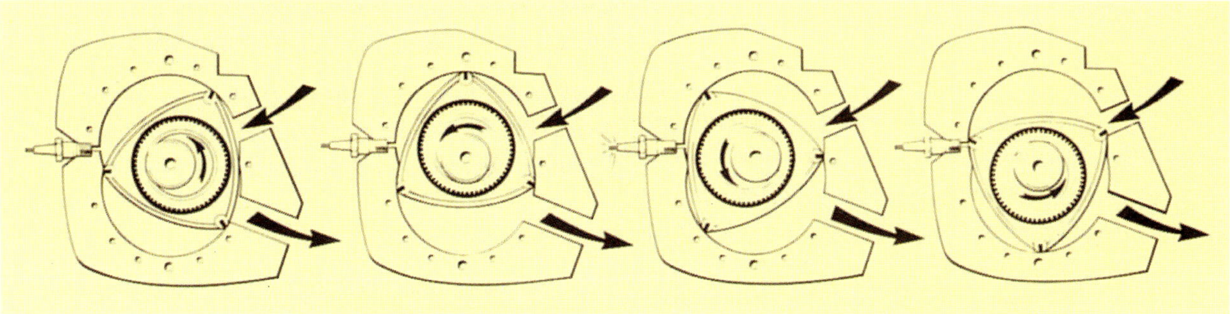

In a rotary engine, circular motion of a rotor provides rotary energy directly to the shaft.

Trochoids to you!

Trochoid is the word used to describe an unusually shaped chamber around which the rotor(s) travel. (This chamber replaces the more conventional cylinder in which the piston travels up and down in a reciprocating engine). A trochoid describes the curve traced by a point in the plane of a circle rolling round another, such as one circle inside another and touching it. If the outer circle is rolled around the inner an epitrochoid can be traced. Conversely, if the inner is rolled within the outer a hypotrochoid is created. So, the ratio of the circle diameters (analogous to the rotor gearing in the engine) determines the number of lobes in the epitrochoid or the number of sides on the hypotrochoid. For a given type of figure, the precise shape is governed by the ratio of radius (from the centre of a rolling circle to tracing point) to eccentric. For example, with a two-lobe epitrochoid the smaller the radius relative to the eccentric, the more pronounced the waist.

Seeing is believing!

When NSU tried to patent the Wankel world wide, some countries rejected the application as being 'impractical'! Others could not understand the sequence of events. After all, not many people had heard of a 'trochoid', even less an 'epitrochoid'. So NSU made a number of working models beautifully crafted from solid Perspex. Vic and I shared an office at the time and I remember him returning from Germany with a model which he proudly demonstrated for years afterwards! The model showed at a glance the simplicity and ingenuity which characterised the Wankel principle and which ultimately was responsible for NSU's successful patent applications world wide.

BSA lead development

Initial development in the UK started in 1968 with a single-rotor Fitchtel and Sachs industrial KM914 Wankel developing a modest 18-bhp. By the time Lawrie came to do his drawing, development had moved on so rapidly that a twin-rotor engine had evolved with power increased by a factor of four. On this unit, a large air filter, the box-like structure at the top of the drawing, collected cool air from below the front of the petrol tank. Directed from the filter (see arrows) into the engine, the air acted as a coolant for the rotors and shaft before it was ducted via the two end plates of the engine to plenum chambers mounted on the carburettor entries. Short inlet tracts led from the carburettors directly to the inlet ports of the engine.

David Garside

Recruited by BSA to its Group Research Centre (GRC) at Kitts Green, Birmingham, in 1969, David Garside was put in charge of Wankel engine development. David had previously worked for three years on a diesel version of the Wankel at Rolls-Royce, Crewe, in the mid 1960's. He had been to NSU several times, had met Herr Wankel and had attended the 1966 Wankel Licencees Technical Conference in USA. David was with the motor cycle project through all its stages, and under various managements, until the mid-80's when he became almost totally involved in the aircraft applications of the engine.

His only regret is that he did not select water cooling in 1969, 'it would have saved a lot of heartache, time and effort!'

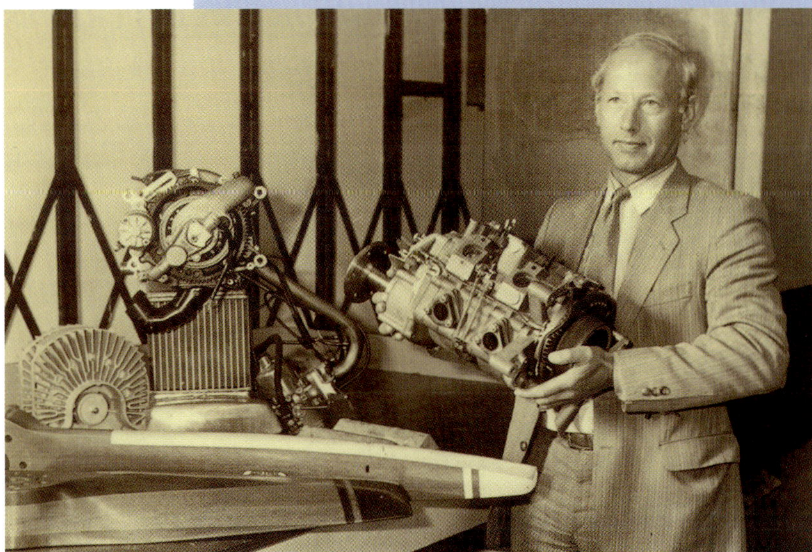

David Garside, Director of Engineering, with a 90-bhp, twin-rotor, liquid-cooled Norton rotary engine.

Head man David

Lawrie remembered doing the drawing but he'd forgotten when and where. I needed to find someone who could complete the picture of this unique engine's UK development. My old friend Norman Hyde suggested David Garside, who had been the engineer in charge of the project, but Norman had lost touch. It was Keith Blair another member of 'The Club' (a club for executive members of the motor cycle industry), who finally put me in touch with David. That he was a mass of information is a huge understatement! He recalled that it was around 1973/4 that 'Lawrie executed his extraordinarily skilful sectioned drawing of a fairly early prototype'. But how, after all this time, did David identify the year of the drawing? 'It has some features, such as the Amal Concentric carburettor, which were certainly not used after 1976', replied David. 'It also shows the left hand engine end plate with an integral primary chaincase which was the 'B' prototype engine. Later developments and production machines, used the third stage 'C' type engine which had the chain case on the right'. Shortly after Lawrie's drawing was done, but before development could be completed and production of the machine commenced, BSA collapsed. Following a report which David Garside had made to Dennis Poore in 1973, the incoming chairman of Manganese Bronze which had taken over what was left of the BSA group in 1974, the Wankel project became the property of NVT International and moved from BSA's development centre at Umberslade Hall to Kitts Green where NVT were then based. Development of prototypes resumed. David recalled that 'the type 'B' was much earlier than the motorcycles we sold to the police'. It's too easy to forget, after all the intervening years, that UK police used state-of-the-art Wankel-engined motor cycles more than 25 years ago, and they were British made!

Early years

In 1976, the project moved to Shenstone, near Lichfield. Of the three major versions constructed during the research stages, the third (C type), was the one subjected to serious development and durability testing before going into low-volume production. It's worth recalling that what had started as a BSA project, a prototype had been sent to Triumph where it was worked on by development engineer Doug Hele and his assistant Fred Swift, but it was as a Norton that the machine finally went on sale to the public. Even stranger, the original design had been German, the engine being fitted by NSU first in their beautifully proportioned Ro80 saloon and then in the Spyder sports car. Fichtel and Sachs had been licensed to make industrial versions of the engine. But it was in a car that the Wankel engine made its first appearance installed in the shapely rear of NSU's Spyder small sports car. An example had been bought by Triumph's research and development department to see what made the Wankel tick. The Motor Industry Research Association (MIRA) also acquired one, pulled it apart, had already pulled one apart and published a very detailed report. BSA based their research on a single-rotor industrial KM914 Fichtel and Sachs engine. Installed in a BSA production C12 frame, the industrial engine came equipped with an axial fan which air-cooled the housings and with an induction-air cooled rotor. Drive was through the existing BSA gear box. This was the project which David Garside inherited when he joined BSA's GRC in May 1969 as project engineer. Using F&S rotor assemblies and ancillary bits, three twin-rotor engines were produced by the following year (the type 'A'). One was installed transversely in a Gold Star frame, a second was for dyno use and the third was lent to Triumph's Doug Hele and Fred Swift for development. With the BSA R&D department moving to the new motor cycle Design and Development Centre at Umberslade Hall in 1971, Wankel development became a significant part of BSA's future. Some external styling was incorporated into the castings of the air-cooled type 'B' engine (Lawrie's drawing). BSA's intensive development of the twin rotor engine soon produced 70-bhp, an impressive output for a production engine and a match for the 68-bhp of Honda's CB750 four cylinder. The Wankel's increased power necessitated beefing up the standard BSA frames as they had not been designed for anything like that output.

BSA collapse

For what reason nobody knew, but Bert Hopwood set up a short-lived rival Wankel design group in 1973 at Norton's base at Kitts Green. It soon collapsed. Unfortunately, BSA also collapsed the same year... Thereafter, the BSA-Wankel became a Norton when it moved firstly to Kitts Green and then, in 1976, to Shenstone, near Lichfield.. Power output had by now reached 75bhp under a new development programme when the rights to the project were re-acquired by Manganese Bronze Holdings from the receivers of BSA. (Triumph remained now as a totally separate co-operative in their factory at Meriden). Over the following eight years, engine development continued until, in 1983 and now in water-cooled format, it reached a very respectable 90-bhp at 9,000 rpm which was the most powerful motorcycle production engine produced by the UK industry until the reborn Triumph marque 15 years later, a landmark still unheralded by the media. By now, many miles of durability testing had been achieved using relays of test riders - some of them off-duty police men--in two shifts, targeting 600 miles per day (and night) in all weathers. Police bought the first production units and their feedback was fed into the programme for civilian production machines. Another 1983 landmark was the return of Doug Hele, after a brief sojourn at British Seagull, as chief development engineer of NVT. Doug's brief was to refine the complete machine for marketing. (Some of this 'refining' came about on 'The Club's' spring and autumn runs when the bike would be subjected to detailed scrutiny by Doug's fellow industry members...).

Mixed joy

Is there a happy ending to this story? Yes and No. Taking the original single rotor Fichtel & Sachs commercial engine and developing it into a twin rotor unit developing 90-bhp in motor cycle form was a technical tour de force. Especially in the face of immense commercial pressure on two fronts; lack of development money from the declining UK motor cycle industry and, secondly, the emergent Japanese industry which was fuelled by successfully capturing the world's market in lightweight machines. People are quick to criticise apparent 'failure', but to put it in context, the firm which was first bold enough to put the Wankel in a car, NSU, went out of business (or, was absorbed into the Audi empire). The Dutch Van Veen company also failed to gain acceptance for their 100-bhp twin rotor bike, which used an oil and water cooled engine straight out of the NSU Ro80 car. A brief ride in Amsterdam impressed me with the bike's potential, though circumstances precluded a ride outside the city. Suzuki also marketed a single rotor 500cc version of the Wankel but it too had a fairly short innings, mainly because conservative motorcyclists failed to accept, or understand, it. Mazda have kept faith with the Wankel since the beginning; I test drove a twin rotor 500cc (a two litre equivalent) coupe at Brands Hatch in 1968. The most impressive feature was the phenomenal smoothness of the 100bhp engine which, naturally, endowed it with a deceptive turn of speed. I had an example of its sporting handling later in the day when out on the track again and Stirling Moss threw the Mazda into a huge four-wheel drift as he slid past me into Druids Hill bend....and I thought I was motoring swiftly in a sporty little S800 Honda....Mazda, after 40 years of continuous production of the Wankel, have garnered themselves a select market but no thanks to anyone for the lack of publicity which they received for their magnificent Le Mans 24 Hours win in 1991 when they finished 1st, 3rd and 6th with specially-tuned four-rotor engines, the new 654cc rotor sizes developing 700-bhp. Of the 57 starters, only about 20% of the reciprocating-engined cars completed that race in spite of being the products of companies with much more racing experience and engineering resources than Mazda. The perfect riposte to the most frequently asked question about Wankel engines; are they reliable? No component in the Wankel engine is as highly stressed than an exhaust poppet valve in a reciprocating engine. Which is why the Wankel has finally found its niche powering expensive and important surveillance-type UAVs.

Mazda's philosophy

Of the manufacturers who have been involved with Wankel only Mazda has stayed the course. Answering those who had criticised Mazda's dedicated pursuit of the Wankel over the preceding 15 years as 'a visionary's pipe dream', the man responsible for Mazda's success, Kenichi Yamamoto, told a meeting of the World Automobile Congress in America that 'The rotary is an engine, not a poem. Why did we say that? What did we mean?'. He explained that Mazda 'pursued the rotary because its virtues as a power plant are rooted in science and technology... but it took more than science to hold us to our path... new ideas, even technological ones, mean crossing established frontiers. The paths are uncharted... but the Wankel engine represented an advance investment in Mazda's future... and Mazda's need for a clear identity'. He described vividly how, over the years, those who worked on the engine's development overcame obstacles which had never before been presented to automobile engineers. As he described it 'Defiance of the old, and essential quality of innovative ideas, is necessary to bring on the new... this was not a totally lonely battle, we received so many letters from people of all stations from all over the world, encouraging letters. This was one of the intangible benefits that came from accepting the challenge of an innovative technology'. He ended by saying

'Like a poem, the rotary engine was the child of our dreams, Like a poem, it put together elements in a way that was startling in the overall effect, And like a poem, the rotary captured a spirit, the pioneering spirit, that spirit has captured our company... We will never be the same.'

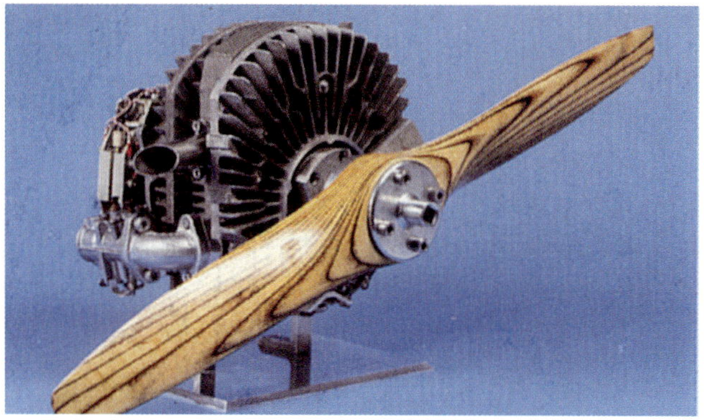

Eventful innings

From modest beginnings, the original BSA-Wankel project had not only survived an eventful 14 years in various ownerships but, more significantly, it had realised much of its early potential. Commercially, it found uses other than in motor cycles and to explore and develop future uses David Garside was put in charge. When bikers are spinning yarns down the pub, more often than not the term 'It was flying mate, absolutely flying…' is prelude to some derring-do… In the case of a Wankel it could literally be true because that's what the engine is most popularly used for today, powering unmanned aircraft. This application started on a small scale in 1985 with the first sale of units for Alvis UAV Ltd (unmanned air vehicles) use. Those engines were still in use in Egypt in 2001! Several thousand of a low cost, low power (38-bhp) versions have been manufactured for target drones. Meantime, some 100 'Classic' air cooled motor cycle engines were made for sale to the public.

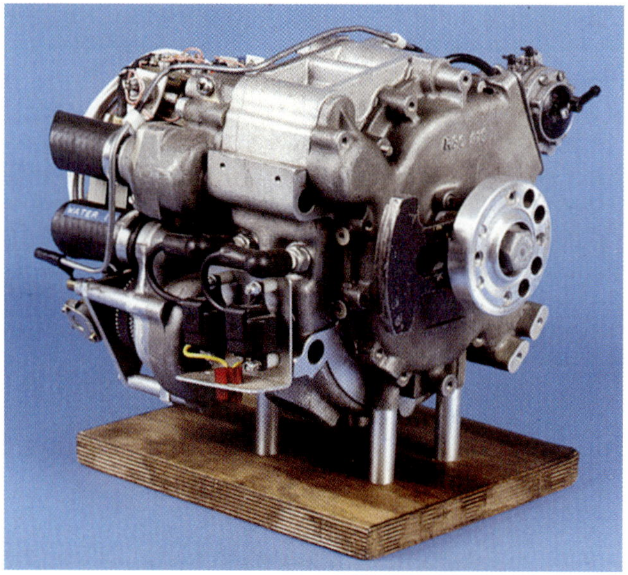

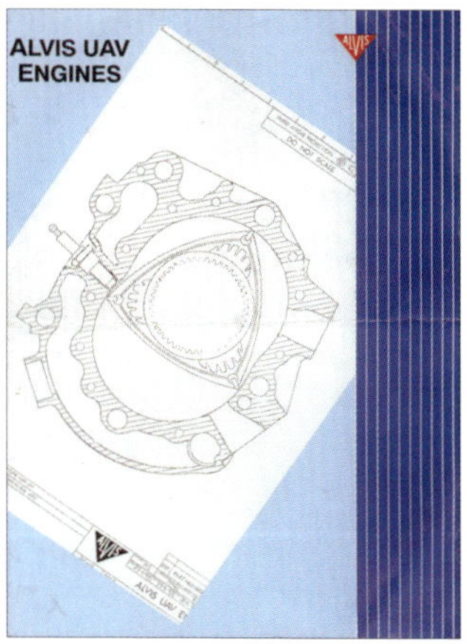

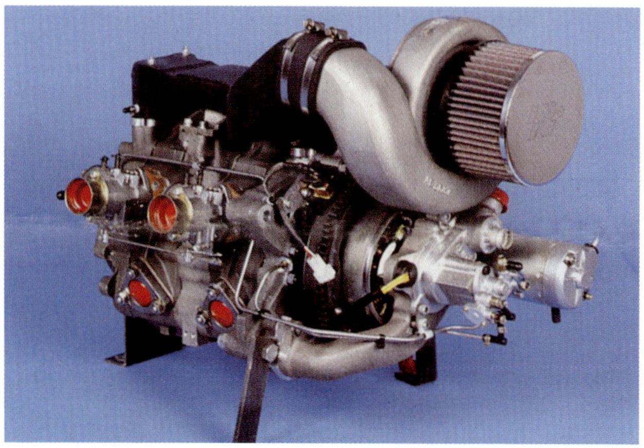

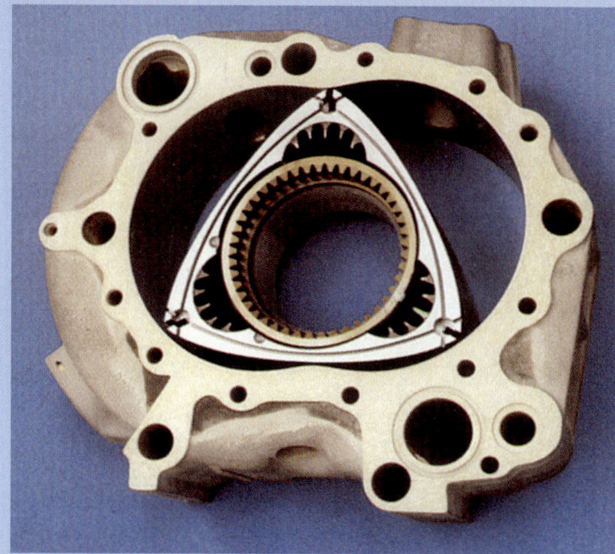

Alvis UAV Ltd.

In December 1992, David Garside led a management buy-out to secure the exclusive rights to rotary technology for use in UAVs. Called Alvis UAV Ltd, the new company was dedicated to the design and production of power systems for Unmanned Air Vehicles (UAV's) and became a 75% owned subsidiary of Alvis plc, a multi-national UK-based engineering group specialising in the design and manufacture of defence and aerospace equipment. Covering 20 to 120-bhp petrol-fuelled rotary engines, the Alvis UAV product range claimed to be the world's leading contender for the supply of UAV propulsion units due to their exceptionally high power to weight ratio. It was claimed that 'Alvis rotary engines were flying in more than 20 types UAV worldwide, ranging from target drones to one-trip lethal UAV's to high-value surveillance-type aircraft'.

SAE paper presented in America

In September 1982, David Garside presented a paper entitled 'Development of the Norton Rotary Motorcycle Engine' to the Society of Automotive Engineers in Milwaukee, Wisconsin. A lucidly written paper capable of being understood by non-technical readers, the paper describes in detail the history of the BSA-Wankel project through to the time it was taken over by Norton. Anyone reading the paper could be forgiven for coming to a frustrating conclusion; if only the suits supposedly running the British industry had realised the potential of the Wankel five years earlier, then the industry would have had a fighting chance of saving its lucrative heavyweight market. For the Wankel achieved not only the targets outlined for it when Lawrie was drawing it, but the engine went on to prove itself inherently reliable at maintained high engine speeds, something which could not be said of most previous British designs. No higher accolade could be awarded to an engine than to be chosen as the motive power for manned and unmanned aircraft, which is where the engine finally found its niche.

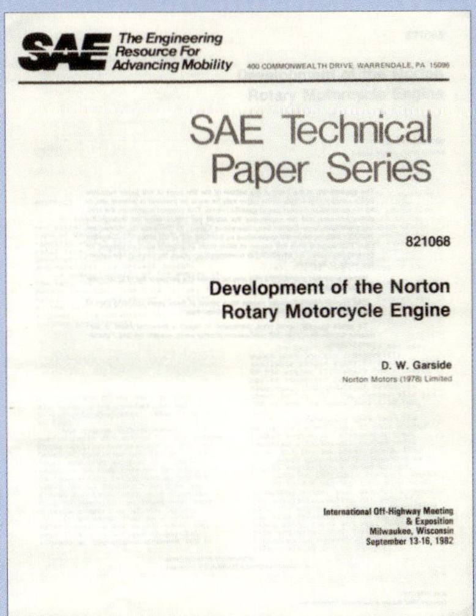
SAE The Engineering Resource For Advancing Mobility 400 COMMONWEALTH DRIVE, WARRENDALE, PA 15096

SAE Technical Paper Series

821068

Development of the Norton Rotary Motorcycle Engine

D. W. Garside
Norton Motors (1978) Limited

International Off-Highway Meeting & Exposition
Milwaukee, Wisconsin
September 13-16, 1982

However, after only two years Alvis sold their shares to Silver Arrow, who are the current owners of UAV Engines Ltd. (NB: Some years later, David Garside heard that Alvis had regretted their decision to sell!). Called 'UAV Engines Ltd', the company is now owned by an Israeli UAV manufacturer and occupies the whole of the Norton site at Shenstone. That company, and Mazda cars in Japan, are the only two companies in the world today still manufacturing the Wankel type engine.

I told Lawrie the story I'd discovered behind his drawing and, especially, how the story had a bittersweet ending. 'I find that quite ironic, and yet comforting', Lawrie ruminated as he lit another evil-smelling cigar. 'I've always been a keen motorcyclist, but my hobby has been flying, for over 60 years. During the many occasions my old mate, Bob Currie, and I chewed the fat over the reasons for the demise of the British motor cycle industry we discovered that most of the missed opportunities which formed part of the industry's downfall were totally wasted opportunities. But, as this Wankel story shows, our engineers proved that they were up there with the best, and when I say the best, I'm borrowing my old mate Vic's words. He felt that German engineering was probably the best in the world. Yet our guys took over development of the Wankel from the Germans, while still co-operating with them, and improved it into something which I'd like to think was world class. That it did not succeed in its chosen use in motorcycles may be blamed on the suits who made commercial decisions probably too late in the day to buck the market trend. So I feel that Mazda deserve enormous credit for sticking with the Wankel and refining it for car use, winning international awards for performance, reliability and owner satisfaction as well as that most remarkable win at Le Mans. But, and this should be remembered, Mazda have an extensive range of other models, with reciprocating engines, which provide the firm with the income which any commercial organisation needs to survive. The RX models, which use Wankel engines, have always been in niche markets. Even after 40 years of Wankel power, the vast majority of car and bike owners haven't a clue what the Wankel engine is about. And, in my opinion, therein lies a big problem. I think that Mazda may have finally realised this and that's why they put the Wankel into their gorgeously styled RX-8 four-door coupe. It's won so many international awards, not only for its looks but also for the way it suits everyday driving, that I wouldn't be surprised if Mazda eased the Wankel into their mainstream saloons. I'd applaud that decision for it means that engineering has won a long-deserved victory'.

Corby Iron and Steel Works

What have farm machinery and an old steel mill in common? Not a lot, except for Lawrie Watts. 'I wanted to include this drawing of the Corby Iron and Steel Works as a tribute to all who worked there and who produced the material which went into many of the cars, bikes and machinery which I drew in the years before the works closed in 1980. I think that this is the only surviving copy of the original drawing,' he explained. 'At the heart of British manufacturing for almost 60 years Corby Iron and Steel was forced out of business because of cheaper foreign imports. I also wanted to include the drawing because it was quite different from any other structure I had drawn at that time, plus which the working conditions could not have been more different - I've not been in anything like it since! As the works has been flattened I'd like to think that this drawing will remind future generations what a British steel mill - at least this part of it - looked like, when real men got themselves very hot, very dirty, and sometimes badly injured, making the raw material from which our cars, household goods and machinery were manufactured.' In its existence, the former Stewarts and Lloyd mill developed and spread over two square miles of Northamptonshire countryside around the little town of Corby which, before the works came in 1933, had a population of 1,500. This had grown to 60,000 when the mill ceased production in 1980, when most of the population were reliant on the works for employment. But before it closed, it had produced 35,813,424 tonnes of steel and 30,879,525 tonnes of iron. Most of this once great works made way for new developments, including the Rockingham Raceway and a Safeway supermarket complex. So how come that Lawrie found such a dramatic change of subject? Well, as a London based artist was not available, and Coventry was not that far from Corby, Lawrie's boss decided that an interlude from drawing farm machinery would be good for Lawrie (though how he squared that with the fresh air down on the farm to the pollution of a steel mill is a thought process peculiar to bosses). But the experience did leave its mark, even 40 years later. 'A truly enormous place,' recalled Lawrie, 'with a mixture of old and new buildings spread all over the place. To connect these and move the raw materials and iron and steel around, the plant had its own railway, with more than 100 miles of track and its own locomotives, and rolling stock, some dating from the 1930s.. Giant cranes, one of which had a 217ft boom and was imported from America, operated drag lines with buckets which held up to 36 cu yards of ironstone dragged from the ground. Steam, smoke, dust and noise was overlaid with the pungent smell of coke, natural gas and oil from the blast furnaces together with fumes from burnt coke dumped outside to cool off before being disposed of. Another cause of air pollution was the coal dust and gas given off from the coke ovens when they were at full stretch, sufficient to cause a heart attack among today's Health and Safety executive! More so if they saw how the staff dressed; hard hats, yes, but sports jacket and flannels for the supervisors? Shirt sleeves and ordinary trousers for those not immediately in front of the furnaces was normal, but no obvious safety rig for those supervising the pouring of molten metal only feet away.... And those fumes! Although extractors inside the plant sucked fumes into an extraction and cleaning plant before being released into the atmosphere, the air still retained quite a whiff, leaving you in no doubt that you were inside a major manufacturing facility! So this was my office for a while around 1964 while I was drawing a five storey building then being constructed to make basic oxygen steel, known as BOS in the industry. It posed me yet another challenge; how to show what was going on inside a 145ft high building, not yet fully operational. No use just showing the machinery in place, I had to get my head around the various functions in making steel so that everything would be positioned correctly. To complicate matters, a three storey building alongside which supplied raw materials and was connected to the main building by a feed pipe, also had to be included. It drew different materials from other machines and sent them to the main hopper on level three of the big building. While thinking about my problem, I had this vision of a little boy cutting Plasticine away from his toys to show what was underneath....so that's how one side of the building was left open to show the interior.'

The size of the building was such that it accommodated five vats containing 260 tons of raw materials - lump lime and ore - on the top floor, 110ft from the ground. This material went to machines on the floor below, there to be reduced to powder form before being metered into a chamber where it was mixed with oxygen as part of the processing before going through the 'blow' procedure in the coke ovens. From the initial drawing, there was nothing to suggest that this was not a Tonka toy chemistry set; the vastness of the plant only became apparent when Lawrie introduced figures of workers on each floor. Look closely, there's a woman with 'her' arms outstretched on the third floor, being approached by a man with an evening paper. In fact, the 'woman' is a man wearing a 'skirt' of heat resistant material as a protection against heat reflected from the molten metal. The 'newspaper' in the other man's hand is the file containing the relevant information and instrument readings of the next ▶

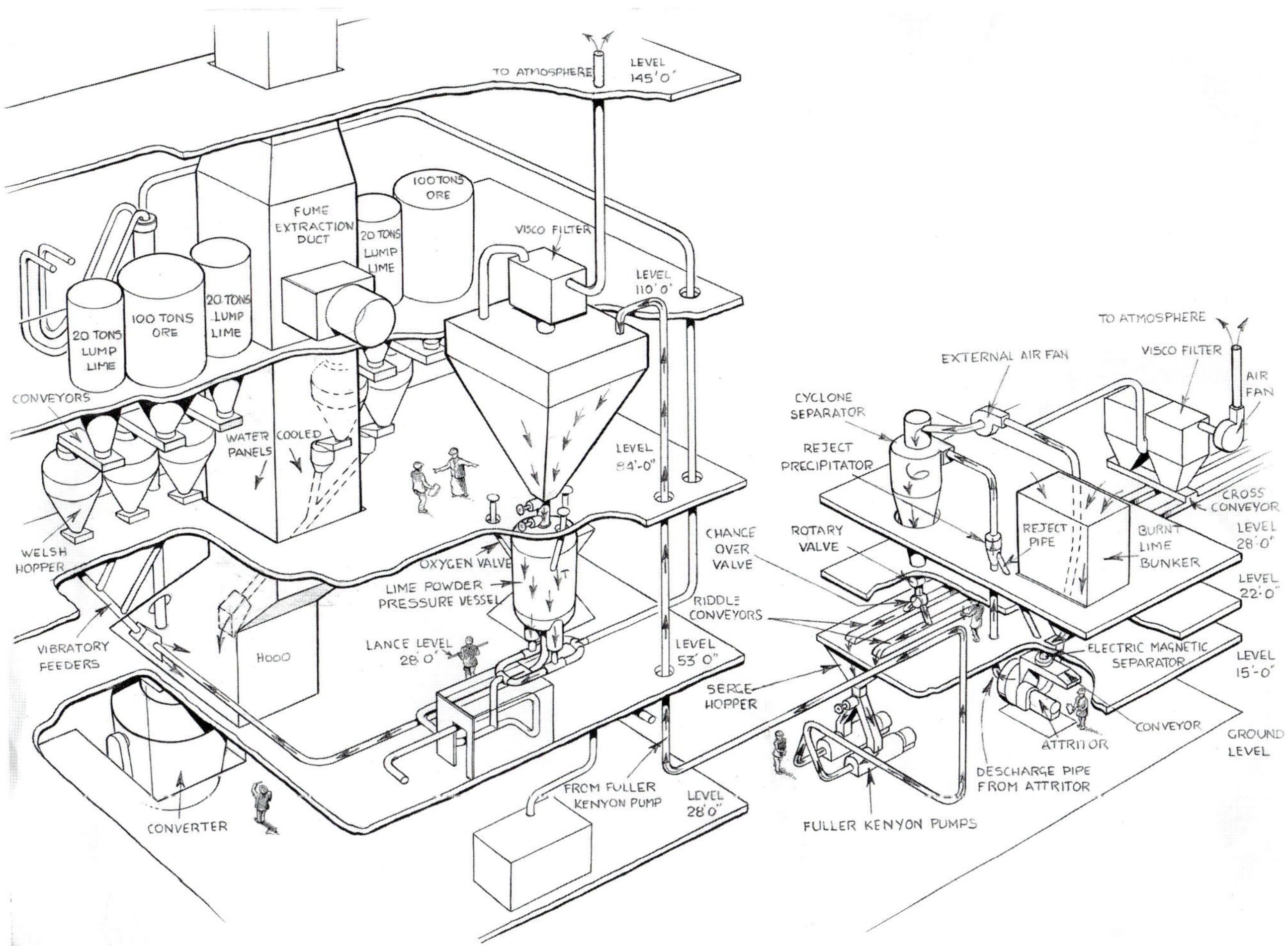

mixture to be fired in the ovens. Such information is vital in controlling the 'cooking' time of each 'blow' (as the mixture is called) and knowing which type of steel is being produced during the eight hours it took to make 100 tonnes. On the floor below you will see a man holding a lance, which is actually a thermocouple for measuring the temperature of the blow in the oven; the normal temperature is around 1,600C. This man is probably taking a break because his work station is on the ground floor near to the oven. Meanwhile, down on the ground floor, the chap apparently looking at television is actually holding up a piece of coloured glass to shield his eyes while inspecting the colour of the flame in the converter or oven. Flame colour is essential information in knowing how the blow is progressing; to check that the metallurgy is right, a sample is extracted, poured into a small cylinder and taken to the laboratory nearby for analysis. In the supply building next door, Lawrie positioned three figures each of whom is monitoring different machines, busy feeding materials to the mill; two men have documents in hand, no doubt keeping accurate read outs from the machines in front of them.

Who would want to work in such an atmosphere? One of the employees who was there when the works finally shut in 1980, Steve Purcell, told me of the strength of loyalty and affection the workers had for Corby Iron and Steel. He felt sufficiently strongly to write and illustrate an absorbing 128 page illustrated book called, unsurprisingly, 'Corby Iron and Steel Works' (£11.99, Tempus Publishing Ltd) to which I'm indebted for much of the above information. According to Steve, a strong bond of comradeship existed among the work force which inspired them to fight a rearguard action to stop the works being closed. In the end, they lost out to market forces, which sounds familiar when applied to much of the UK's former industries.

Alpha 250cc disc valve

Another drawing that Lawrie found, and could remember only vaguely, was the very promising 250cc Alpha Centuri twin two stroke racer. 'I believe it was designed by a man called Frank Cutler who was one of the two brothers behind the Alpha Bearings company,' recalled Lawrie.'Can't remember much else about it except it had an ingenious feature whereby a single carburettor fed twin rotary discs which were mounted in the centre of the crankcase. As the discs were assembled at the same time as the crankshaft it was difficult for me to get a clear indication of where everything went. Unlike other designs that I'd drawn, which had twin carburettors feeding the engine through rotary discs on either end of the crankshaft, the Centuri's single inlet tract entered the crankcase below the rear of the cylinder block. A bifurcated section at the inner end of the tract directed the in-going mixture into the path of the two discs alongside the central flywheel and thence to transfer ports in the cylinder walls, as shown by the directional arrows in my drawing. Look closely and you'll just be able to see two arrows at the top of the left hand transfer ports indicating that the charge is exiting from a split port. There were two unequal size openings; the smaller one directed cooling gas across the piston crown. The engine I drew had positive oiling to the main bearings, apart from the centre bearing, but this was later dropped as unnecessary in favour of a 20:1 Castrol R and petrol mix. Instead of re-inventing their own gear box, Alpha took the expedient course of using an existing Albion box which bolted to the flanges I have drawn on the rear of the Centuri's crankcase. Altogether, a compact engine with some very clever ideas. It deserved to succeed.'

Later I was to test the bike at Silverstone as part of the 'Racer Test' series initiated by Bruce Main-Smith of 'Motor Cycling' before that paper amalgamated with 'The Motor Cycle' in September 1967. For one which was intended to become a 'clubman's affordable production racer' according to Frank Cutler, the Centuri was very impressive. His feeling that it would be on a par with the production Yamahas which were then making their presence felt in the 250cc class was close to the mark, judging by our lap times that day at Silverstone. When tested in October 1967, the Centuri was in its fourth year but yet to go into production; previous versions had borrowed DMW and Royal Enfield frames. For a production version to be viable, Alpha needed to make their own frame; this was the one I tested. Because it fitted me perfectly, I was able to concentrate on extracting the best from the machine. Fred Hadley, later known as 'Mr Piston' because of the highly-successful Omega piston company he developed and which he still runs, was then on a one year contract to develop the Centuri. The frame had been his first major contribution. 'Until now, Frank had been forced to use what bits he could find to make a complete development machine', recalled Fred. 'At one stage he used Albion race gear boxes from Royal Enfields, cylinder barrels and heads from the Velocette Viceroy scooter, Dellorto carburettors, Asso pistons and Oldani brakes from Italy, modified frames from DMWs and Enfields. Frank had only ever intended to make engines, that had been the family business for many years. Now that he'd made a competitive race engine, and fixed a price of £200, he thought that riders would be queuing up to buy it to fit in their own frames. But life in the '60's was not like that; riders wanted complete machines. Frank decided that if he wanted to make a commercial success of his new engine, he'd have to offer it in a complete machine. So he gave me a year to sort out a new frame. And that's the machine you tested. However, it was not the complete job because after your test we decided that you'd been too comfortable and we lowered the riding position by nearly three inches! Obviously, this made a lighter and more compact machine with improved road holding and various engine modifications gave even more power. This increased from 37-bhp at 10,500rpm to 43-bhp at 12,000rpm. The Centuri was flying so much so that our race rider, Dave Browning, scored lots of wins across the country'.

Fred recalled later how fate stepped in and killed the project. 'About the time the Centuri was ready for tooling up, the parent company, Alpha, was taken over by a bigger concern, E & H.P. Smith, who also owned Royal Enfield, and the racer project was put on ice. I believe that the intention to produce it for private owners was implicit in the take-over plan but the new company's heart was not in racing and they let the project die. Which was a great pity, because the hard work had already been done and with just a little extra effort the Centuri could have been marketed at a competitive price to private owners. Instead, the market was gifted to products of foreign manufacturers. How often have we heard that story before?'

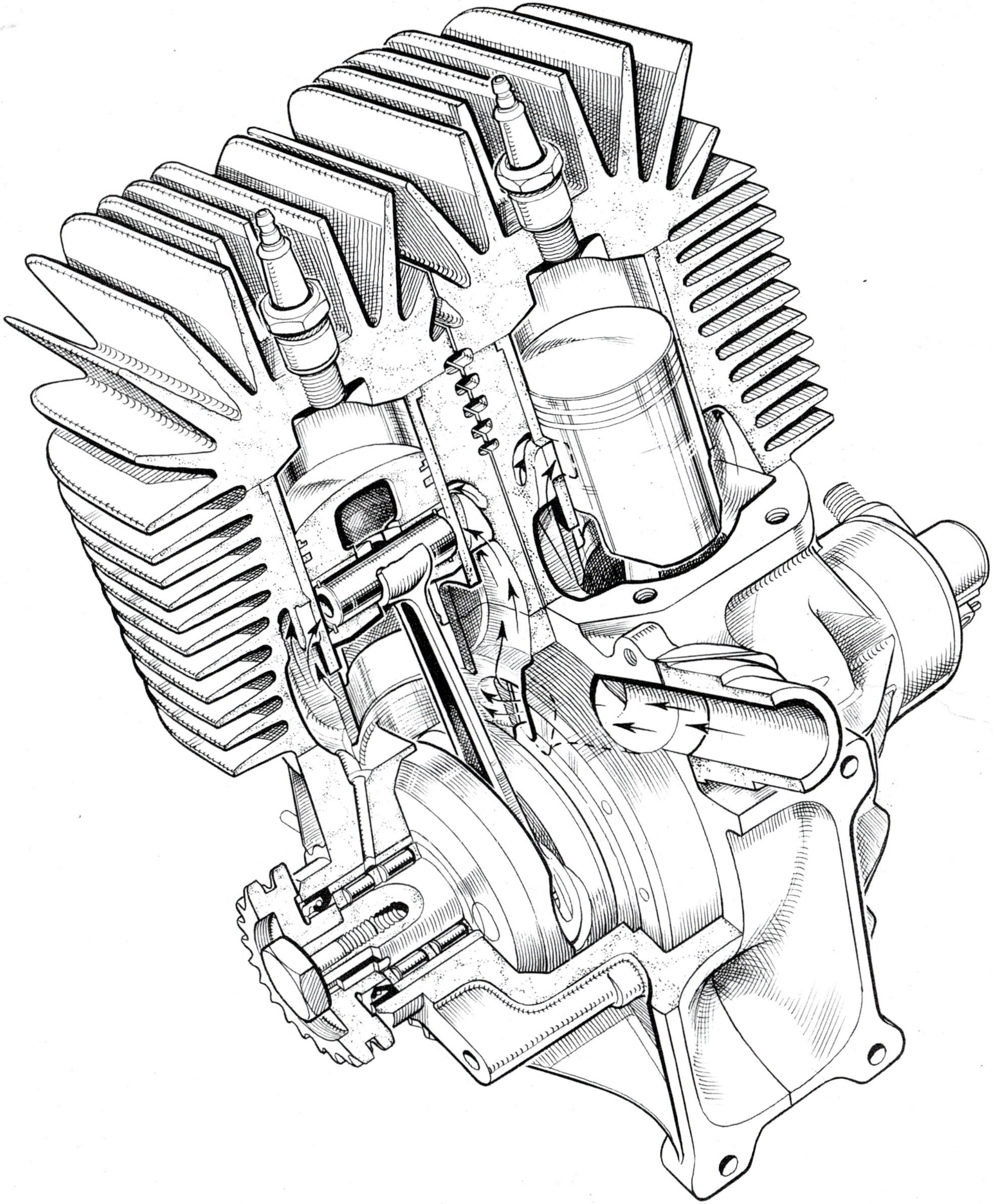

Tester's size matters

One of the problems of being a race tester is that the bikes I tested were almost invariably modified to fit their regular rider. More often than not, he was a different build from my 6ft frame. The greatest discrepancies lay in 125 and 250cc machines; the smaller the capacity the more critical the size issue, because lightweight bikes required lightweight jockeys, smaller the better. It was a rare occasion to jump on a machine and find that it fitted me like a glove. But the Centuri did. So at ease did I feel that I was able to concentrate on doing it justice, lapping the Silverstone Club circuit 0.77s slower than Don Wolfindale, the factory's development rider at that time.

Pants on fire

Jack Williams and Lawrie became close friends. Jack would drop in on a Thursday morning for a chat, usually a deep discussion about aerodynamics. To aid concentration, he would get out his trade mark pipe… then his tobacco pouch… charge the pipe… produce the matches… all the while gesticulating with his pipe. He'd then strike a match, take a couple of puffs from the pipe, resume gesticulating, then a pause to strike another match… repeating the procedure throughout the conversation. 'He never, ever, got the pipe going,' recalled Lawrie, 'There was a time, elsewhere, when the pipe finally did ignite towards the end of a lecture. Jack absent-mindedly stowed it in his pocket, strode out of the lecture room… followed by a trail of smoke from his trouser pocket!'

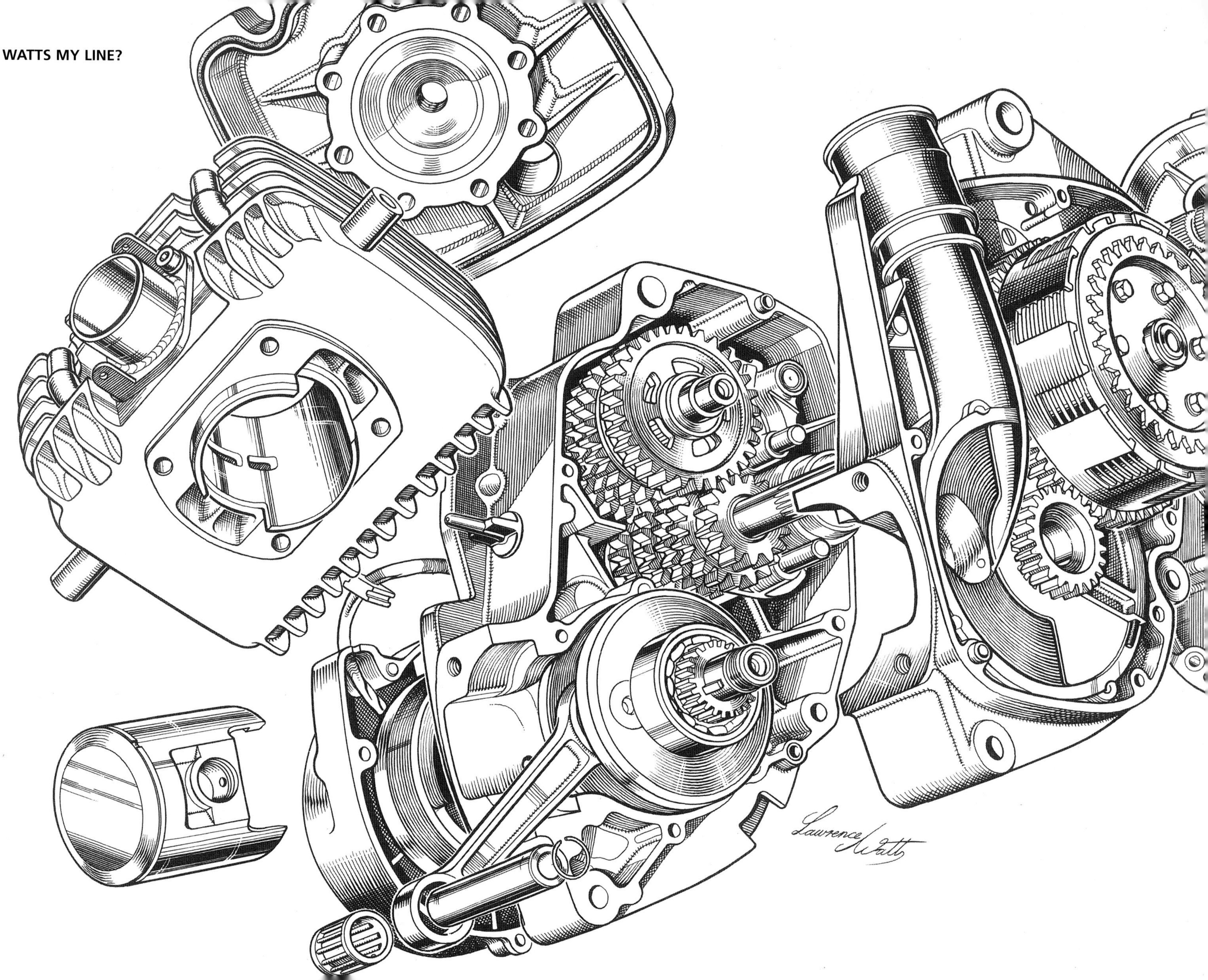
Lawrence Watts

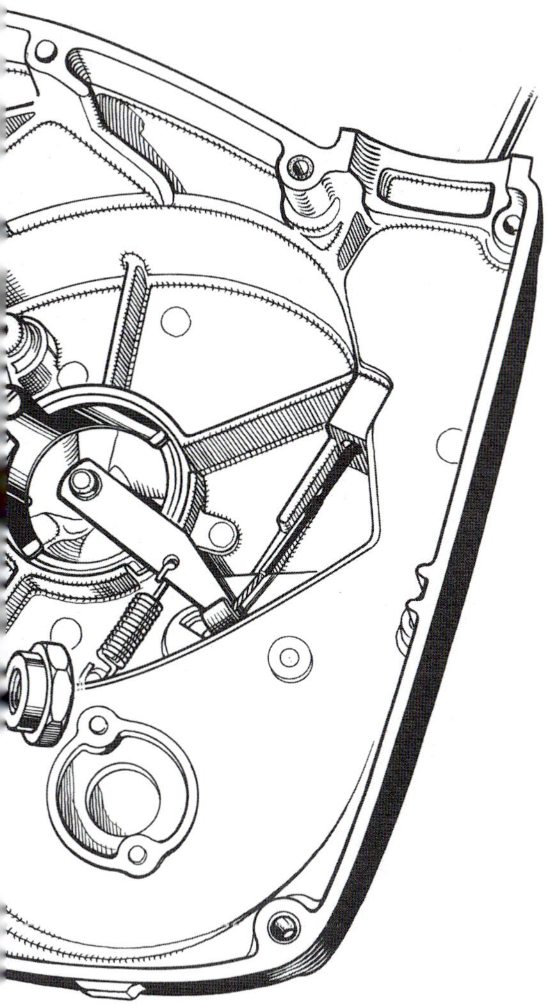

250cc Rotax

Rotax made some of the strongest engines in competitions and from these derived this 250cc design for roadster use. By showing the underside of the cylinder head, and turning the cylinder on its side, Lawrie shows as many features as possible. For instance, the porting layout and, unusually, the piston has only one ring and a slightly bevelled crown which matches with a squish band inside the cylinder head. Although the cylinder has fins, these cover a water jacket; the projections either side of the lower part of the cylinder carry water, are part of the cooling system. No fewer than eight bolts secure the head to the cylinder and a rubberised ring sits in an annular groove to ensure water tightness between the head and cylinder. The disc rotary valve on the left-hand end of the crankshaft would, on the race version, have the carburettor bolted as close as possible to the disc but on the road version, to protect the disc from damage and dirt, a moulded intake carries mixture from the carburettor, which is mounted behind the cylinder barrel, to the disc.

The gear change shaft is co-axial with the kick start shaft; Lawrie shows the two serrated shafts running one within the other (visible below, and to the rear, of the clutch assembly). On the right-hand end of the crankshaft (shown on the left of the drawing) can be glimpsed part of the lighting generator and output cable, a necessity for road use. A number of smaller manufacturers were believed to have been interested in using this engine in British-built chassis but, probably on the score of cost and the increasing competitiveness of Japanese imports, nothing of significance was produced.

The Hooper-Favill 500cc Norton Wulf

This drawing, about which Lawrie could remember very little, started out fairly normally with an instruction from 'The Blue Un' office to 'go and see this new engine...seems a bit unusual for Norton...get the salient features... let's have it tomorrow...old chap...' He soon found himself looking at a remarkable piece of kit, outwardly innocuous but inwardly something very special, as befitted a designer of the calibre of Bernard Hooper, a man well known to Lawrie and for whom he had the greatest admiration. Little did Lawrie know at the time, 30 years ago, that the engine which he was drawing was destined to make its name not in a Norton Wulf, as intended (because Norton was to collapse shortly afterwards), but in the world of marine, automotive, industrial and defence applications. But that's jumping the gun a bit....

Bernard Hooper achieved great acclaim for his design of the highly successful Villiers Starmaker engine. However, he left Villiers after becoming disillusioned with the direction the motorcycle industry was taking. During this time he started work on the stepped piston engine concept. When Dennis Poore took over AMC and Villiers he invited Hooper back to the newly-formed Norton Villiers as chief engineer and Norton Villiers became the first licensee for the SP (stepped piston) engine. However, the first challenge was to develop a new product for the NV Group and, as head of the design team under Director of Engineering Stefan Bauer, Bernard Hooper started the design of the revolutionary Norton Commando with its Isolastic anti-vibration system. On completion of the Commando, Bernard and his colleague, gear box specialist John Favell, were able to start work on single cylinder stepped piston engines and the Norton Wulf project was born.

The Wulf was ultimately intended to be sold under the name of Matchless, also part of the Norton Villiers group, in engines sizes of 350, 500 and 500cc.The cylinder centres of the 500cc unit, which was intended for the Norton Wulf and shown in Lawrie's drawing, were designed wide enough to accommodate 750cc cylinders on the same crankcase. This compromise was at the insistance of Dennis Poore because it made economic sense, but from an engineering perspective it made the 500cc engine wider than it needed to be. This soon became of academic interest because the NV Group collapsed in 1975. Attempts were made at the Wolverhampton plant in Marston Road to save the industry by showing the world, and especially the Government, for the first time what was at stake in the form of the new Norton Wulf and a new Norton Commando. Government assistance was instead focussed on the Chrysler plant, just down the road at Ryton, which was also in serious financial trouble.

With no help forthcoming, Bernard Hooper set up his own consultancy in 1977 and further developed the Wulf as a liquid-cooled Wulf II. Both machines are on display at the National Motor Cycle museum at Solihull courtesy of Peter Hooper, Bernard's son, who now runs Bernard Hooper Engineering Ltd.

THE STEPPED PISTON PRINCIPLE

Uniquely, what Bernard Hooper and John Favell were doing, as shown in Lawrie's drawing of the earlier air-cooled design, was using a form of supercharging by utilising a 'Top Hat', or two-diameter 'stepped piston', to pump mixture into the neighbouring cylinder. In-going mixture passed from the carburettor via reed valves. But instead of entering the crankcase as on a normal two stroke, the mixture entered through ports into the lower cylinder of each stepped piston; from there it was pumped upwards and outwards through transfer ports in the front wall of the cylinder casting, to the opposite cylinder, and thence into the combustion chamber. (Arrows on the drawing indicate the direction of the gas flow). As the piston moved upwards, it compressed and released the mixture into the upper cylinder section via ports which were cast transversely into the cylinder block. From there the mixture was compressed by the upper piston into the combustion chamber.

THE STEPPED PISTON PHILOSOPHY

The use of stepped pistons for charge transfer and combustion allows the key advantages of two and four-cycle engines to be combined and elimination of disadvantages inherent in each of these engine types. As in the case for four-cycle engines, the charging and lubrication processes are physically separate in the stepped piston design. This eliminates a major source of problems that is common in crankcase scavenged two-cycle engines. Isolation of the fresh charge from the crankcase is made possible by the provision of normal four-cycle engine compression and oil control rings on the larger diameter part of the piston. The crankcase, freed from any gas exchange functions, is well lubricated; the working processes being sealed above the piston.

Isolation of the crankcase also permits a full pressure lubrication system to be used, as in four-cycle units. This simplifies fuelling needs with operation on neat fuels, avoiding the battlefield supply problems, uncertainty and risk of error inherent in the oil and fuel mixture required for conventional two-cycle UAV(Unmanned Air Vehicle) engines. Furthermore the inherent piston cooling characteristics of the stepped piston design offer a major durability advantage over conventional two-cycle engines, allowing much leaner fuel delivery than can be sustained with crankcase ▶

A measure of supercharging is achieved by means of a stepped piston, with a smaller diameter piston on top, ('top hat' effect), running in different size bores, the larger one forcing the incoming mixture through passages in front of the cylinder into the top of the neighbouring cylinder where the mixture is ignited. (A similar procedure is effected in the left-hand cylinder, with the mixture being routed through passageways at the rear into the right hand cylinder).

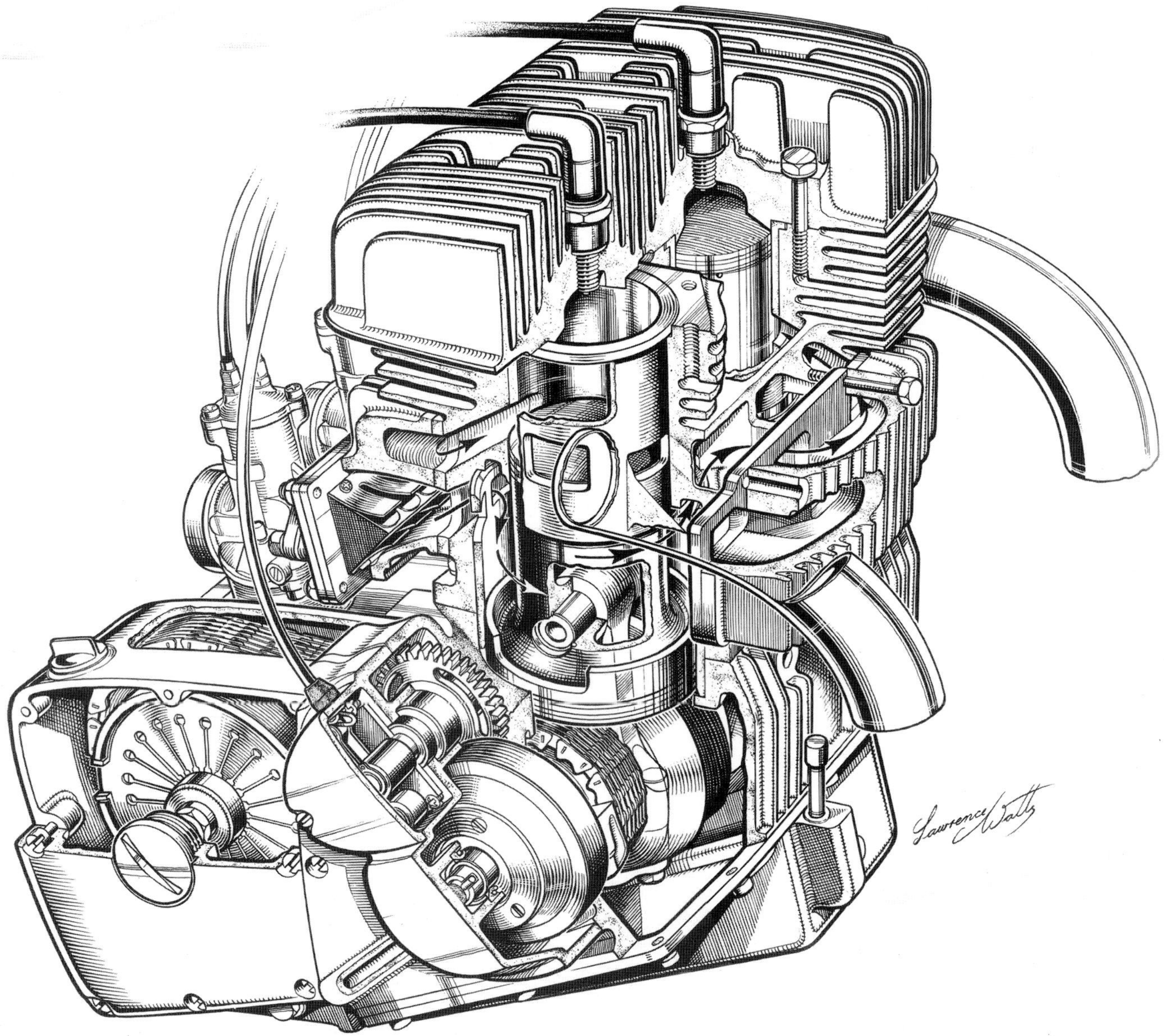

scavenging. Piston overheating and consequent seizure normally occurs with very lean air fuel mixture operation in crankcase scavenged engines.

Technical advantages of the stepped piston engine were gained not only in the greater amount of mixture entering each cylinder and therefore higher volumetric efficiency, but by the fact that the piston can be cooled far more effectively than is the case for conventional crankcase-scavenged engines. Although the piston mass is approximately 20% higher than for a conventional looped scavenged two stroke piston, the step of the piston provides a flange-stiffening effect which means that thinner skirt sections can be used. As the engine uses the two stroke cycle, and therefore with the absence of load reversals, piston mass is not as critical as it is for four stroke engines. The Hooper stepped piston also provides improved load bearings and guidance which results in low ring wear and reduced piston noise. Another advantage was the discovery, following detailed studies by large commercial automotive groups, that there were no serious production issues building the engine in quantity. With no valve gear, plain bearings only, manufacturing costs were down 20/25% per unit over comparable four stroke engines. Low thermal loading of the pistons, low emissions, fast warm up times, extended oil changes (because the lubricant did not degrade), compact size and low weight, coupled with durability, were operational attributes which later attracted so much commercial attention. In short, this Hooper design was a notable break-through in two stroke design, heralding a much more efficient method than traditional two stroke engines. Even more satisfying, from a British industry point of view, it all preceded the days when rotary disc valves would become the norm on Japanese production two strokes. ▶

Villiers stepped-piston V-four diesel

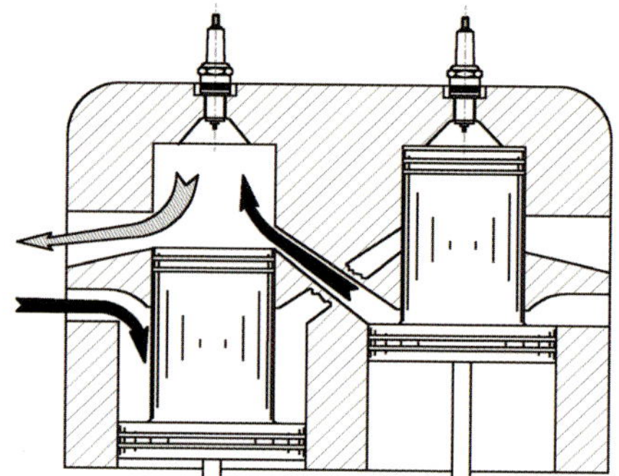

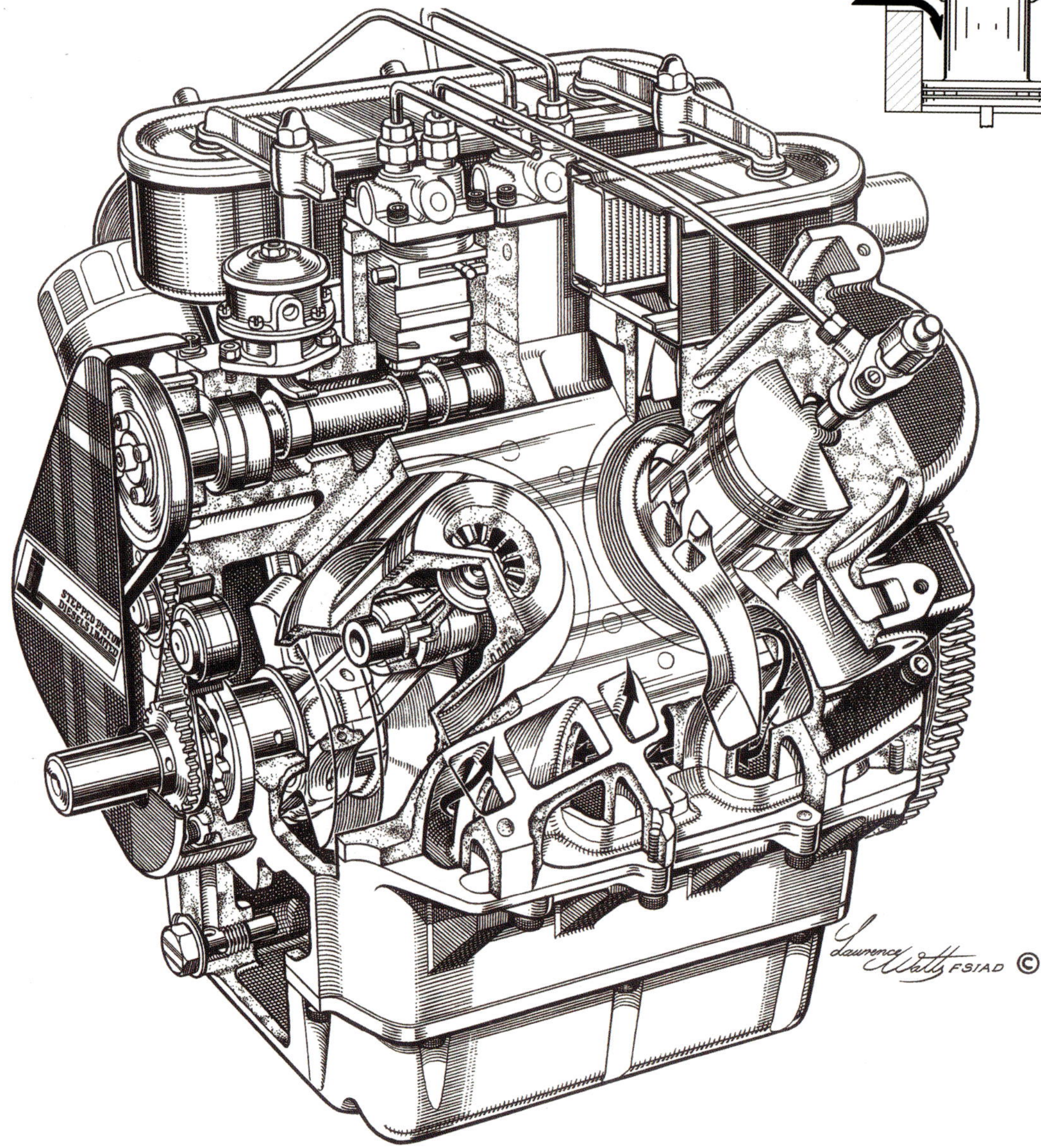

This seemingly complex stepped piston design by Bernard Hooper initially caused Lawrie difficulty due to the internal layout of the crossover porting between the four cylinders. His task was complicated by the need to show the water-cooling passageways.

Look closely at Lawrie's drawing and you will see that the transfer ports carrying the 'supercharged' mixture across the front of the engine pass through detachable castings bolted to the front and rear of the block. Mixture destined for the left hand cylinder passes through an S-shaped passageway in the top of the block while mixture destined for the right hand cylinder travels in a similarly-shaped passage underneath. Making the block detachable eased manufacturing while providing a measure of control over port shapes. An advantage of having the transfer passages at the front ensured optimum cooling of the in-going mixtures. Another innovative feature was the early use of a Morse chain for the primary drive, replacing the ubiquitous roller drive chain. Why? Because the Morse was quieter, stronger and did not require adjustment.

So the engine died the death with the demise of Norton and the motorcycle industry? In fact, the opposite was the case! This heralded the start of the stepped piston's career. In water-cooled form, the MkII Wulf engine found its niche not in motorcycling but in UAVs. The stepped piston principle also formed the basis of Bernard Hooper's ingenious V4 cylinder diesel engine. 'Although I knew that Bernard had come up with another very ingenious design, I had no reason to suspect that I was looking at something which so soon outgrew the Norton for which it was intended and go on to earn an enviable reputation in fields far removed from motorcycling,' remarked Lawrie when I relayed to him Peter Hooper's account of the progress of his father's invention through the years...

This one is highly unusual, a stepped piston V four cylinder diesel engine. The story which Loll related concerned the late Bernard Hooper. 'Bernard, a very talented and under-estimated man, designed this as a static diesel engine for Villiers, the former industrial giant, better known as a supplier of two strokes to the motor cycle industry for many years. Hooper's design was an injected diesel, but an ingenious touch was to use transfer ports more normally found on two strokes, a rather complicated way of supercharging. Very compact, quiet in operation, the engine must have been a nightmare to make and it was difficult for me to even understand how the damned thing worked. Then I had to draw it! Some of it was extremely difficult, requiring many sketches to get it right, especially those convoluted transfer ports which were very, very, difficult to draw. Bernard was very keen on the aerodynamics of the gas flow into the engine, but you couldn't see any of this because it was all cast inside the engine. Suddenly, I was faced with a dilemma, how to illustrate something which I found quite difficult to fathom for myself. It took some time, many little sketches, before I could get the perspective correct, because the shapes Bernard had evolved were very specific. His method of supercharging was ingenious; he used the bottom, stepped part, of one piston to compress the incoming mixture to supercharge the other cylinder. This was done via the transfer ports, which explains why he had opted for a crankshaft throw of 180 degrees. But to illustrate this was not easy, believe me. Everything was out of sight due to the transfer ports' convoluted passage through the crankcase casting. Adding to my problem was the fact that the engine was water cooled and I had to illustrate this too. A nightmare! By much trial and error, I managed to get it right but it was certainly one of the most challenging jobs I've undertaken. Even now, many years later, I still have to look hard at the drawing to remember how the engine actually worked, ...but work it certainly did, as illustrated dramatically in the many variations developed by Bernard's brilliant son, Peter. Bernard, an extremely nice man, never got his due recognition as a designer.'

Versatile Hooper designs

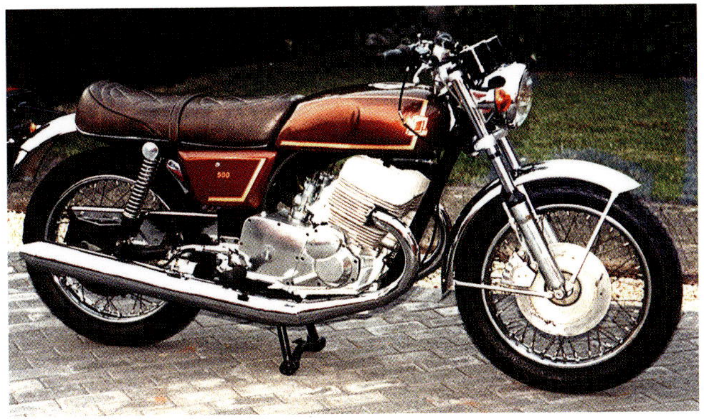

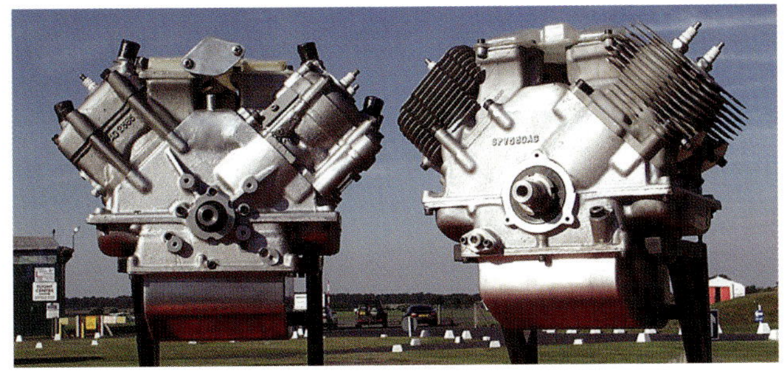

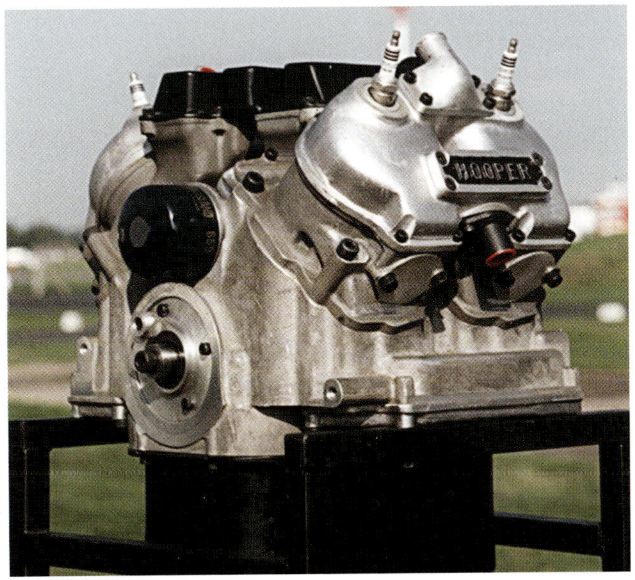

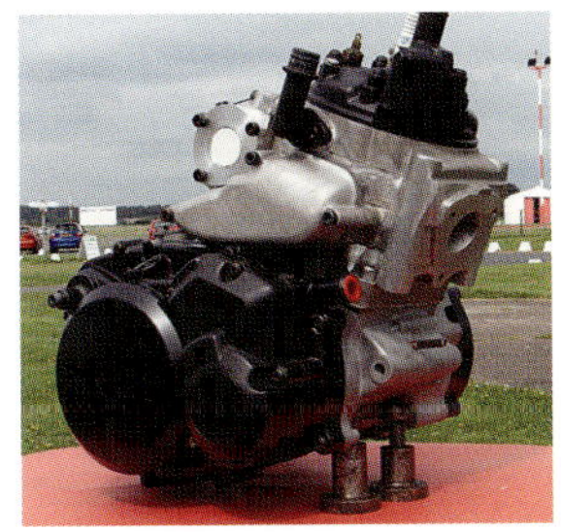

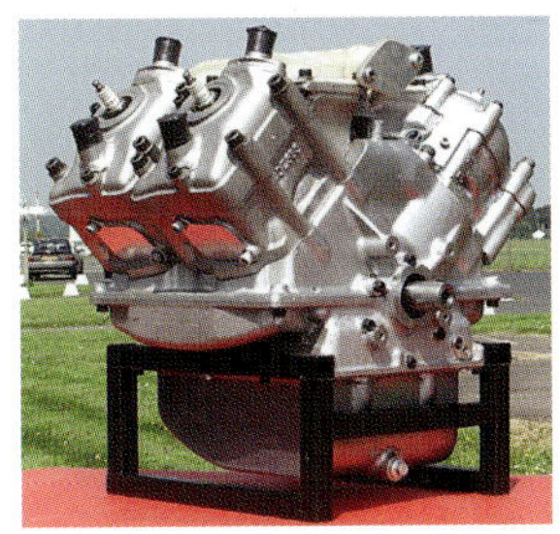

Top left, JGP Norton Wulf prior to delivery to the National Motorcycle Museum.
Top right, air and liquid cooled engines.
Above, SPV1000 V4 cylinder engine.
Right top, SP150cs, 150cc single motorcycle engine.
Right, SPV580LC engine.

Index

Other titles from Redline Books - www.redlinebooks.co.uk

Mick Walker's British Racing Motorcycles
ISBN 09531311 0 6.
Softback 208 pages,
280 x 210mm,
over 800 pictures
in colour and b/w.
£20

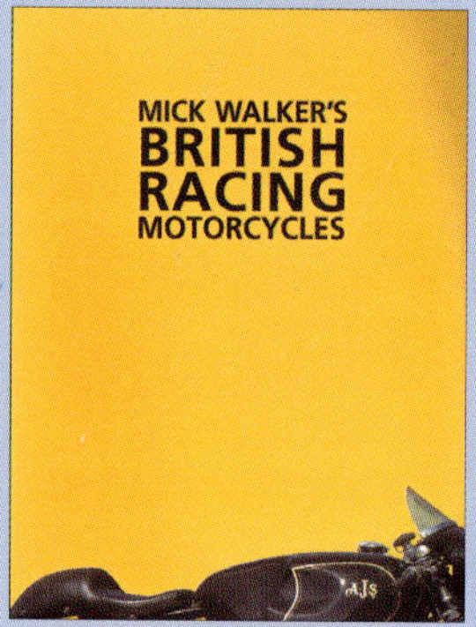

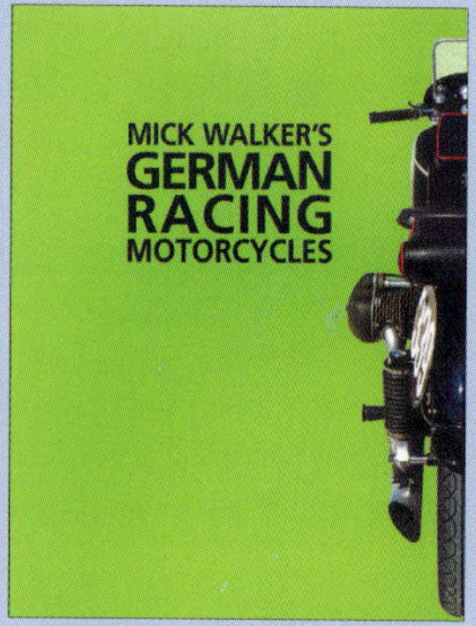

Mick Walker's German Racing Motorcycles
ISBN 09531311 2 2.
Softback 224 pages,
280 x 210mm,
over 500 pictures
in colour and b/w.
£20

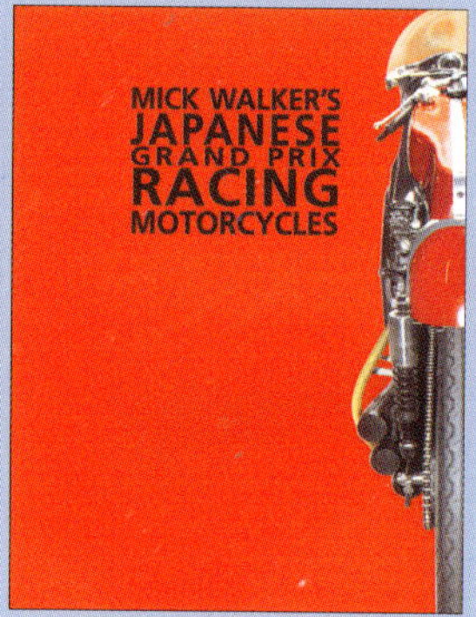

Mick Walker's Japanese Grand Prix Racing Motorcycles
ISBN 09531311 8 1.
Softback 203 pages,
280 x 210mm,
over 500 pictures
in colour and b/w.
£20

Mick Walker's Italian Racing Motorcycles
ISBN 09531311 1 4.
Softback 256 pages,
280 x 210mm,
over 800 pictures
in colour and b/w.
£20

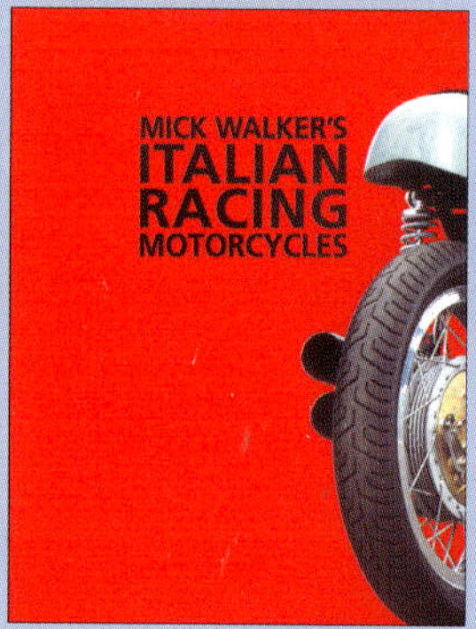

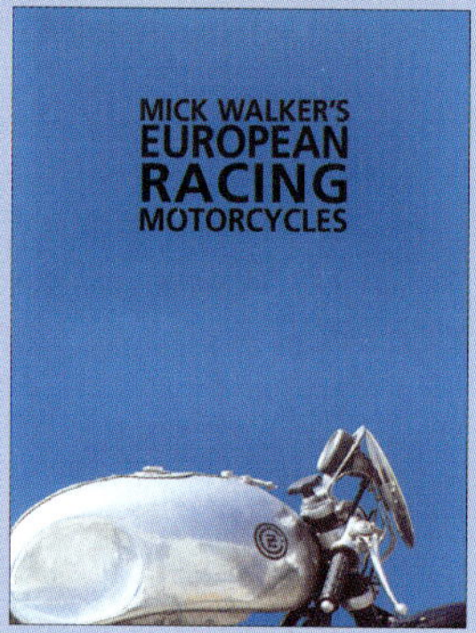

Mick Walker's European Racing Motorcycles
ISBN 09531311 3 0.
Softback 232 pages,
280 x 210mm,
over 500 pictures
in colour and b/w.
£20

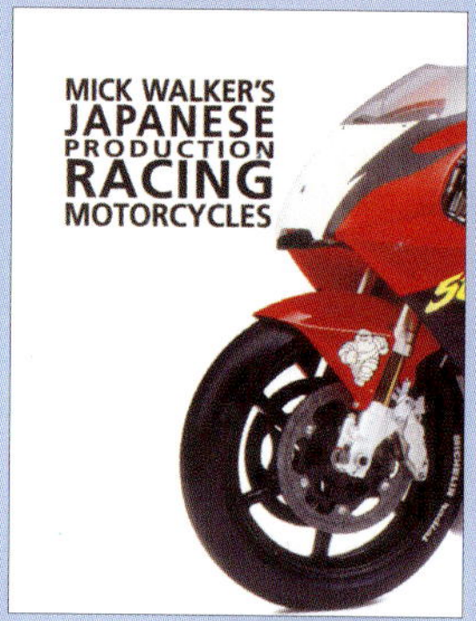

Mick Walker's Japanese Production Racing Motorcycles
ISBN 09544357 0 2.
Softback 336 pages,
280 x 210mm,
over 800 pictures
in colour and b/w.
£25

Bill Lomas World Champion Road Racer
ISBN 09531311 9 X.
Hardback, 304 pages
280 x 210mm,
£30

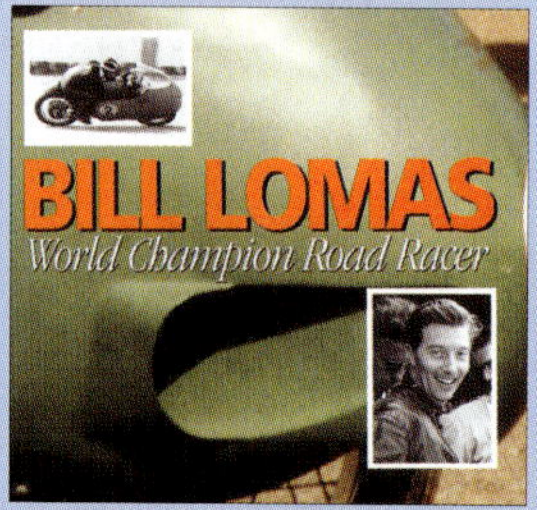

The AJS 7R
205 x 210mm. 223 pages.
ISBN 09531311 7 3.
Hardback
£29
ISBN 09531311 6 5
Softback
£20

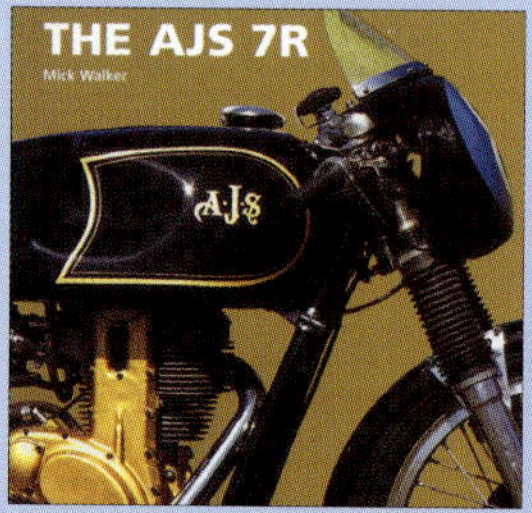

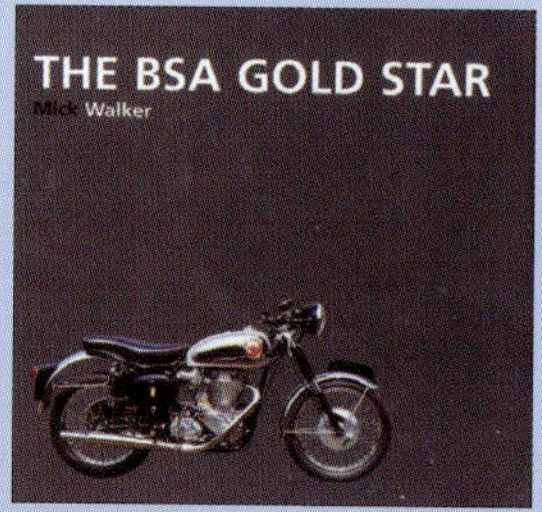

BSA Gold Star
ISBN 09544357 3 7.
Hardback, 264 pages
full colour
205mm x 210mm,
£29.95.

The Manx Norton
ISBN 09544357 9 6
Softback, 312 pages
full colour
205 x 210mm
£29.95

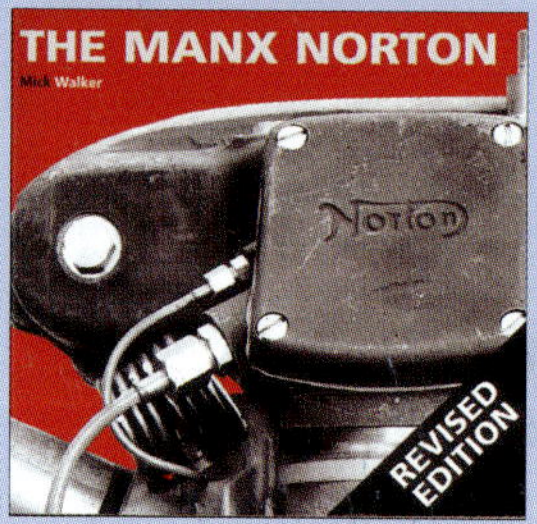

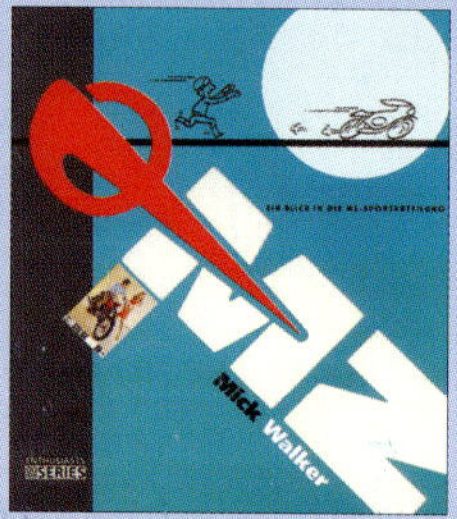

MZ
ISBN 09544357 4 5.
Softback, 144
pages
full colour
220 x 193mm
£17.95

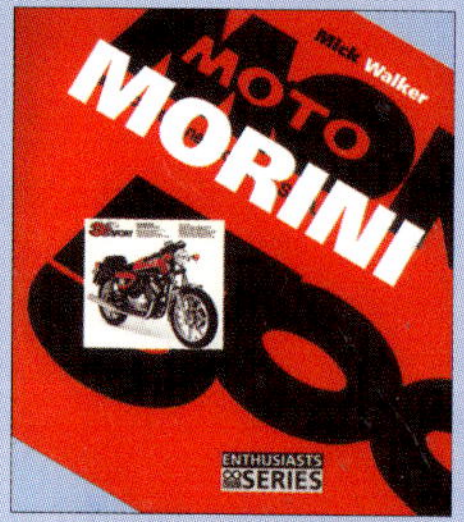

Moto Morini
ISBN 09544357 2 9.
Softback, 144 pages
full colour
220 x 193mm
£17.95

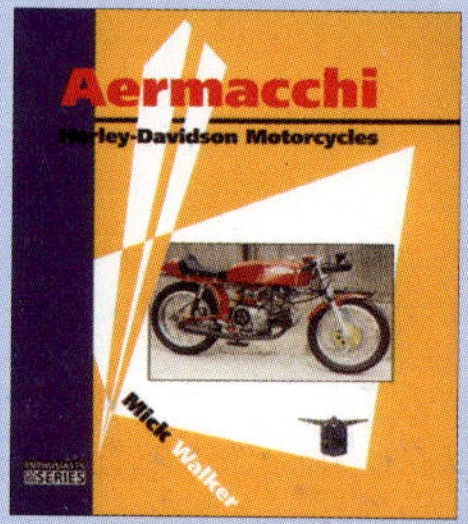

Aermacchi Harley-Davidson Motorcycles
ISBN 09544357 6 1
Softback, 216 pages
full colour
220 x 193mm
£19.95

ISBN: 0 9544357 7 X
ISBN: 978 0 9544357 7 6

Printed in Italy by Printer Trento

REDLINE BOOKS

Designed and published by Redline Books. **www.redlinebooks.co.uk**

www.redlinebooks.co.uk